HYDRAULIC AND ELECTRIC-HYDRAULIC CONTROL SYSTEMS

Hydraulic and Electric-Hydraulic Control Systems

Second Enlarged Edition

by

R.B. WALTERS
Engineering Consultant, Wembley, U.K.

SPRINGER-SCIENCE+BUSINESS MEDIA, B.V.

Library of Congress Cataloging-in-Publication Data

Walters, R. B. (Ronald B.)
    Hydraulic and electric-hydraulic control systems / by R.B. Walters.-- 2nd enl. ed.
    p. cm.
    Rev. ed. of: Hydraulic and electro-hydraulic control systems. c1991.
    ISBN 978-94-015-9429-5   ISBN 978-94-015-9427-1 (eBook)
    DOI 10.1007/978-94-015-9427-1

    1. Hydraulic servomechanisms. I. Walters, R. B. (Ronald B.) Hydraulic and
electro-hydraulic control systems. II. Title.

    TJ857 .W3 2000
    629.8'323--dc21

                                                                    00-060528

    ISBN 978-94-015-9429-5

Printed on acid-free paper

The first part of this book has been reprinted from the first edition.

All Rights Reserved
© 2000 Springer Science+Business Media Dordrecht
Originally published by Kluwer Academic Publishers in 2000

Preface to the First Edition

Force and motion control systems of varying degrees of sophistication have shaped the lives of all individuals living in industrialized countries all over the world, and together with communication technology are largely responsible for the high standard of living prevalent in many communities. The brains of the vast majority of current control systems are electronic, in the shape of computers, microprocessors or programmable logic controllers (PLC), the nerves are provided by sensors, mainly electromechanical transducers, and the muscle comprises the drive system, in most cases either electric, pneumatic or hydraulic.

The factors governing the choice of the most suitable drive are the nature of the application, the performance specification, size, weight, environmental and safety constraints, with higher power levels favouring hydraulic drives. Past experience, especially in the machine tool sector, has clearly shown that, in the face of competition from electric drives, it is difficult to make a convincing case for hydraulic drives at the bottom end of the power range, specifically at fractional horsepower level. A further, and frequently overriding factor in the choice of drive is the familiarity of the system designer with a particular discipline, which can inhibit the selection of the optimum and most cost-effective solution for a given application.

One of the objectives of this book is to help the electrical engineer overcome his natural reluctance to apply any other than electric drives. Another difficulty often encountered among all types of engineers is the unwillingness or inability to tackle the dynamics of hydraulic control systems in view of their relative complexity as compared with electric drives. Owing to the compressibility of the working fluid and the non-linear characteristics of hydraulic control devices, dynamic system modelling involves the manipulation of non-linear, high order differential equations. This fact can have a daunting effect on all but the more analytically inclined engineers, and has contributed to the wide gap that exists between the control engineer and the average hydraulic application engineer. It has often led to the oversimplification of hydraulic system identification, which has frequently necessitated costly re-design and even resulted in litigation.

It is hoped that this book will help in bridging the gap between the academic and the application engineer and thereby make some contribution towards the wider application of hydraulic control systems.

Partly due to its relative complexity, hydraulic control system analysis is an ideal hunting ground for the mathematically biased engineer. Several analytical methods have been developed over the years and every specialist in this field has his own preference. The conventional methods can be briefly summarized as:

(1) Non-linear analysis in the time domain.
(2) Linearized small perturbation analysis using the root-locus (pole-zero) approach.
(3) Linearized small perturbation analysis using the frequency response approach.

The approach adopted in this book is based on method (3), with an extension into the time domain, facilitating the modelling of system transient response to any given duty cycle. This concept, which permits system optimization in the frequency domain, has been developed and successfully applied over more than 15 years, with close correlation between predicted and actual performance over a wide range of applications, which is, in the final analysis, the ultimate criterion of credibility.

This book is based on an earlier version, *Hydraulic and Electro-hydraulic Servo Systems*, published in 1967. The original publication was aimed at an engineer using a slide rule, graph paper and other manual aids, whereas the present edition is focused on an engineer with a personal computer at his disposal. This underlines the considerable advance in communication technology that has taken place over the past twenty years. To bring the contents into line with this monumental change in analytical capability, now readily available to every engineer, the bulk of the text had to be re-written. It will become apparent to the reader that a meaningful dynamic analysis of a complex electro-hydraulic control system is not feasible without the aid of a computer.

In that context it is of interest to note that whereas the manuscript for the original edition was laboriously handwritten and subsequently typed and all graphs were manually drawn, the text for the latest edition was typed directly into a word processor and all graphs were generated and plotted by means of a specially adapted computer simulation program.

Thus time marches on!

Preface to the Second Enlarged Edition

A 'hands-on' exercise to familiarise the reader with the programs contained in the enclosed disk is a logical sequence to Part 1, which is an in depth study of component design and system analysis.

It will have become apparent to a reader of Part 1 that a meaningful dynamic analysis of a complex electro-hydraulic control system is not feasible without the aid of a computer. This leaves us with two alternatives, i.e. to write our own set of programs tailored to our requirements, or to use a suitable proprietary software package, preferably PC compatible, dedicated to the analysis and performance prediction of hydraulic and electro-hydraulic control systems.

It is of interest to list the required expertise to undertake these tasks. Considering first the user of a proprietary software package, minimum requirements can be summarised as:

1. Basic knowledge of hydraulic control concepts.
2. Ability to formulate a Performance Specification.
3. Access to a PC.

The Engineer choosing to write his own software programs would have to satisfy the following additional requirements:

4. Detailed knowledge of hydraulic control concepts.
5. Working knowledge of control theory.
6. Ability and time to write and debug software.

The majority of Application Engineers working for hydraulic equipment manufacturers, distributors or end users would fall into the first category, whereas some System Designers would have the capability to write their own programs., although frequently, in view of the considerable amount of effort required to write and debug programs,pressure of work would preclude this.

The objective of presenting part 2 in the form of an Operating Manual is two-fold, on the one hand to provide some guidelines to those Engineers prepared to write their own programs, and on the other hand to highlight the features and versatility of a proprietary dedicated knowledge-based performance prediction software package with special emphasis on the presentation of performance characteristics in the graphical form of single and multiple plots.

In contrast to Part 1 of this textbook, Part 2 does not contain any equations or algorithms, underlining the earlier stated requirement for a user of a suitable software package, which does not include the need for any analytical or programming skills.

Acknowledgment

The author wishes to express his thanks to Vickers Systems Ltd, a Trinova Company, and Flotron Ltd for permission to use some of the material and illustrations.

Contents

1

Introduction

All control systems can be reduced to a few basic groups of elements, the elements of each group performing a specific function in the system. The division into groups of elements can be carried out in a number of different ways, but selecting the following four groups forms a convenient structure for the definition of hydraulic and electro-hydraulic control systems.

(1) The power source.
(2) The control elements.
(3) The actuators.
(4) The data transmission elements.

The power source consists invariably of a pump or combination of pumps and ancillary equipment, e.g. accumulators, relief valves, producing hydraulic energy which is processed by the control elements to achieve the required operation of the actuator.

The control elements can be valves of one type or another, variable displacement pumps or variable displacement motors. Some control systems contain a combination of all or some of these control elements.

The actuator converts the hydraulic energy generated by the power source and processed by the control elements into useful mechanical work. An actuator producing linear output is referred to as a cylinder, jack or ram, whilst an actuator giving continuous rotation is a hydraulic motor and an actuator giving non-continuous rotation is usually called a rotary actuator.

The control elements act on information received from the data transmission elements; in a 'simple' hydraulic control system the data transmission elements are mechanical linkages or gears, but in 'complex' systems data transmission can take many forms, i.e. electrical, electronic, pneumatic and optical, or combinations of these types of data transmission. Although 'simple' or mechanical–hydraulic control systems are

1

still in use, they are being progressively replaced by the more versatile and flexible electro-hydraulic control system, using electronic data transmission.

Control systems can be subdivided into two basic types: on–off, or 'bang-bang', and proportional. A typical example of the former is an electro-hydraulic system controlled by solenoid-operated directional valves, where actuator velocity is pre-set but not controlled, whereas an example of the latter would be a velocity control system controlled by a solenoid-operated proportional valve controlling the flow to the actuator and hence its velocity. This book will confine itself to proportional systems, i.e. to control systems where a functional relationship exists between the controlled output quantity and the demand signal.

2

Hydraulic Power Source

In hydraulic control systems the pump supplies fluid either at substantially constant pressure which is independent of the external load acting on the actuator or, alternatively, at a supply pressure which is a function of the external loading.

In systems which have the supply pressure maintained at a constant level, the hydraulic power source can be either a fixed or variable displacement pump. In either case, as pressures normally encountered in power hydraulic systems are relatively high, the pump would be one of the three positive displacement types, that is gear, vane or piston. Piston pumps can have either axially or radially mounted pistons. The choice of pump depends mainly on the maximum pressure and rate of flow required for the operation of the system.

Three typical arrangements of a constant pressure supply are shown in Figs 2.1–2.3. In its simplest form, shown in Fig. 2.1, the power unit consists of a fixed displacement pump, a pressure relief valve and a reservoir. As the

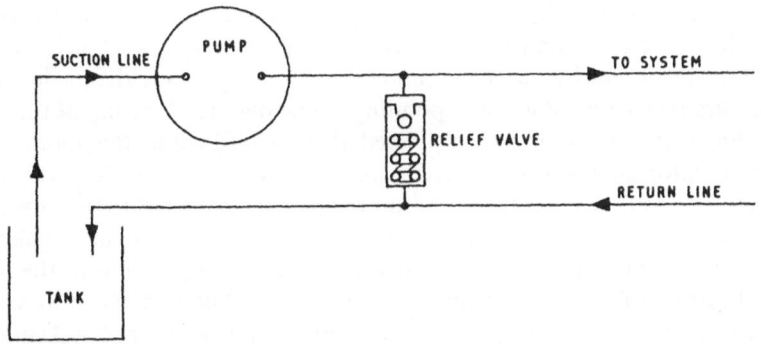

Fig. 2.1 Fixed displacement pump with relief valve.

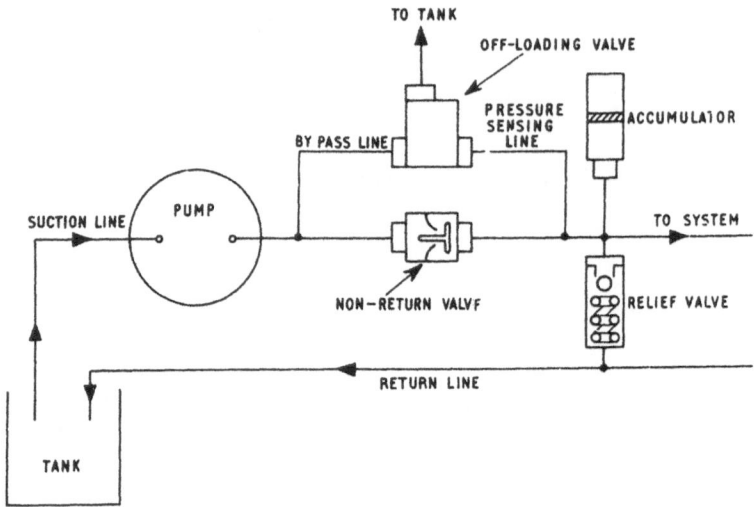

Fig. 2.2 Fixed displacement pump with accumulator and off-loading valve.

pump delivers fluid at a constant rate, the supply pressure is determined by the setting of the pressure relief valve which diverts pump flow in excess of system demand back to tank. The energy dissipated through the relief valve is not recoverable and is converted into heat energy causing a rise of temperature of the fluid in the system, which can be counteracted to some extent by increasing the amount of fluid circulating in the system, although normally, other than for very low power levels, some form of cooling would have to be provided.

A more efficient supply system, employing some additional components, is shown in Fig. 2.2. The additional components are: an accumulator, an off-loading valve and a non-return valve. On start-up the bypass line is closed and, as the pump charges the accumulator, system pressure rises; when the pressure reaches a value corresponding to the high level setting of the off-loading valve the bypass line is opened, thereby off-loading the pump. The accumulator now maintains system pressure with the non-return valve in its closed position. Since the flow required by the system has to be provided by the accumulator, the supply pressure will drop as the accumulator discharges. When the pressure reaches a value corresponding to the low level setting of the off-loading valve, the bypass line is blocked and the pump again supplies the system; the accumulator is re-charged and supply pressure rises. The sequence of operations is then repeated, the relief valve

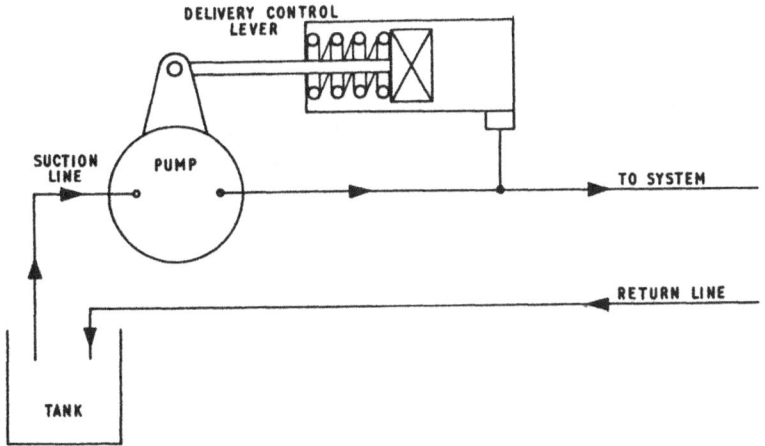

Fig. 2.3 Pressure-compensated variable displacement pump.

acting solely as a safety device and remaining seated under normal operating conditions.

This arrangement is particularly suitable for systems operating for extended periods under idling or low flow conditions. The rating of the pump and size of the accumulator depend on the duty cycle; therefore, if in addition to long idling periods the system is required to operate for extended periods at peak or near peak demand, the pump should be rated at the maximum flow, enabling a small accumulator to be used to maintain supply under idling conditions. Energy dissipation and the consequent cooling problem is considerably reduced by using a power unit of this type rather than the simpler arrangement previously described.

An alternative solution, that overcomes the inefficiency of the first arrangement, is to use a double pump unloader system incorporating a sequence valve to unload the larger pump.

The most efficient hydraulic power source is a variable displacement pump. Figure 2.3 shows a diagrammatic arrangement of a pressure compensated variable displacement pump which supplies fluid to the system at constant pressure and varying flow. In a variable displacement pump the rate of flow is controlled by operating the delivery control lever which in turn varies the displacement of the pump from nominal zero to its maximum rating. Different methods are used to obtain variable displacement, but most designs use either the principle of variable swash-plate angle or provide means for varying the angular position of the cylinder block assembly in relation to the drive-shaft axis.

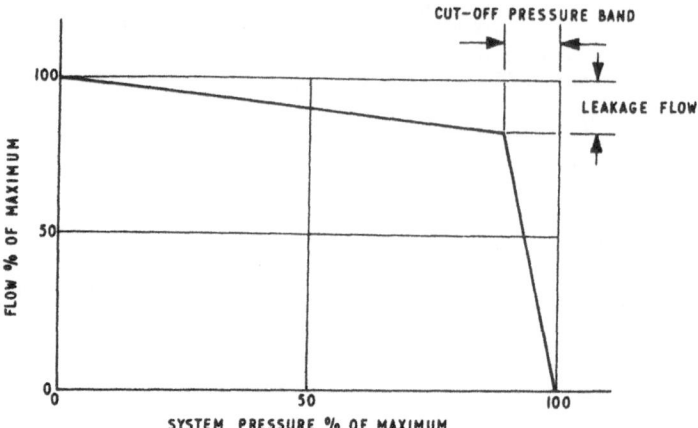

Fig. 2.4 Pressure-flow characteristics of a pressure-compensated variable displacement pump.

The pre-load of the bias spring in the pressure control unit determines the supply pressure. If the pump delivers fluid in excess of that required by the system, supply pressure will momentarily rise, unbalancing the load acting on the piston of the delivery control actuator which will then reduce the displacement of the pump until flow equilibrium is restored. If the delivery control lever overshoots the equilibrium position, supply pressure will drop below the nominal setting, unbalancing the control piston and thereby increasing pump flow until equilibrium is regained. Typical pressure flow

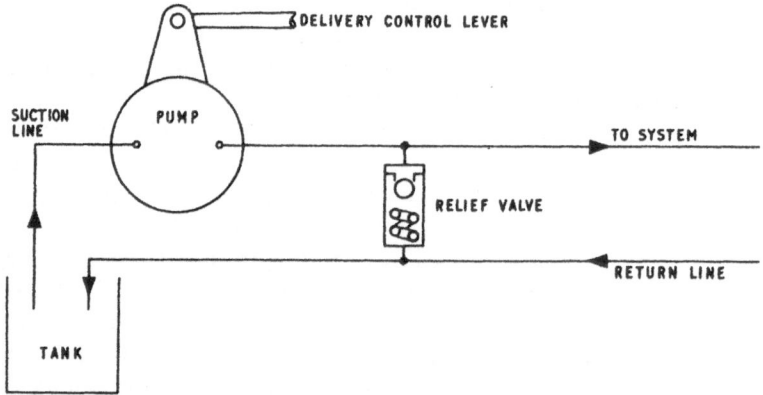

Fig. 2.5 Variable pressure supply system.

characteristics of a pressure-compensated variable delivery pump are shown in Fig. 2.4. Unless the pump is allowed to saturate, working pressure is always within the cut-off pressure band, the design of the pressure control unit determining pressure flow characteristics. Normally supply pressure at maximum flow is 90% or more of the pressure at zero flow.

In hydrostatic transmissions the velocity or position of the actuator is directly controlled by varying the flow of a variable displacement pump. In such a system the supply pressure does not remain constant but varies as a function of the external loading. A schematic arrangement is shown in Fig. 2.5. Under normal operating conditions the relief valve is seated and acts as a pressure limiting safety valve.

3

Working Pressures

Working pressures in hydraulic control systems cover a wide range, from as low as 5 bar to 300 bar or higher. Hydraulic presses are in a special category, sometimes working at pressures as high as 700 bar. The advantage of higher working pressures are reduced size and weight of components, which is a factor of particular importance for aerospace applications. For industrial applications the major consideration is often installed cost, and since the pump is frequently the most expensive single item in a control system its choice can have an important bearing on total system cost. Three distinct types of positive displacement pump are available: gear pumps, vane pumps and piston pumps.

Up to a few years ago pressure ratings for gear and vane pumps were well below those of piston pumps; this is, however, no longer the case. High pressure gear and vane pumps with maximum pressure ratings of 200 to 250 bar are now available. As the majority of hydraulic control systems operate within this range, the determining factor in choosing a pump will often be a compromise between volumetric efficiency and cost. Gear pumps are, as a rule, cheaper than comparably rated piston pumps, vane pumps occupying the middle ground. The high volumetric efficiency of piston pumps cannot normally be matched by either gear or vane pumps.

As the cost of the pump increases with size, higher working pressures at lower flow rates will result in a cheaper unit than a pump of similar power rating operating at lower pressure and higher flow. There are, however, other factors which have to be taken into consideration in choosing a working pressure suitable for a given application.

The pressure range over which control systems give a satisfactory performance is limited, at both ends of the range, by the compressibility of the fluid. At low pressures there is a greater danger of air inclusions in the fluid, which can considerably reduce its effective bulk modulus, causing excessive elasticity in the system.

Let us now examine the effect of working pressure on fluid compressibility. Let the maximum working pressure $= P$ and the volume of fluid under compression $= \delta V$, then the energy stored in the fluid

$$E = \tfrac{1}{2} P \delta V \qquad (3.1)$$

and the useful work done

$$W = PV \qquad (3.2)$$

where V is the volume delivered.

Also the bulk modulus of the fluid

$$N = PV/\delta V \qquad (3.3)$$

Dividing eqn (3.1) by eqn (3.2), $E/W = \tfrac{1}{2}\delta V/V$ and combining with eqn (3.3),

$$E/W = \tfrac{1}{2} P/N \qquad (3.4)$$

Hence the energy stored in the fluid due to its compressibility is directly proportional to the working pressure.

The effect of pressure on the elasticity of the system sets the upper limit to the generally acceptable working pressure in a hydraulic control system, and this, together with the additional hazards introduced by very high pressures, restricts maximum pressure ratings of the majority of practical applications to around 300 bar.

4

Hydraulic Actuators

The function of the actuator in hydraulic control systems is to convert the hydraulic energy supplied by the pump and processed by the control elements into useful work. Actuators have either a linear or rotary output and can be classified into three basic types:

(1) Cylinders or jacks.
(2) Motors.
(3) Rotary actuators.

Rotary actuators, which are essentially non-continuous motors, consist of a cylindrical body to which one or two vanes are rigidly attached. The output shaft carries a moving vane or, in the case of a double-vane actuator, two vanes, the torque output of a double-vane actuator being twice that of a single-vane unit. Maximum angle of rotation is limited to approximately 150° for double-vane and 300° for single-vane actuators. This type of actuator is particularly suitable for applications requiring accurate position control at high output torque, as it obviates the need for reduction gearing, which can introduce elasticity and backlash, thereby adversely affecting system performance.

Hydraulic motors are essentially hydraulic pumps in which the sense of energy conversion has been reversed. While a pump converts mechanical energy supplied to its drive shaft by a prime mover into hydraulic energy, the motor reconverts the hydraulic energy provided by the pump into mechanical energy at its output shaft. Because of this basic similarity they are, except for some minor differences, identical in construction. Hydraulic motors fall into two categories:

(1) High-speed, low-torque.
(2) High-torque, low-speed.

High-speed, low-torque motors can be subdivided into three groups:

(1) Gear motors.
(2) Vane motors.
(3) Axial piston motors.

High-torque, low-speed motors are usually of the radial piston-type construction.

Once the required output power of the motor has been established, it has to be decided whether a high-speed, low-torque or a high-torque, low-speed unit will be the most suitable for the given application. Industrial high-speed, low-torque motors are considerably cheaper than high-torque motors of equivalent power rating; but since the former almost invariably necessitates the provision of reduction gearing, which for closed loop systems has to be of high grade, it is often advantageous to select a direct-drive high-torque unit. System stability and performance considerations will also affect the choice of motor type and capacity.

The most commonly used hydraulic actuator is the cylinder or jack. Cylinders can be either single-acting or double-acting. Single-acting cylinders are power driven in one direction only, whereas double-acting cylinders are power driven in both directions. Cylinders can be constructed as single-ended or double-ended, as shown in Fig. 4.1. Double-ended symmetrical cylinders, Fig. 4.1(a), are frequently used for high performance servo systems, but have greater overall length and are more expensive than single-ended actuators of similar work output. Because of their lower cost

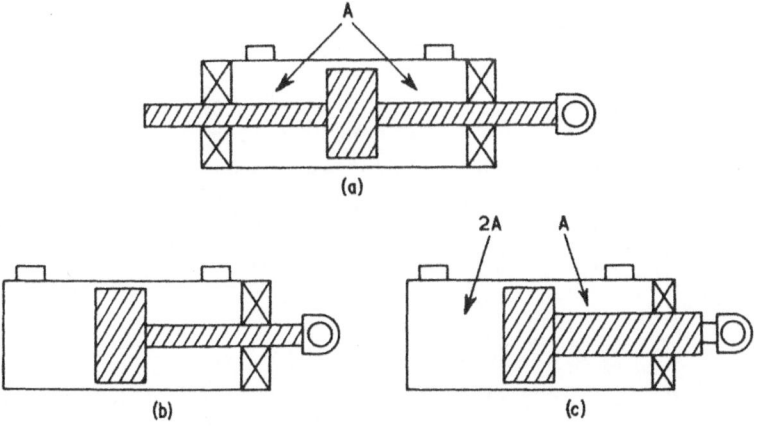

Fig. 4.1   (a) Symmetrical. (b) Single-ended. (c) Differential.

and smaller size, single-ended cylinders are widely used for both industrial and aerospace control system applications. Although single-ended cylinders are asymmetrical, they are perfectly suitable for closed loop systems, provided the effect of the asymmetry on performance and stability is taken into account when carrying out a system analysis.

Standard commercial cylinders are quite adequate for many control applications, but their relatively high seal friction can give rise to slip–stick effects at low speeds, which can be overcome by using cylinders incorporating low-friction seals, e.g. nylon, tufnol, metal piston rings, etc.

5

Control Elements

The function of any hydraulic control device is to control one or more of the following actuator output variables: direction, velocity, acceleration, deceleration, position or force. The control element controls the mechanical output variables by manipulating the hydraulic variables, pressure and flow. The input variables to the control element are usually in the form of mechanical, pneumatic, hydraulic or electrical signals. In modern proportional hydraulic control devices the input variable is most likely to be a low-power electrical analogue or digital signal, the former providing infinitely variable, the latter discrete control of either flow or pressure.

We can therefore reduce the control elements to two basic component groups:

(1) Electro-hydraulic pressure controls.
(2) Electro-hydraulic flow controls.

Since there is little commonality, functional or constructional, between the two groups, it is more practicable to consider each group separately.

5.1 PRESSURE CONTROLS

A cross-sectional drawing of a proportional pressure relief valve is shown in Fig. 5.1. It is essentially a single variable orifice controlled by a proportional solenoid. A magnetic force proportional to the applied input current is generated and reacted by the pressure force. Typical steady-state performance curves are shown in Fig. 5.2. Since the valve is a single stage device, flow rates are limited to around 5 litres/min. To control higher flow

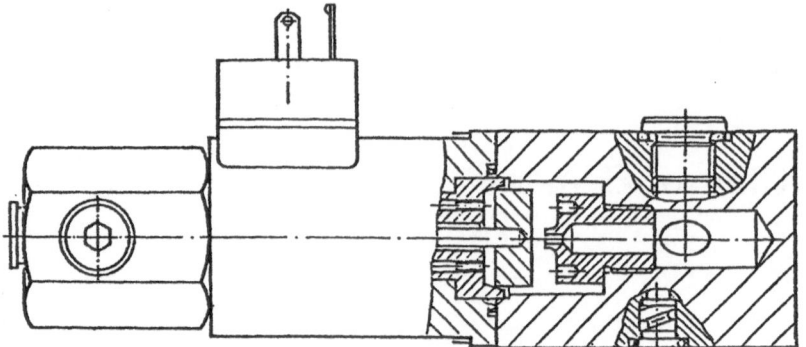

Fig. 5.1 Proportional pressure relief valve.

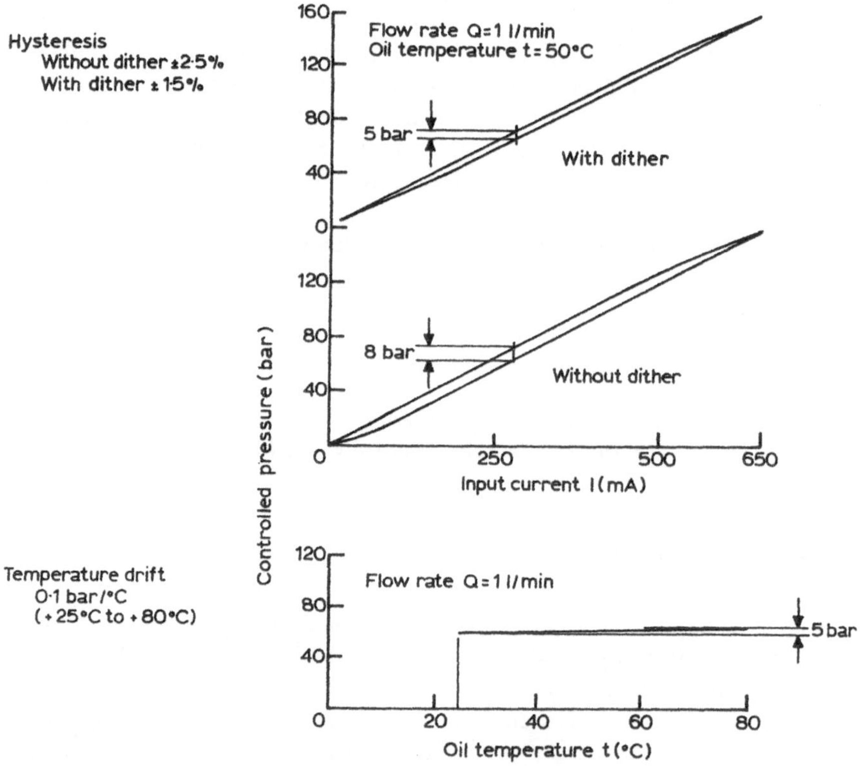

Hysteresis
Without dither ±2·5%
With dither ± 1·5%
Flow rate Q=1 l/min
Oil temperature t=50°C
5 bar
With dither
8 bar
Without dither
Controlled pressure (bar)
Input current I (mA)
160
120
80
40
0
250
500
650
Temperature drift
0·1 bar/°C
(+25°C to +80°C)
Flow rate Q=1 l/min
5 bar
20
60
Oil temperature t (°C)

Fig. 5.2 Steady-state characteristics.

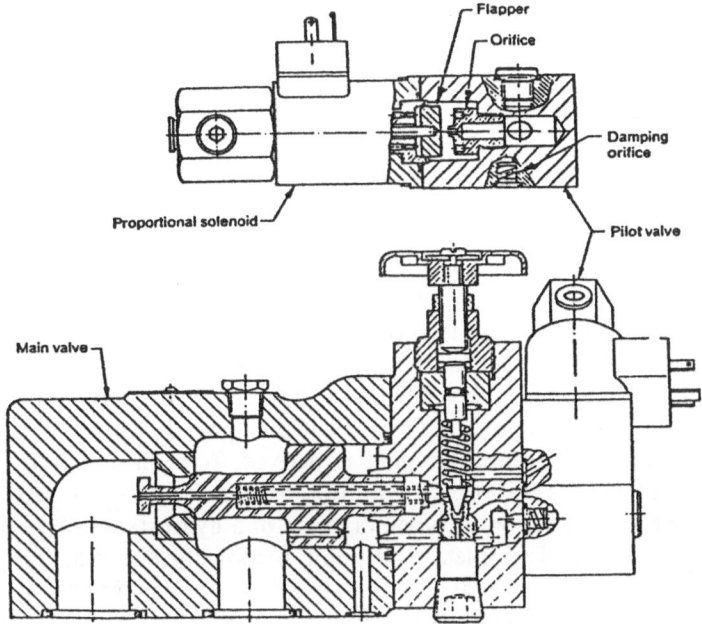

Fig. 5.3 Two-stage proportional pressure relief valve.

rates, the valve can be used as a pilot to control the vent flow of a hydro-cone type relief valve as shown in Fig. 5.3. Typical steady-state performance curves at 100 litres/min flow are shown in Fig. 5.4.

Another type of proportional pressure control is a pressure-reducing valve, facilitating electro-hydraulic proportional control of the reduced pressure.

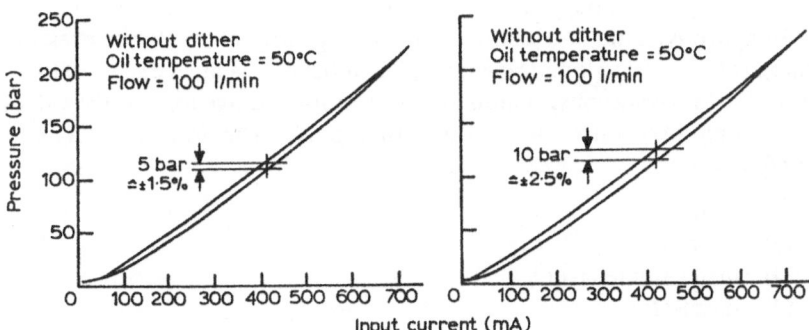

Fig. 5.4 Two-stage valve, steady-state characteristics.

Proportional pressure controls are used to limit the force exerted by an actuator or to produce in-cycle pressure profiles to improve system efficiency.

5.2 FLOW CONTROLS

Flow control can be achieved by one of two quite distinct methods: hydrostatic or hydrodynamic. In a hydrostatic transmission flow is controlled by varying the swash angle of a variable displacement pump or motor. An electro-hydraulic servo valve is often used to perform this function, with positional feedback taken from the pintle. In a velocity control system, a signal derived from an actuator-driven tacho can be used either as an alternative or additional feedback signal. In a position control system, the major feedback signal is derived from an actuator-driven positional feedback transducer, with the swash angle feedback providing the minor feedback loop. A block diagram of a hydrostatic transmission employing a variable displacement pump is shown in Fig. 5.5.

For unidirectional velocity control systems, a one side of centre swash angle control is sufficient, whilst a bidirectional velocity control system requires an additional directional valve or an over-centre swash angle control. A hydrostatic position control system invariably requires an over-centre swash angle control.

In a hydrostatic transmission where the control element is a variable displacement motor, the directly controlled variable is output torque, which can be utilized to achieve motion control by installing the appropriate feedback transducer.

The majority of variable displacement pumps and motors are of the vane or piston type, although units of different design are on the market.

In hydrodynamic flow control a throttling valve converts the pressure energy at the inlet to kinetic energy at the outlet port. Valves can be of two-, three- or four-way construction, the conventional usage to describe valves containing two, three or four external ports. The four basic circuit configurations:

(1) meter-in
(2) meter-out
(3) meter-in/meter-out
(4) bleed-off

are shown in Figs 5.6 to 5.9. Since most actuators require velocity control in

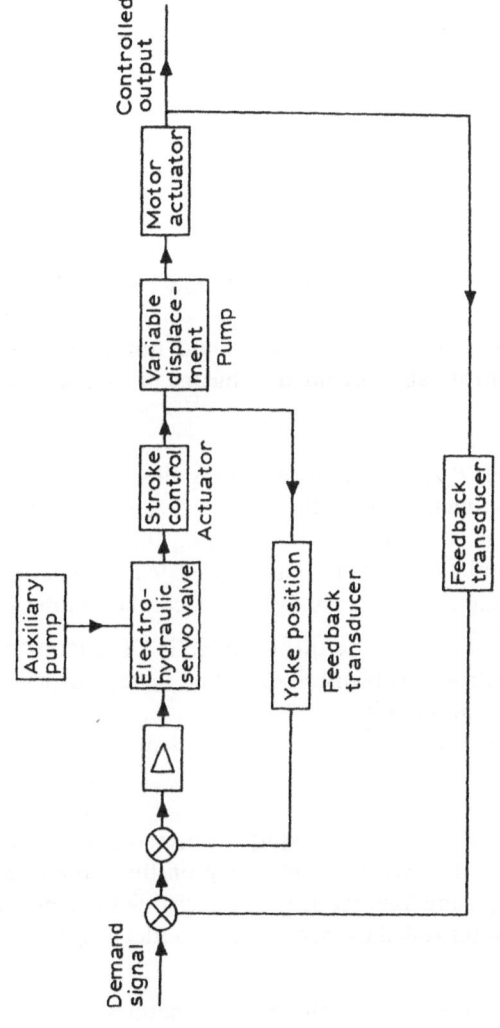

Fig. 5.5 Variable delivery pump block diagram.

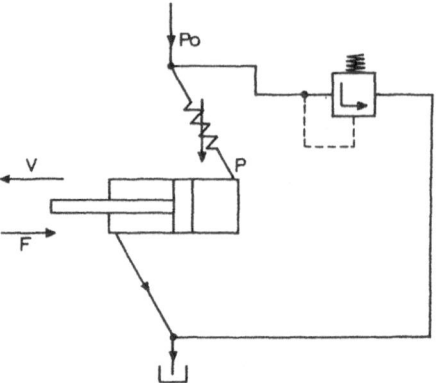

Fig. 5.6   Meter-in flow control.

both directions, the above circuits would normally include additional directional control valves or integral means of directional control.

Meter-in

Supply pressure, P_0, is pre-set by adjusting the pressure relief valve. Load pressure, P, is a function of the external load, F, acting on the actuator: $P = F/A$. Velocity, v, is controlled by throttling the inlet flow $Q = vA$ to the cylinder. The outlet flow from the actuator is fed straight back to tank. The salient features of the above configuration can be summarized as follows:

(1)   Energy wasted by the throttling action results in lower efficiency than hydrostatic transmission, but better efficiency than meter-in/meter-out circuit.
(2)   Only opposing loads can be controlled.

Meter-out

Supply pressure, P_0, is set by adjusting the pressure relief valve. Pressure, P, is determined by the external load acting on the actuator and the supply pressure, an opposing load reducing, an assisting load increasing pressure, P. Velocity, v, is controlled by throttling the outlet flow from the actuator to tank.

(1)   Efficiency is similar to the meter-in configuration.
(2)   Both opposing and assisting loads can be controlled.
(3)   Care has to be taken to avoid excessive pressure build-up due to pressure intensification when encountering overrunning loads with asymmetrical cylinders.

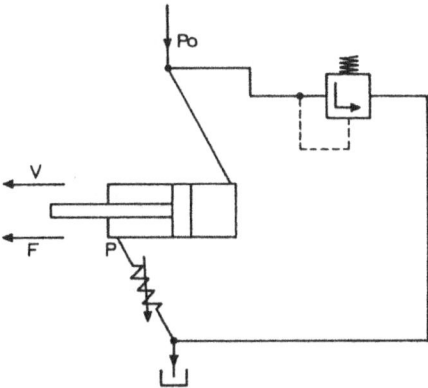

Fig. 5.7 Meter-out flow control.

Meter-in/meter-out

Supply pressure, P_0, is again pre-set by adjusting the pressure relief valve. Velocity, v, is controlled by throttling both the inlet to and the outlet from the actuator. Cylinder chamber pressures are a function of supply pressure setting and external loading of the actuator.

(1) Because of the double throttling action, this is the least efficient configuration.
(2) Both opposing and assisting loads can be controlled.
(3) Trapped oil volumes in both inlet and outlet cylinder chambers improves stiffness by reducing effects of oil compressibility on system stability and performance.

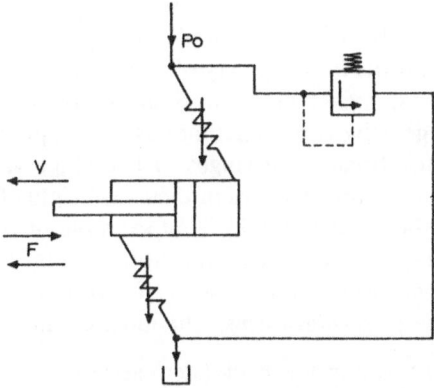

Fig. 5.8 Meter-in/meter-out flow control.

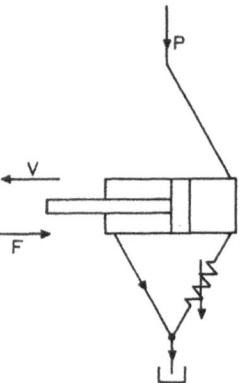

Fig. 5.9 Bleed-off flow control.

Bleed-off
Other than for safety reasons a pressure relief valve is not required, as operating pressure, P, is determined by the external loading of the actuator. Velocity, v, is controlled by bleeding off flow from the fixed displacement pump, which in this arrangement acts as a source of constant flow. The difference between pump flow and throttled flow is the effective flow fed to the actuator.

(1)   This is the most efficient circuit, since operating pressure, P, is directly related to the external load.
(2)   Only opposing loads can be controlled.
(3)   Particularly attractive for applications requiring relatively constant velocity at varying actuator loading, e.g. conveyors, lifts.

Control valves can be of either single- or multi-stage construction. Bernoulli forces limit the flow capacity of single stage valves to around 10 to 15 litres/min, most higher flow valves are of two-stage construction, although three-stage valves are sometimes used for applications requiring very high flow rates. Some pilot stages of two-stage valves can also be employed as low-flow control valves in their own right. The pilot stage of the control valve shown in Fig. 5.11 is an example of this. Some typical examples of two-stage control valves are shown in Figs 5.10 to 5.14. Most two-stage control valves incorporate some sort of internal feedback loop, which can take one of several forms. The most common are:

(1)   Mechanical position feedback (stem servo).
(2)   Electrical position feedback.

(3)   Force feedback.
(4)   Flow feedback.

An example of construction (1) is shown in Fig. 5.10. The three-way pilot valve is housed co-axially inside the main spool, thereby providing positional feedback between the two stages. The main spool is controlled by pilot control pressure acting on an annular control chamber at one end of the spool counteracted by supply pressure acting on a bias piston of half the control chamber area at the opposite end of the spool. The pilot valve is operated by a torque or linear force motor.

An example of construction (2) is shown in Fig. 5.11. In this particular design electrical positional feedback is taken from both the pilot and main stage; in alternative designs the pilot stage feedback is omitted. The pilot valve is operated by a single solenoid acting against an opposing spring; alternatively two opposing solenoids can be used, only one being activated at a time. In the construction shown, a pressure reducing module is interposed between the two stages. This provides constant reduced pilot supply pressure, irrespective of main supply pressure variations, providing improved valve performance characteristics.

An example of construction (3) is shown in Fig. 5.12. Oil is supplied at full system pressure to a pair of fixed orifices which discharge into two control pressure chambers at either end of the main spool. Oil from these two chambers is allowed to spill to tank via the two flapper-controlled variable orifices. The end covers at either end of the body are directly vented to tank return. A current passed through the force motor coils will tend to

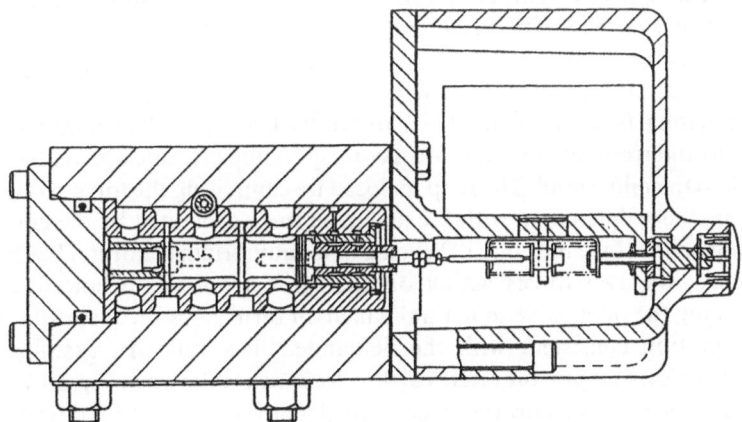

Fig. 5.10   Mechanical position feedback two-stage valve.

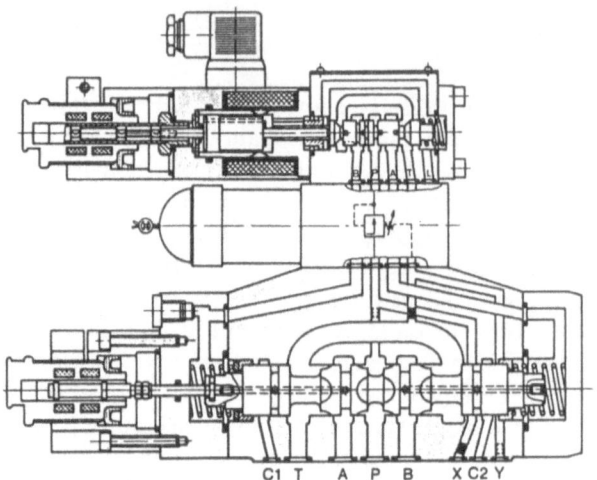

C1 T   A P B   X C2 Y

Fig. 5.11   Electrical position feedback two-stage valve.

unbalance the flapper which in turn will set up a differential pressure in the control chambers, displacing the main spool in the opposite direction to that of flapper movement. The feedback springs housed inside the main spool convert spool displacement into a proportional force acting on the flapper and force motor armature, thus balancing the magnetic force produced by the input current. When the new equilibrium condition has been reached, the flapper will be re-centred between the two nozzles. A current will therefore produce a proportional spool displacement.

An example of construction (4), employing hydraulic flow feedback, is shown in Fig. 5.13. The valve consists of three basic elements: a main control stage comprising a four-way meter-in/meter-out spool, a force motor operated pilot valve and a flow sensing feedback element. The bidirectional flow sensor situated in one of the two service lines converts the flow to and from the actuator into a corresponding pressure differential fed back to the pilot spool. The loop is closed by comparing the force set up by this pressure differential with the magnetic force generated by the current acting on the force motor coil. When a new equilibrium condition has been reached, the two forces acting on the pilot spool are balanced, thus returning the pilot valve to neutral which, in turn, locks the main spool in the position consistent with the demanded flow rate. An example of construction 4 employing electrical flow feedback is shown in Fig. 5.14. The valve incorporates a flow transducer which converts the output flow into a corresponding electrical feedback signal, which is fed to a summing

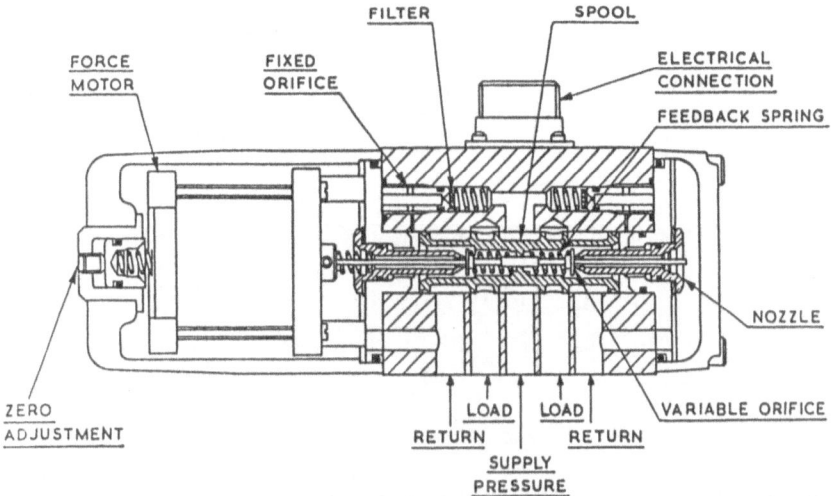

Fig. 5.12   Force feedback two-stage valve.

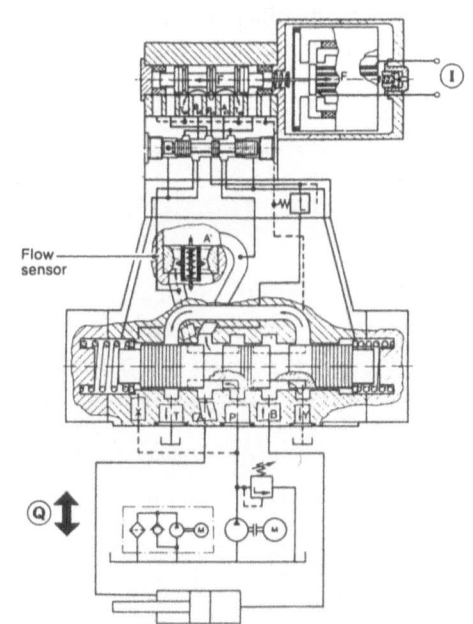

Fig. 5.13   Hydraulic flow feedback two-stage valve.

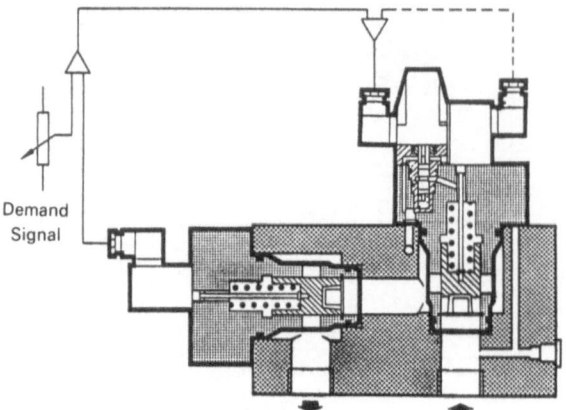

Fig. 5.14   Electrical flow feedback two-stage valve.

junction to provide an error voltage signal to a drive amplifier controlling the solenoid operating the pilot stage, thus closing the loop.

The effect of the choice of the control element on overall system performance, together with a clarification of the current terminology, including an assessment of the differences and similarities of proportional and servo valves, will be highlighted in Chapter 7, 'The Control System'.

6

Data Transmission Elements

The function of the data transmission element is to sense the controlled output quantity and to convert it to a signal which can be used to either monitor the output or to act as a feedback device in a closed loop control system. The controlled output variable in a hydraulically operated force–motion control system can be:

(1) Force.
(2) Velocity.
(3) Position.
(4) Acceleration/deceleration.
(5) Pressure.
(6) Flow.

Although the output signal produced by the data transmission element can take many forms, the most common type used in modern control applications will be a transducer creating an electrical signal, usually at a low power level. Transducers measuring any of the six listed output variables are commercially available.

1 Force
Force transducers are termed load cells. They directly sense the load exerted by the actuator or a load-bearing member of the system. Most commercially available load cells are designed to measure loading acting in one direction only.

2 Velocity
Velocity transducers are usually referred to as tachos. the most readily available having a rotary input which is converted into an AC or DC

output voltage directly proportional to input spindle velocity. DC tachos are more suitable for closed loop control applications. Tachos can have a wide operating range but have a threshold which limits the minimum running speed. In applications where linear motion has to be sensed, some form of rack and pinion gearing or cable drive has to be employed, and care has to be taken that backlash is eliminated, since the presence of backlash can have an adverse effect on system stability and performance.

3. Position

Position transducers can have a linear or rotary input and either an AC or DC output signal. The output signal can be a voltage or current, usually at low power level. DC position transducers, or potentiometers, are variable resistances of wire-wound or plastic film construction. The resolution of wire-wound potentiometers is limited by the pitch of the windings, whereas plastic film potentiometers have infinite resolution. The resistance value of the potentiometer has a bearing on the linearity of the input–output characteristics; $1\,k\Omega$ is a fairly typical value. Since potentiometers are contacting devices, their life and reliability are limited, particularly if they are used as feedback sensors. It is common practice to use potentiometers as reference or demand signal generators and non-contacting devices as feedback sensors.

Non-contacting transducers can produce either an analogue or a digital output signal, analogue devices providing an infintely variable output related to an absolute datum, digital devices providing a discrete incremental output without an absolute datum. Typical examples of non-contacting analogue transducers are LVDTs, RVDTs, synchros and resolvers; examples of digital transducers are encoders and proximity sensors.

Inductive devices producing an AC output signal are termed LVDTs, i.e. linear variable displacement transducers or RVDTs, i.e. rotary variable displacement transducers. Some typical circuits are shown in Fig. 6.1. Packaged units incorporating integrated oscillator-demodulator circuitry, thus providing a DC output, are readily available.

Synchros and resolvers are rotary AC transducers generating a sinusoidal ouput. An example of a circuit employing two synchros, one acting as a demand transducer, the other as a feedback element, is shown in Fig. 6.2.

Non-contacting transducers have a virtually unlimited life and infinite resolution, but their linearity is usually inferior to that of potentiometers.

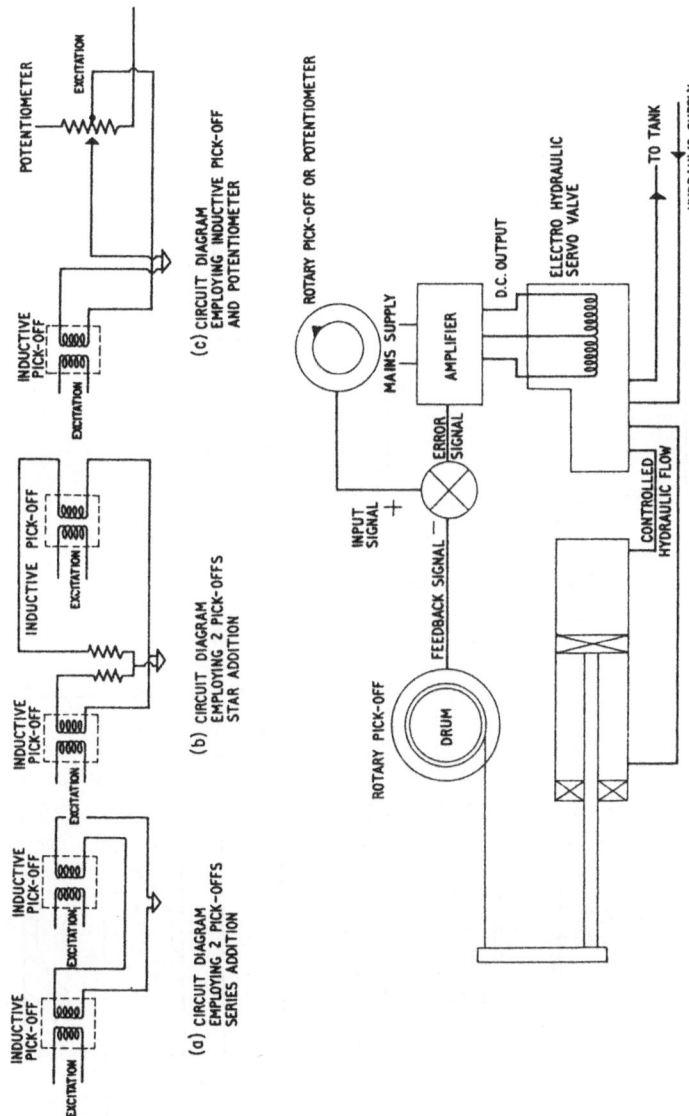

Fig. 6.1 Positional control system using LVDT/RVDT.

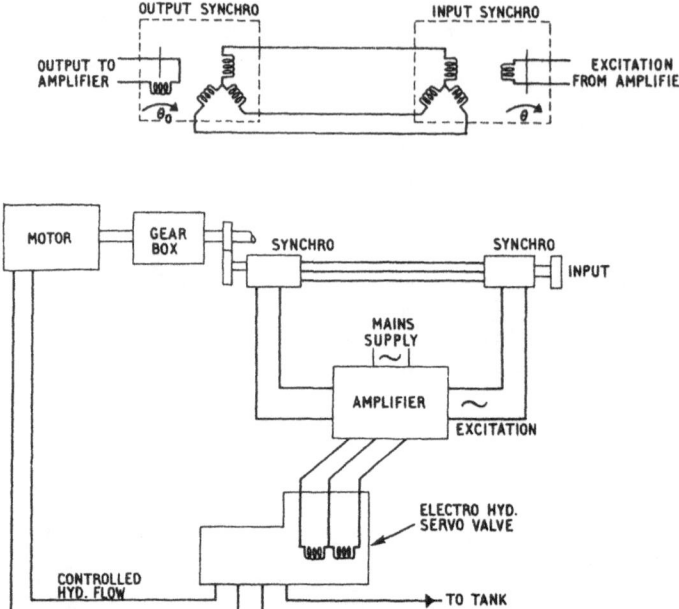

Fig. 6.2   Positional control system using synchros.

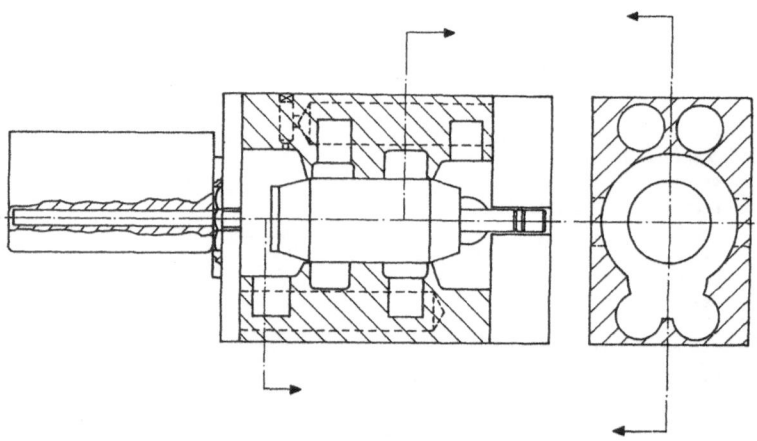

Fig. 6.3 Bidirectional flow transducer.

4. Acceleration

Although acceleration sensors are on the market, they are not widely employed in hydraulic control systems. Since acceleration is the first time derivative of velocity and the second time derivative of displacement, both acceleration and deceleration are easily handled in motion control systems by controlling the rate of change of velocity. Various methods of accomplishing this will be discussed in subsequent chapters.

5. Pressure

Pressure transducers fall into two distinct categories, i.e. single and differential pressure versions. In a single pressure transducer, pressure is sensed by means of a diaphragm, bellows or spring arrangement and the resulting displacement converted to an electrical signal by using a position transducer, giving a linear relationship between pressure and output signal.

Differential pressure transducers measure the pressure difference at two different pressure tappings and are required to provide a signal proportional to pressure differential unaffected by base pressure variations. Construction of differential pressure transducers is similar to that of single pressure units.

6. Flow

Three types of flow transducers can be used: positive displacement units, piezo-electric venturi meters, and variable resistance flow meters. The most suitable for hydraulic control applications are positive displacement units and variable resistance flow meters. The former are essentially hydraulic motors using either tachos or rotary position transducers, while the latter make use of the principle of hydro-kinetic flow to convert flow to pressure which in turn is converted into a corresponding displacement and hence an electrical output signal. An example of such a device is shown in Fig. 6.3.

Another function of the data transmission elements is to generate and process the input signal. In this sense they constitute the element generically described as the controller. Various types of controllers will be discussed in the following chapter.

7

The Control System

A block diagram of a hydrostatic transmission described in Chapter 5 was shown in Fig. 5.5. A block diagram of a valve operated control system is shown in Fig. 7.1. The major feature distinguishing the two systems is the part played by the pump. In the hydrostatic transmission the pump performs the dual role of power source and control element, whereas in the valve-operated system it acts as the power source to the control element. Both systems configurations can be represented by a single block diagram, Fig. 7.2. Since in the majority of cases the controller will produce an electrical low-power signal, an additional element is required to process this signal in order to provide a signal compatible with the control element input characteristics. This element is designated as the electro-hydraulic interface in the block diagrams, Fig. 7.2.

The manipulated variable, i.e. the output from the control element, would normally be pressure for force control and flow for motion control applications. In either case, open or closed loop control can be used. In defining closed loop control we have to differentiate between major and minor feedback loops, bearing in mind that most control systems contain a combination of feedback loops. It is impracticable to rigidly classify feedback loops, but feedback emanating from the actuator, as shown in Fig. 7.2, would usually be the major loop, while spool feedback as shown in Fig. 5.11, if used in a closed loop system, would be a minor loop. By treating the major feedback loops included in the three block diagrams as optional features, Fig. 7.2 can be taken as basically representing all types of control system. As a rule a major feedback loop will enhance system performance, but care has to be taken to avoid instability. Accurate and repeatable position control can only be achieved by employing a positional feedback element.

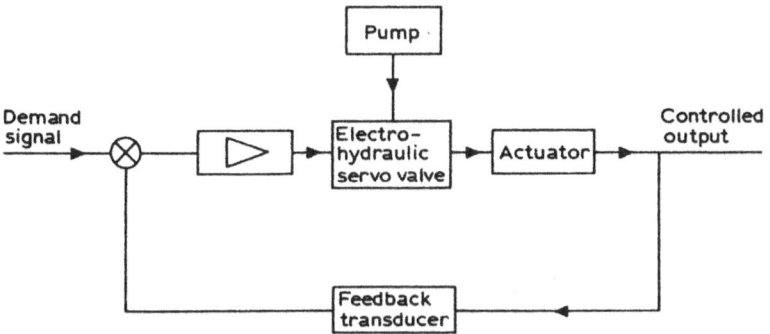

Fig. 7.1 Valve-operated control system block diagram.

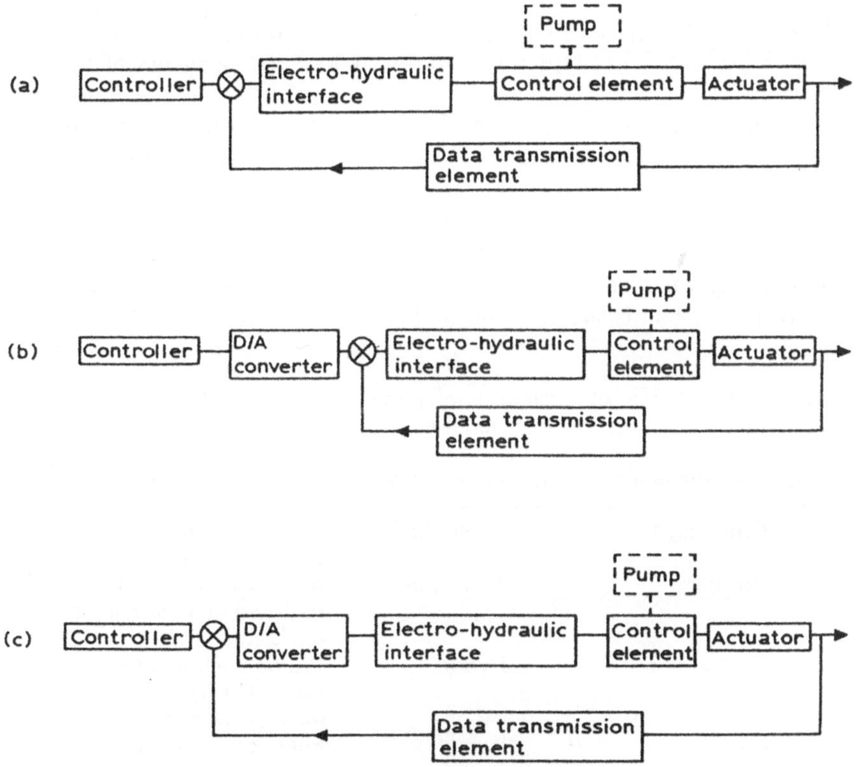

(a)

(b)

(c)

Fig. 7.2 System block diagrams.

7.1 THE CONTROLLER

Controllers cover a wide range, from human operators to highly sophisticated computer or microprocessor controls. The control requirements of the system can be reduced to four fundamental groupings:

(1) Manually controlled or pre-set demand.
(2) Manually controlled or pre-set demand with automatic correction.
(3) Discrete in-cycle demand changes.
(4) In-cycle contour changes.

Control systems fall into three basic classes:

(1) Analogue.
(2) Digital.
(3) Hybrid, i.e. a combination of (1) and (2).

Block diagram Fig. 7.2(a) represents a pure analogue or pure digital system, Figs 7.2(b) and 7.2(c) represent two alternative arrangements of hybrid systems.

Control requirements (1) and (2) can be accomplished by analogue, control requirements (3) and (4) by digital or hybrid control. Pure digital control systems require digital control elements. Although digital electro-hydraulic control devices are on the market, they are currently not widely used. Stepper-motor-driven hydraulic valves were popular for a period, particularly for machine tool drives. A typical arrangement of a pulse controlled system is shown in block diagram Fig. 7.3. Table 7.1 links the

Table 7.1 Choice of system arrangement and controller for different requirements

Control requirements	System arrangement	Controller
Grouping 1	Analogue	Potentiometer, synchro, pre-set resistor
Grouping 2	Analogue	As above plus level detector, gyro compass, inertial navigator
Grouping 3	Hybrid	Programmable logic controller (PLC), microprocessor
Grouping 4	Hybrid	Function generator, microprocessor, computer

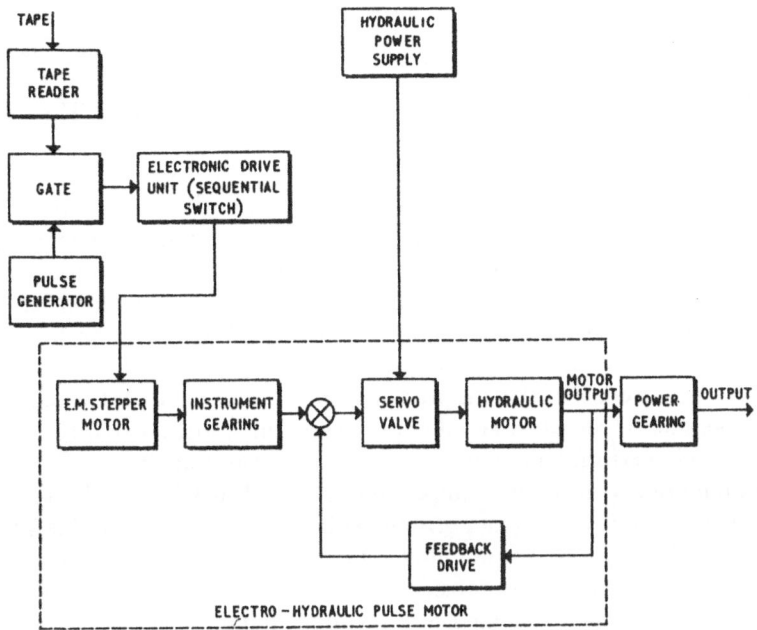

Fig. 7.3 Pulse-controlled system block diagram.

control requirements to the most likely system arrangement and type of controller.

7.2 THE CLOSED LOOP OPTION

Before a decision on whether to close the loop can be taken, a number of questions have to be addressed:

(1) Effect on system performance.
(2) Installed cost.
(3) Life and reliability.

The answers to these questions will be greatly influenced by the nature of the controlled output variable. Referred to the actuator, three alternative outputs can arise: (1) force or loading, (2) velocity, (3) position. An important point to remember is that only one variable can be controlled at any one time, although other variables can act as overrides which can either inhibit or replace the original output.

Normally a clear distinction can be drawn between force and motion control systems, the former being controlled by a proportional electro-hydraulic pressure control valve of the type described in Section 5.1, the latter by one of the flow control elements covered in Section 5.2. There are, however, exceptions to this rule, where pressure control devices are used to control motion, and flow control devices to control force.

The main justification for closing the loop in force control systems is the presence of frictional forces such as seal friction in cylinders. An open loop pressure-controlled system cannot differentiate between external and frictional forces, whereas in a closed loop system the external force can be identified and fed back by means of a load cell. Closing the loop by employing a pressure transducer in addition to the inherent pressure feedback loop contained in the pressure control valve has to be treated with caution, since it can cause stability problems and, particularly with regard to dynamic performance, could even be counter-productive.

For motion control it is important to establish whether the critical output variable is velocity or position. Although on the face of it this might appear to be obvious, it is not always the case. Two typical examples will illustrate this. Let us first consider an application requiring the motion control of a cylinder-operated reciprocating sliding table. In its simplest form the system could comprise a proportional flow control valve controlled by a potentiometer or pre-set resistor. The demand voltage determines the output flow from the valve and hence cylinder velocity. The distance travelled by the table is specified as one unit length and the

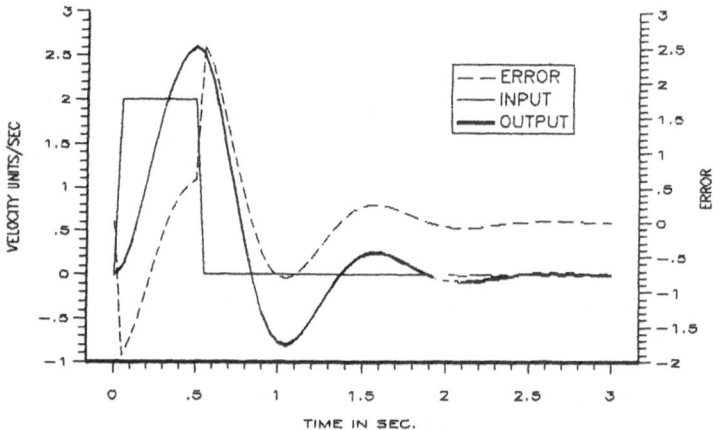

Fig. 7.4 Velocity profile open loop control system.

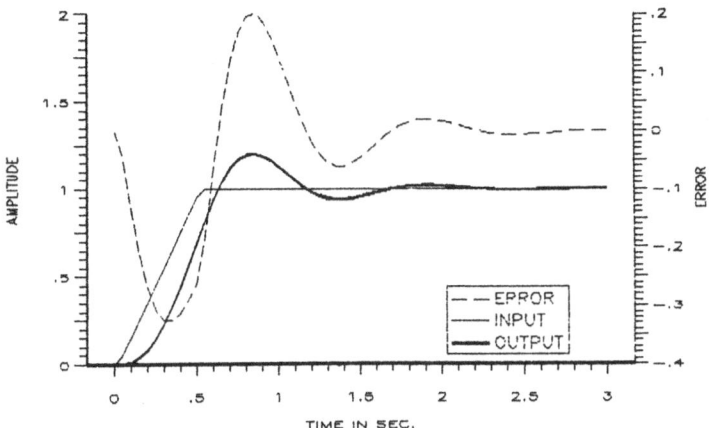

Fig. 7.5 Displacement profile open loop control system.

dynamic performance of the system is equivalent to a second order system of 1 Hz natural frequency and 0·35 damping factor. Let us initially select a velocity/time profile as shown in Fig. 7.4. A step demand corresponding to a velocity of two length units per second is maintained over a period of half a second. Cylinder velocity will over- and undershoot until the table comes to rest after approximately two seconds. The corresponding displacement profile can be obtained by calculating the time integral of the velocity profile, as shown in Fig. 7.5. It can be seen that after one over- and undershoot the table settles at the specified displacement of one unit length at zero error. This would indicate that accurate positioning can be achieved by velocity control; in fact this is far from being the case. The reason for this is that we have been analysing an ideal system, which presupposes a fixed steady-state relationship between demand and output velocity. In practical systems the steady-state input/output characteristics are subject to errors due to several causes:

(1) Valve output flow affected by pressure and temperature variations.
(2) Actuator loading.
(3) Seal friction.
(4) Actuator leakage.
(5) Performance scatter of components due to production tolerances (effect of scatter can be eliminated by accurate calibration).
(6) Valve hysteresis.

We can demonstrate the effect of steady-state errors by introducing a 10% velocity deviation to our ideal system. The resulting velocity profile is

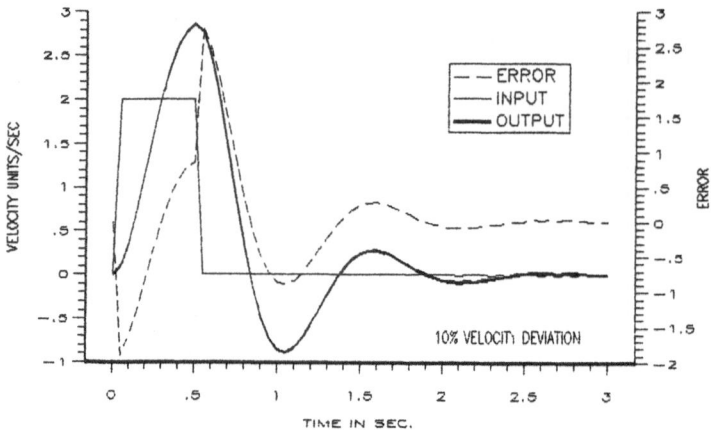

Fig. 7.6   Velocity profile open loop control system with velocity deviation.

shown in Fig. 7.6, and the corresponding displacement profile in Fig. 7.7. It can be seen from Fig. 7.6 that, apart from altering the amount of over- and undershoot, system response is almost identical to that of the ideal system, but that a 10% displacement error has now been introduced, as shown in Fig. 7.7. So far we have only investigated unidirectional motion. By extending the duty cycle to include a complete reciprocating table movement, as shown in Figs 7.8 and 7.9, we can see that a velocity deviation

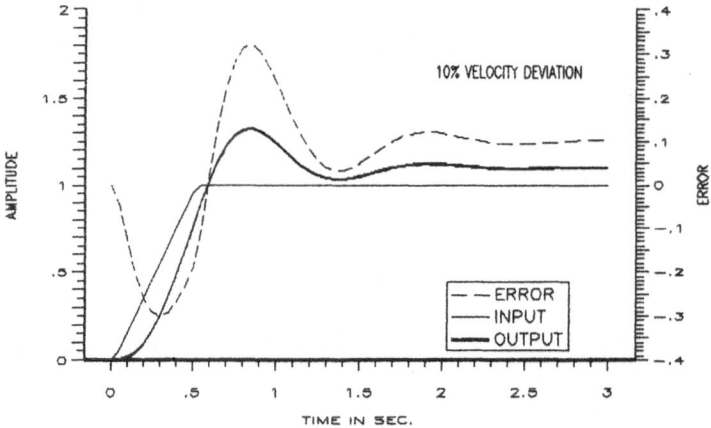

Fig. 7.7   Displacement profile open loop control system with velocity deviation.

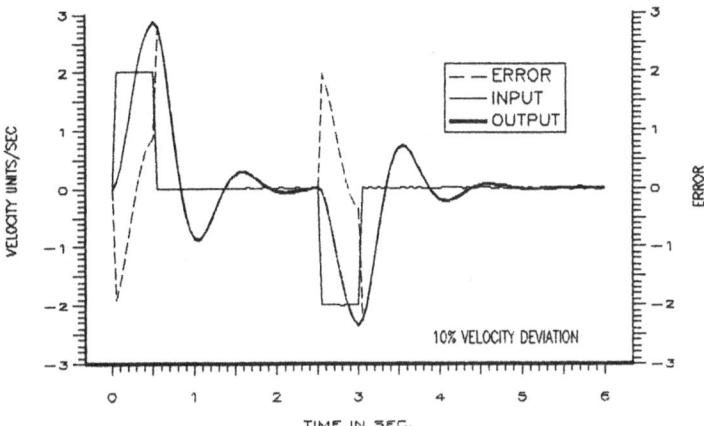

Fig. 7.8 Reversing velocity profile open loop control system with velocity deviation.

introduces a cumulative displacement error to an open loop system. The only practical way to eliminate these errors and to achieve accurate and repeatable position control is to close the loop by utilizing an output-driven position feedback element. An alternative approach which is suitable for systems where absolute positioning accuracy is less critical is to reduce positional overshoots by changing the velocity profile and to eliminate cumulative errors by introducing output-activated limit switches.

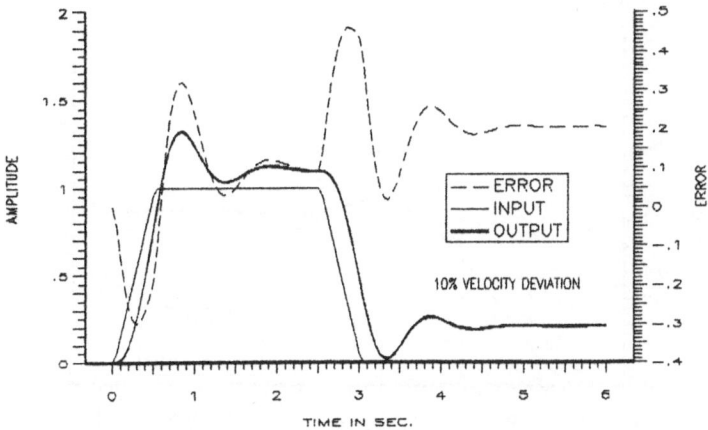

Fig. 7.9 Reversing displacement profile open loop control system with velocity deviation.

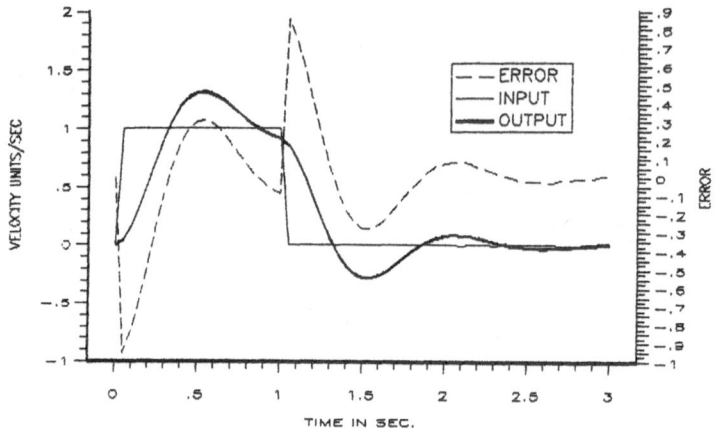

Fig. 7.10 Velocity profile open loop control system.

In Fig. 7.10 the velocity profile of our ideal system, Fig. 7.4, has been changed to a reduced velocity step demand of one length unit per second maintained over an increased period of one second. This has reduced velocity over- and undershoots and decreased position overshoot from 20% to 7%, as shown in Fig. 7.11. Similar results can be obtained by ramping the velocity demand as shown in the velocity profile, Fig. 7.12, and the corresponding displacement profile, Fig. 7.13. Many amplifier drive

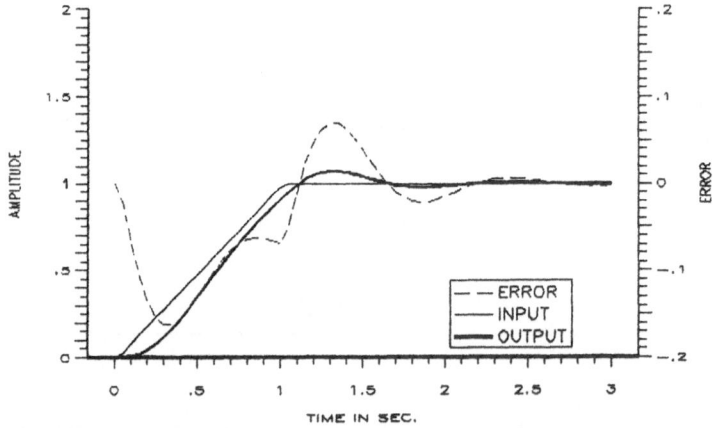

Fig. 7.11 Displacement profile open loop control system.

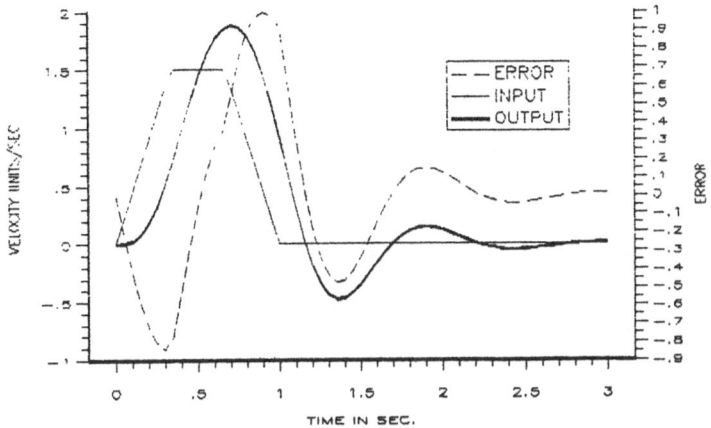

Fig. 7.12 Ramped velocity profile open loop control system.

cards for proportional control valves incorporate means for selecting ramps, either from a pre-set integral potentiometer or from an external signal.

For our second example we will examine an application requiring the synchronization of two cylinders. For our first attempt we will use an open loop control system identical to that of our first example, but this time duplicated to give us individual control of each cylinder. Obviously an ideal

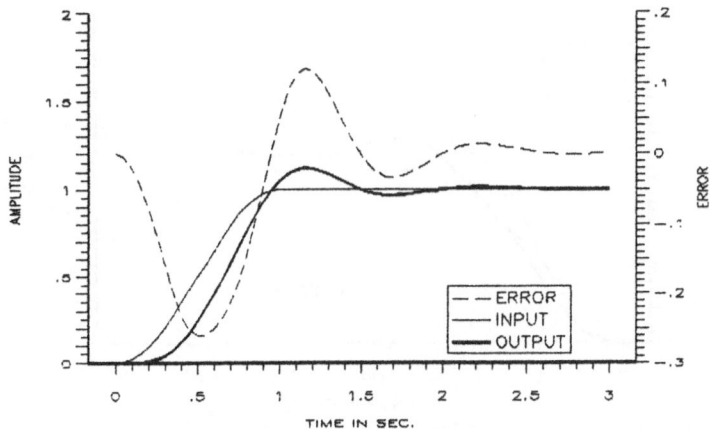

Fig. 7.13 Ramped displacement profile open loop control system.

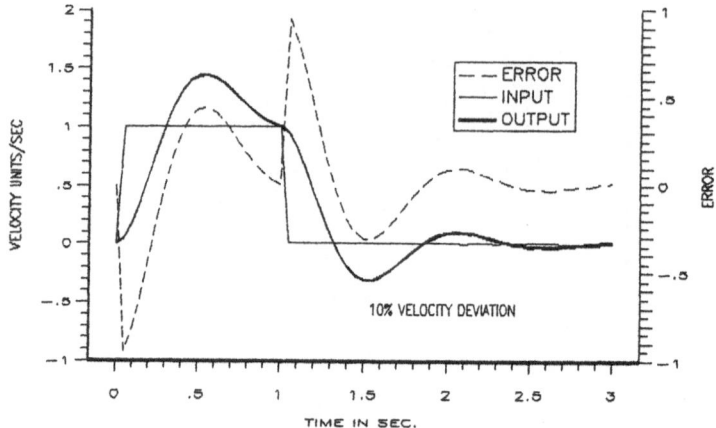

Fig. 7.14 Velocity profile open loop control system with velocity deviation.

system as depicted in Figs 7.10 and 7.11 will present us with no problems, since both cylinders will follow an identical path, as shown in Fig. 7.11. If, however, we again consider an actual system with a 10% velocity deviation, the velocity and displacement profiles will alter to those shown in Figs 7.14 and 7.15. The effect of a 10% velocity deviation between the two control systems during one reciprocating cycle is to again introduce a cumulative error, as shown in Fig. 7.16.

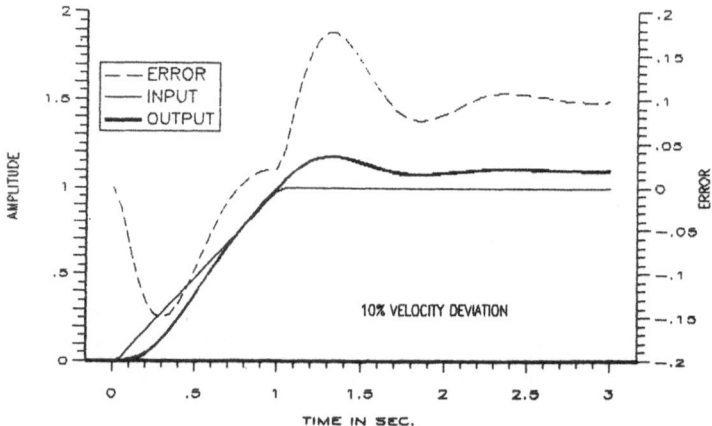

Fig. 7.15 Displacement profile open loop control system with velocity deviation.

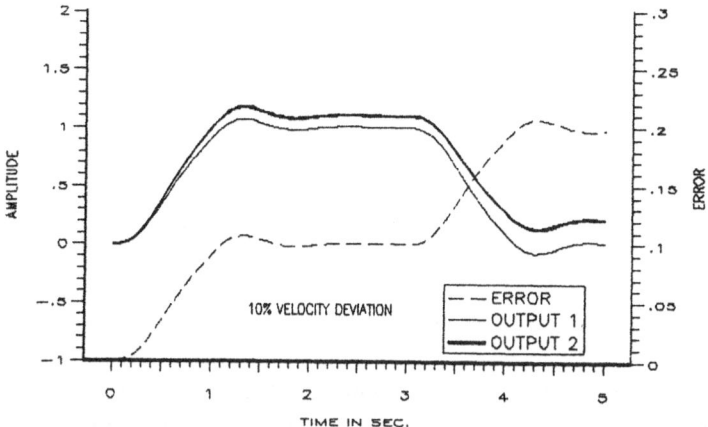

Fig. 7.16 Displacement profile dual open loop control system.

We can therefore conclude that effective synchronization of two or more actuators can only be achieved by closed loop position control.

8

Control Concepts

The majority of closed loop systems are liable to become unstable or inadequately damped if the critical parameters are not maintained within suitable limits. One of the major objectives of a theoretical system investigation is to determine parameters compatible with adequate stability.

Two distinct methods can be used in the theoretical assessment of control systems: analysis or synthesis. The essential difference between these two methods can best be illustrated by expressing the design specification in terms of three sets of requirements, as shown in the two sequence diagrams Figs 8.1 and 8.2.

The three sets of requirements which make up the complete system specification are:

(1) General performance requirements, e.g. loads, velocities, power ratings.
(2) Steady-state performance, e.g. static positional accuracy, repeatability, velocity errors, hysteresis, load errors.
(3) Dynamic characteristics, e.g. transient response to step input or given duty cycle, resonant frequency and system damping, frequency bandwidth.

Referring to Fig. 8.1, the first step is the selection of components complying with the general performance requirements laid down in the design specification. The system is then rationalized and presented in a form suitable for a theoretical assessment. By applying suitable stability criteria, all relevant parameters are then established. Finally a steady-state and dynamic performance analysis is prepared and compared with the requirements of the design specification. If the calculated results are not within limits of the specification, the sequence of operations has to be

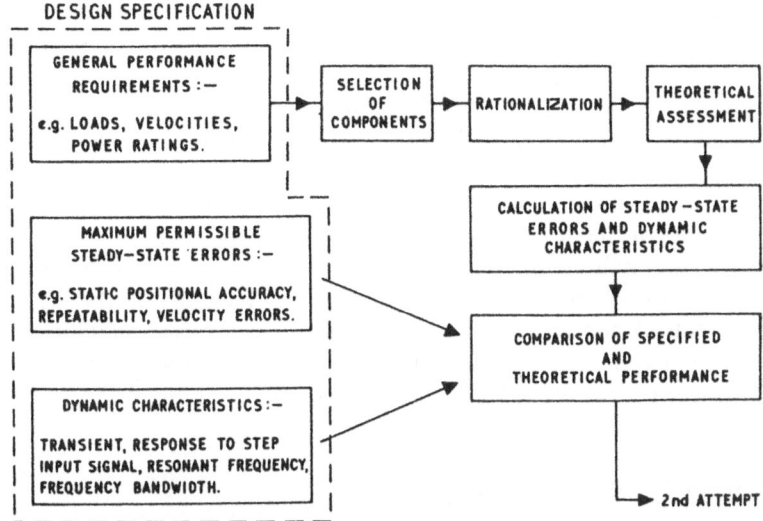

Fig. 8.1 System analysis sequence diagram.

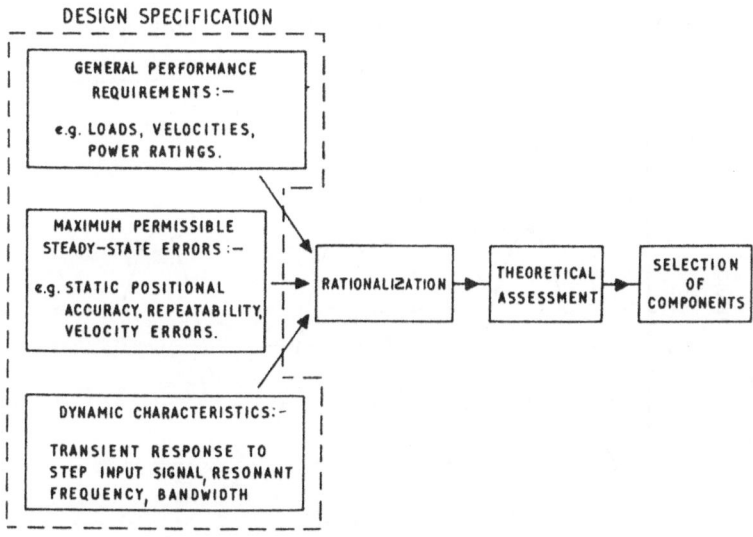

Fig. 8.2 System synthesis sequence diagram.

repeated. Although this is essentially a trial-and-error solution, it is not always practicable or possible to adopt a more rational approach. At first sight the analysis method appears to be long-winded and time-consuming; the use of computer programs does, however, reduce the problem to manageable proportions. This will become apparent in subsequent chapters.

In the more direct synthesis method, as shown in Fig. 8.2, all the requirements of the design specification are taken into account when the system is rationalized in preparation for the analytical assessment, in practice, however, it is often not possible to feed in all the design requirements. A compromise solution is sometimes adopted in which the mandatory requirements are used as a basis for synthesis, all other performance parameters being determined by the analysis method.

8.1 DEFINITION OF TERMS

The principles of automatic feedback control, which are applicable to all control systems, are extensively covered in a number of books, technical papers and articles in technical journals; readers are referred to publications on general control theory listed in the Bibliography. For the benefit of those readers not familiar with the subject, a brief summary of the more important concepts and terms generally used will now be given.

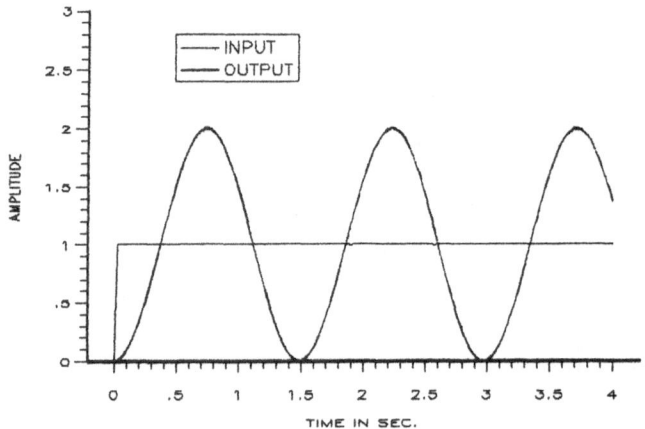

Fig. 8.3 Transient response—absolute stability.

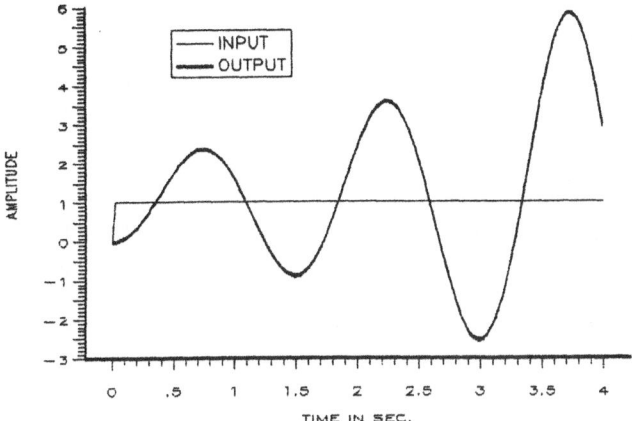

Fig. 8.4 Transient response–instability.

8.1.1 Stability

This is the property of a control system whose response to a stimulus dies down if the stimulus is removed or dies down. Absolute stability is defined as the property of a system whose response to a stimulus remains constant, i.e. the absence of both convergent and divergent oscillations. Adequate stability is the property of a system with sufficient damping to rapidly eliminate divergent oscillations, say after less than two or three over- or undershoots. Typical transient response plots of systems subjected to a step input signal are shown in Figs 8.3, 8.4 and 8.5.

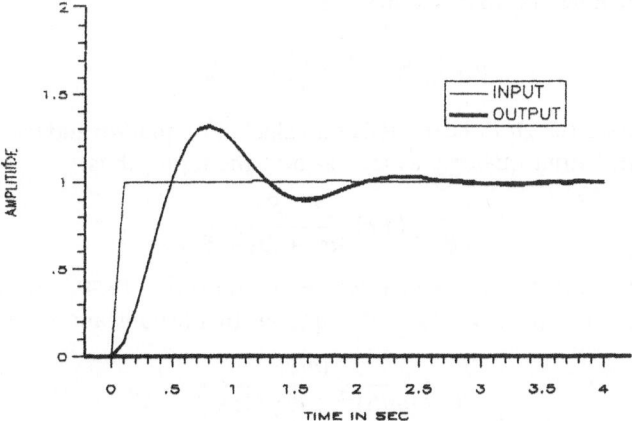

Fig. 8.5 Transient response—adequate stability.

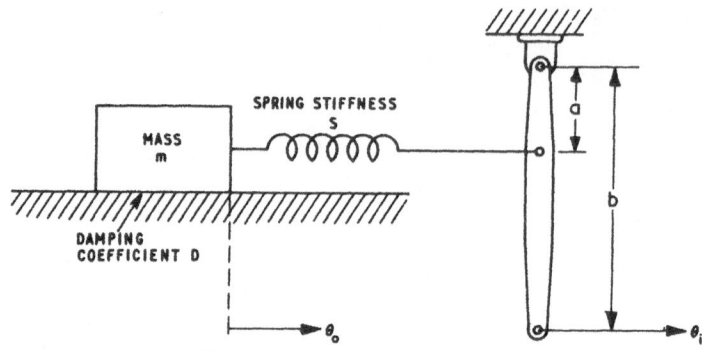

Fig. 8.6 Spring mass system.

8.1.2 Transfer Function

The transfer function of an element or system is a differential equation which uniquely defines its dynamic behaviour. To illustrate how a transfer function is derived, consider the simple spring mass system shown in Fig. 8.6. The equation of motion of the system is given by the expression:

$$m\frac{d^2\theta_o}{dt^2} + D\frac{d\theta_o}{dt} = \left(\frac{a}{b}\theta_i - \theta_o\right)S \qquad (8.1)$$

and making the substitution,

$$\frac{d}{dt} = s$$

where s denotes the Laplace operator,

$$ms^2\theta_o + Ds\theta_o = \left(\frac{a}{b}\theta_i - \theta_o\right)S \qquad (8.2)$$

The expression can now be treated as an algebraic equation and transposed to give the output quantity as the ratio of the input, thus

$$\frac{\theta_o}{\theta_i} = (a/b)\frac{S}{ms^2 + Ds + S} \qquad (8.3)$$

In order to express the characteristic equation $ms^2 + Ds + S$ in the form $as^2 + bs + 1$, eqn (8.3) is divided throughout by the constant S, and hence

$$\frac{\theta_o}{\theta_i} = \frac{(a/b)}{(m/S)s^2 + (D/S)s + 1} \qquad (8.4)$$

$$= KG(s)$$

where the constant $K = a/b$ and the time-dependent portion of the transfer function,

$$G(s) = \frac{1}{(m/S)s^2 + (D/S)s + 1}$$

8.1.3 Steady-State Gain
The steady-state gain is the ratio of the output quantity to the input quantity of an element or system after all transients have died down. The steady-state gain of the spring mass system of Fig. 8.3 is given by the ratio of the lever arm a/b.

8.1.4 Loop Gain
The loop gain is the parameter which ultimately determines the performance of the system, under both steady-state and transient conditions. It is the product of the steady-state gains of the individual elements making up the loop.

The dynamic characteristics of the elements in the loop determine the limiting value of the loop gain of the system; if this value is exceeded, the system is liable to become unstable or insufficiently damped, while a reduction of loop gain will result in a deterioration of both steady-state and dynamic performance.

8.1.5 Frequency Response
System analysis in the frequency domain is a convenient method of establishing stability margins. By considering the frequency characteristics of the elements in a loop, i.e. their response to a sinusoidal input stimulus, the characteristics of the overall system can be determined without the necessity of solving differential equations, frequently of higher orders.

The response of an element or system to a sinusoidal input stimulus is shown in Fig. 8.7. The phase shift ϕ between the input and output and the ratio of the output to input amplitude $|\theta_o/\theta_i|$ fully define its characteristics at any particular frequency. The frequency response characteristics of the elements in a loop can only be utilized for system modelling if the frequency spectrum over a sufficiently wide range of frequencies is known.

The frequency characteristics can be obtained either from the transfer function by making the substitution for the Laplace operator $s = j\omega$ and thence treating the transfer function as a function of a complex variable or, empirically, by measuring the phase angle and amplitude of the output relative to the input. As an example, consider the transfer function of the

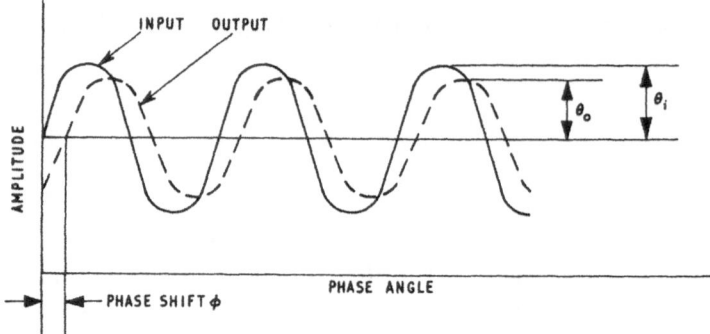

Fig. 8.7 Response to sinusoidal input stimulus.

spring mass system given in eqn (8.4) as a function of s. Expressing the coefficients of the characteristic equation in terms of the steady-state gain, K, natural frequency, ω_0, and the damping factor, ζ

$$\frac{\theta_o}{\theta_i}(s) = KG(s) = \frac{K}{(s^2/\omega_0^2) + (2\zeta/\omega_0)s + 1} \tag{8.5}$$

where $K = a/b$, $\omega_0 = \sqrt{(S/m)}$ and $\zeta = D/(2\sqrt{(S/m)})$.

Then making the substitution, $s = j\omega$,

$$\frac{\theta_o}{\theta_i}(j\omega) = KG(j\omega) = \frac{K}{(1-(\omega^2/\omega_0^2)) + j(2\zeta(\omega/\omega_0))} \tag{8.6}$$

The phase shift, ϕ, is given by the argument of the function of the complex variable, or

$$\phi(j\omega) = -\tan\frac{2\zeta(\omega/\omega_0)}{1-(\omega^2/\omega_0^2)} \tag{8.7}$$

and the ratio of output to input amplitude is given by the modulus, or

$$|\theta_o/\theta_i(j\omega)| = \frac{K}{\sqrt{((1-(\omega^2/\omega_0^2))^2 + (2\zeta(\omega/\omega_0))^2)}} \tag{8.8}$$

It is generally more convenient to represent the transfer function logarithmically, and since communication engineers played a leading part in the development of feedback control theory, the decibel was adopted as the logarithmic unit for the modulus. Equation (8.8) then becomes

$$L_m KG(j\omega) = 20\log\left|\frac{\theta_o}{\theta_i}(j\omega)\right| \tag{8.9}$$

Open and closed loop transfer functions of some typical hydraulic transmissions are tabulated in Table 8.1. Overall system transfer functions are usually of a higher order, since they contain additional elements which influence the performance of the electro-hydraulic control system. More complex systems will be analysed in subsequent chapters.

When the open loop transfer function is represented in its general form as $\theta_o/\varepsilon = KG(s)$, the closed loop transfer function is given by the expression

$$\frac{\theta_o}{\theta_i} = \frac{KG(s)}{1 + KG(s)} \tag{8.10}$$

System (i) is an idealized velocity control system in which all dynamic effects have been omitted. It can be seen from eqn (8.12) in Table 8.1 that when the loop is closed, the output quantity approaches the reference input, provided the loop gain K is substantially larger than unity.

System (ii) is a velocity control system in which dynamic effects have been taken into account.

In the position control system (iii) the actuator is assumed to be a pure integrator, whereas in system (iv) dynamic effects have been included.

All the transfer functions given in Table 8.1 can be expressed as a function of a complex variable by making the substitution $s = j\omega$. In Fig. 8.8 the transfer functions are represented on a complex plane over the entire frequency range, and in Fig. 8.9 the phase angle and log modulus are plotted separately versus the frequency. The former frequency response contours are called Nyquist diagrams, the latter Bode diagrams.

At low frequencies, the log modulus of the G function of both the open and closed loop transfer functions of the velocity control system (ii), represented by eqns (8.13) and (8.14) respectively, approach zero. As the Bode diagrams plotted in Fig. 8.9 include the loop gain K, the log modulus of the open loop system approaches this value at the low end of the frequency spectrum, while the closed loop system approaches a constant value $K/(1 + K)$. At the high end of the frequency spectrum, both open and losed loop contours are asymptotic to a slope of $-12\,\mathrm{dB}$ per octave, i.e. !oubling of frequency.

At the resonant frequency of $\omega = \omega_0$ for the open loop and $\omega = \sqrt{(1 + K)}\omega_0$ for the closed loop system, the phase shift is $-90°$, reaching a maximum value of $-180°$ at high frequencies.

In the position control system (iii) which represents a pure integrator, the open loop transfer function attenuates at $6\,\mathrm{dB}$ per octave at a constant phase shift of $-90°$. When the loop is closed, the log modulus at low frequencies approaches $0\,\mathrm{dB}$ while at high frequencies it is asymptotic with

Table 8.1 Table of transfer functions

System	Open Loop Transfer Function		Closed Loop Transfer Function	
(i) Velocity Servo	$\dfrac{\dot\theta_o}{\varepsilon} = K$	(8.11)	$\dfrac{\dot\theta_o}{\theta_i} = \dfrac{K}{1+K} = \dfrac{1}{1+(1/K)}$	(8.12)
(ii) Velocity Servo	$\dfrac{\dot\theta_o}{\varepsilon} = \dfrac{K}{\dfrac{s^2}{\omega_0^2} + \dfrac{2\zeta s}{\omega_0} + 1}$	(8.13)	$\dfrac{\dot\theta_o}{\theta_i} = \dfrac{K}{\dfrac{s^2}{\omega_0^2} + \dfrac{2\zeta s}{\omega_0} + K + 1} = \dfrac{K}{(1+K)}\left[\dfrac{1}{\dfrac{s^2}{(K+1)\omega_0^2} + \dfrac{2\zeta s}{(K+1)\omega_0} + 1}\right]$	(8.14)
(iii) Position Servo	$\dfrac{\theta_o}{\varepsilon} = \dfrac{K}{s} = \dfrac{1}{\tau s}$	(8.15)	$\dfrac{\theta_o}{\theta_i} = \dfrac{1}{\tau s + 1}$	(8.16)
(iv) Position Servo	$\dfrac{\theta_o}{\varepsilon} = \dfrac{K}{s\left(\dfrac{s^2}{\omega_0^2} + \dfrac{2\zeta s}{\omega_0} + 1\right)}$	(8.17)	$\dfrac{\theta_o}{\theta_i} = \dfrac{K}{\dfrac{s^3}{\omega_0^2} + \dfrac{2\zeta s^2}{\omega_0} + s + K} = \dfrac{1}{\dfrac{s^3}{K\omega_0^2} + \dfrac{2\zeta s^2}{K\omega_0} + \dfrac{s}{K} + 1}$	(8.18)

POSITION SERVO

VELOCITY SERVO

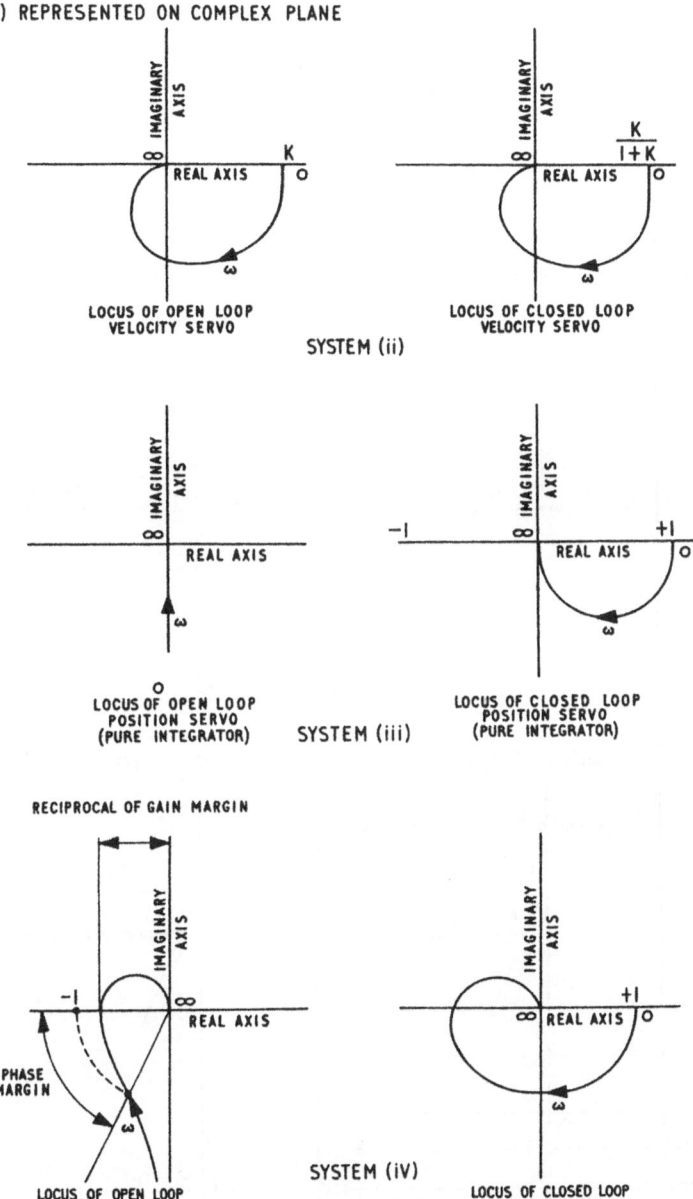

Fig. 8.8 Nyquist diagrams.

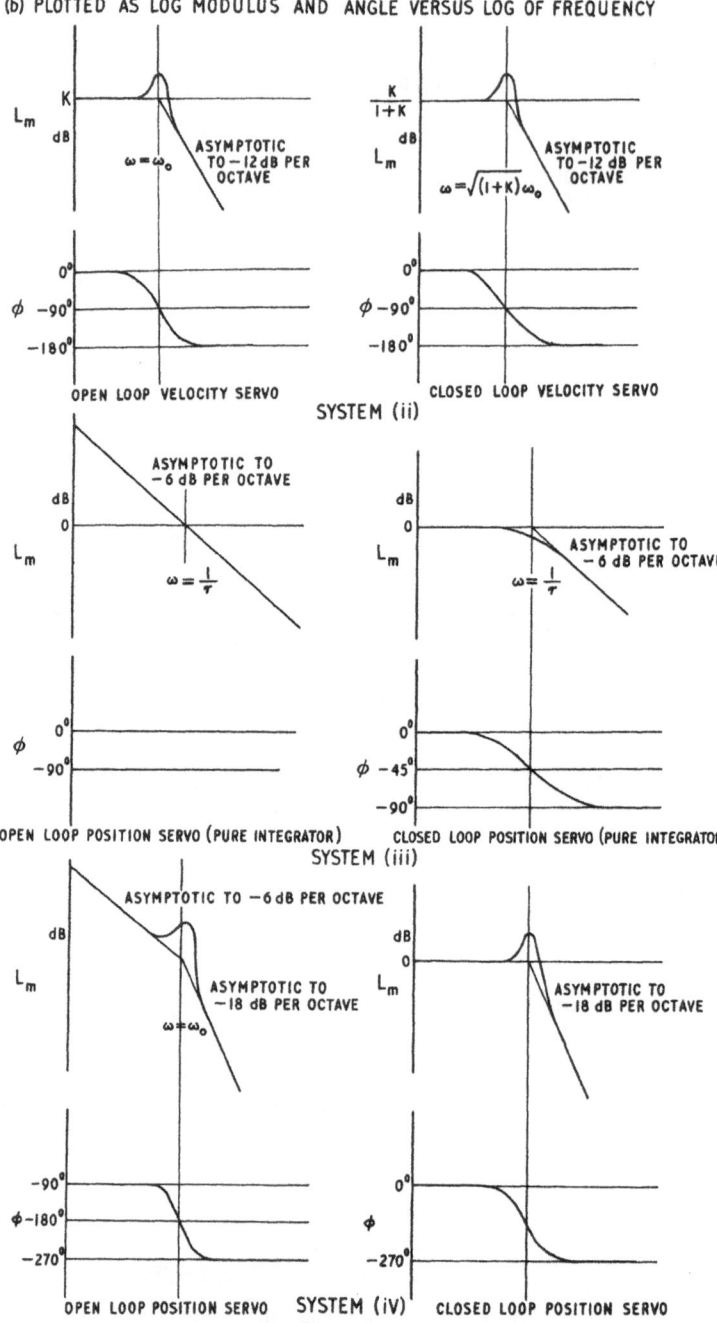

(b) PLOTTED AS LOG MODULUS AND ANGLE VERSUS LOG OF FREQUENCY

Fig. 8.9 Bode diagram.

a $-6\,dB$ per octave slope. At the break or corner frequency of $\omega = 1/\tau$, the phase lag is 45°, reaching a maximum value of 90° at high frequencies.

The open loop log modulus of the position control system (iv) approaches a low frequency asymptote attenuating at $6\,dB$ per octave and a high frequency asymptote attenuating at $18\,dB$ per octave. The two asymptotes intersect at the natural frequency ω_0, corresponding to a phase lag of 180°. At low frequencies, the integration in the loop causes a phase lag of 90° which increases to a maximum value of 270° at high frequencies. Closing the loop removes the effect of the integration at low frequencies. The frequency at peak amplitude is called the resonant frequency of the system.

8.1.6 Stability Criteria

The objective of any system analysis is to determine suitable parameters compatible with the performance specification without exceeding acceptable limits of stability.

Several analytical methods have been developed to determine absolute stability; these are of academic interest but are not practicable for analysing and synthesizing actual systems. The open loop frequency contours plotted on a complex plane as Nyquist diagrams in Fig. 8.8 are a convenient method of establishing stability margins. For feedback control systems which consist of a single loop or which can be reduced to a single loop, the criterion for absolute stability is related to the encirclement of the $-1 + 0j$ point. Enclosure of this point by the $KG(j\omega)$ locus indicates instability, i.e. the presence of divergent oscillations.

By referring to Fig. 8.8 it can be seen that the only plot which can enclose the $-1 + 0j$ point is that of the position control system (iv). It will by now be apparent to the reader that absolute stability criteria can only give a qualitative indication of stability and do not adequately define a satisfactory system.

Quantifying an adequately damped system can be obtained by a number of means. The most widely accepted stability criteria associated with the $KG(j\omega)$ locus, or Nyquist diagram, are the gain margin, the phase margin and the closed loop amplitude ratio. The gain margin is defined as the reciprocal of the modulus $KG(j\omega)$ when the phase angle $\phi(j\omega)$ is $-180°$. The phase margin is defined as the phase difference between the phase shift and $-180°$ when the modulus $KG(j\omega)$ is unity. Gain and phase margins for the position control system (iv) are indicated in Fig. 8.8.

It is usually more convenient to plot the $KG(j\omega)$ locus on a Log modulus versus phase angle chart. This logarithmic representation is known as a Nichols chart. As an illustration the position control system (iv) is plotted

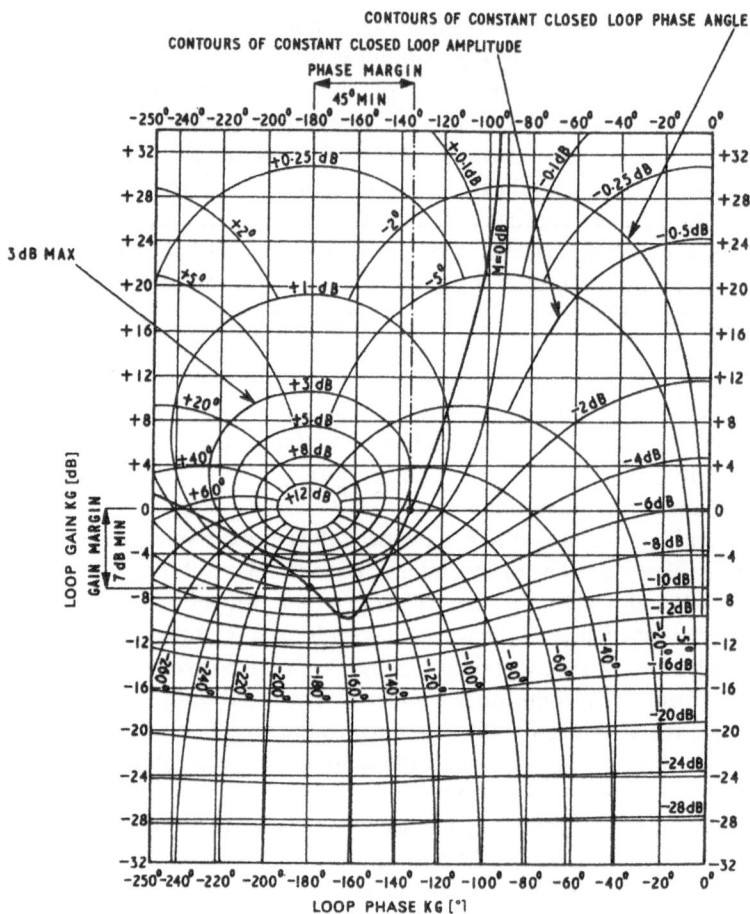

Fig. 8.10 Nichols chart.

on a Nichols chart in Fig. 8.10. The chart includes contours of constant closed loop amplitude ratio and phase shift. The gain margin, which is invariably expressed in decibels, can now be defined as the minus log modulus at $-180°$ phase shift. Typical limiting values of gain and phase margin for an adequately damped system are 7 dB and 45° respectively. Maximum value for the closed loop amplitude ratio $|\theta_o/\theta_i|$ would normally be around 1·4 or 3 dB.

An alternative technique of system analysis is the pole-zero approach developed by Evans as the root-locus method for the investigation of closed loop systems.

9

Principles of Flow Control for Valve-Operated Systems: Part 1

In hydraulic control systems of this type, flow is controlled by throttling the fluid passing through the variable orifice or orifices of a proportional control valve, thus converting pressure energy into kinetic energy. Provided the flow demand is within its capacity, the pump will act as a source of constant or substantially constant pressure.

If the controlled flow is primarily turbulent, it is mainly dependent on the pressure differential, the density of the fluid, the coefficient of discharge and the orifice area, and is substantially independent of fluid temperature. When flow is laminar, it is highly sensitive to viscosity and hence fluid temperature variations. The variable orifices in proportional control valves are specifically designed to be sharp-edged to avoid sensitivity to fluid temperature variations.

The general relationship between valve opening, pressure drop and controlled flow in a throttle valve is

$$q = C_d a \sqrt{(2/\sigma)} \sqrt{(\delta P)} \qquad (9.1)$$

where q is the flow through the orifice, a the orifice area, δP the pressure drop across the orifice, σ the density of the fluid and C_d the coefficient of discharge.

When the four-way control valve shown in Fig. 9.1 is displaced by an amount y, the cylinder will move at a velocity v against an opposing external force F. The controlled flow is then given by the equation

$$q = C_d a \sqrt{(2/\sigma)} \sqrt{(P_0 - P_1)} \qquad (9.2)$$

or by the equation

$$q = C_d a \sqrt{(2/\sigma)} \sqrt{(P_2 - P_R)} \qquad (9.3)$$

55

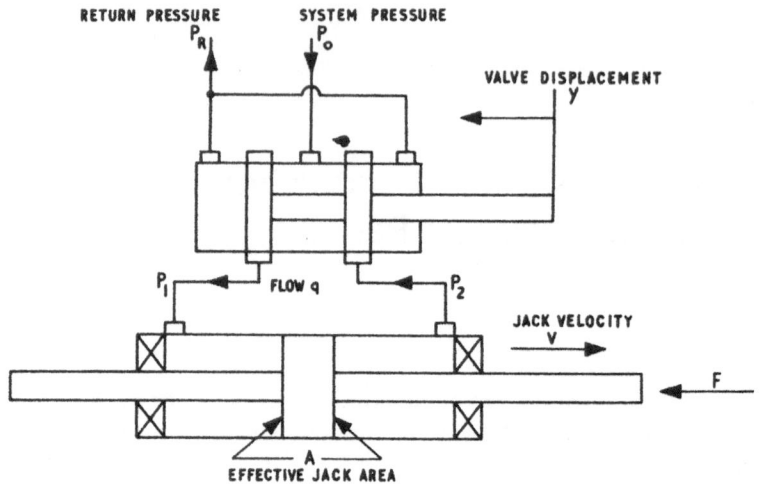

Fig. 9.1 Four-way valve controlling symmetrical cylinder.

where system and return pressure are denoted by P_0 and P_R respectively, and P_1 and P_2 are the pressures in the two cylinder chambers.

Let the load pressure drop across the cylinder $= P_L$, then

$$P_L = F/A = P_1 - P_2 \qquad (9.4)$$

and

$$q = vA \qquad (9.5)$$

Then by combining eqns (9.2), (9.3) and (9.4)

$$q = \frac{C_d}{\sqrt{\sigma}} a \sqrt{(P_0 - P_R - P_L)} \qquad (9.6)$$

The total pressure drop across the valve,

$$P_V = P_0 - P_R - P_L \qquad (9.7)$$

Equation (9.6) can be simplified to

$$q = ka \sqrt{(P_0 - P_R - P_L)} \qquad (9.8)$$

where $C_d/\sqrt{\sigma}$ is replaced by the valve constant k.

For a valve with rectangular orifices, the orifice area $a = wy$, where w is the total width of one set of orifices, and eqn (9.8) then becomes

$$q = kwy \sqrt{(P_0 - P_R - P_L)} \qquad (9.9)$$

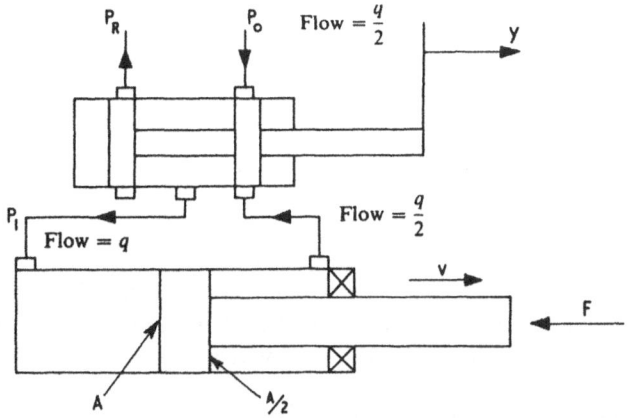

Fig. 9.2 Three-way valve controlling differential cylinder.

The maximum flow rating for a given valve spool diameter, D, is obtained when the orifices consist of annular grooves; eqn (9.9) can then be expressed as

$$q = k\pi Dy\sqrt{(P_0 - P_R - P_L)} \qquad (9.10)$$

When q is in litres/minute, P_0, P_R and P_L in bars, w, y and D in centimetres, and taking the specific gravity of a typical mineral oil and corresponding coefficient of discharge as 0·83 and 0·80 respectively, the valve constant $k = 75·5$.

An alternative system arrangement, comprising a differential (2:1 area ratio) cylinder controlled by a three-way valve, is shown in Fig. 9.2. The corresponding equations are

$$q = C_d a\sqrt{(2/\sigma)}\sqrt{(P_0 - P_1)} \qquad (9.11)$$

where q is the flow from the control port of the valve and q/2 the flow supplied by the pump.

$$P_L = \frac{F}{A} = P_1 - \frac{P_0}{2} \qquad (9.12)$$

and by combining eqns (9.11) and (9.12),

$$q = C_d a\sqrt{(2/\sigma)}\sqrt{\left(\frac{P_0}{2} - P_L\right)} \qquad (9.13)$$

eqn (9.13) can be simplified to

$$q = ka \sqrt{\left(\frac{P_0}{2} - P_L\right)} \qquad (9.14)$$

or for valves with rectangular orifices to

$$q = kwy \sqrt{\left(\frac{P_0}{2} - P_L\right)} \qquad (9.15)$$

The valve constant, k, for metric units now becomes 106·8. For round orifices a coefficient of discharge of 0·625 is normally used, so that valve constants for four-way and three-way valves are reduced to 59 and 83·4 respectively.

It should be noted that the above equations are equally applicable if valve travel and hence actuator movement is reversed, provided the external force remains positive, i.e. opposing motion. In the presence of a negative force, i.e. an assisting or overrunning load, P_L has to be taken as negative, which changes the valve pressure drop for a four-way valve to $P_0 + P_L$ and for a three-way valve to $(P_0/2) + P_L$.

Since return pressure, P_R, is usually insignificant compared with supply pressure P_0, P_R has been omitted from equations subsequent to eqn (9.10).

Pressure flow characteristics of a four-way valve at valve openings varying from 25% to 100% are shown in Fig. 9.3. The useful power output from the valve

$$h = P_L q \qquad (9.16)$$

and by combining eqns (9.16) and (9.8)

$$h = ka P_L \sqrt{(P_0 - P_L)} \qquad (9.17)$$

This gives a maximum value for h when $P_L/P_0 = 2/3$ while at $P_L = 0$ and $P_L = P_0$ no useful power is transmitted.

The maximum power rating of the valve is given by the expression

$$h_m = \frac{2}{(3\sqrt{3})} ka P_0^{1·5} \qquad (9.18)$$

The power rating of a four-way throttle valve, as given by eqn (9.17), is plotted non-dimensionally in Fig. 9.4.

Three parameters which have an important bearing on the performance of closed loop systems are the flow gain, the pressure factor and the output

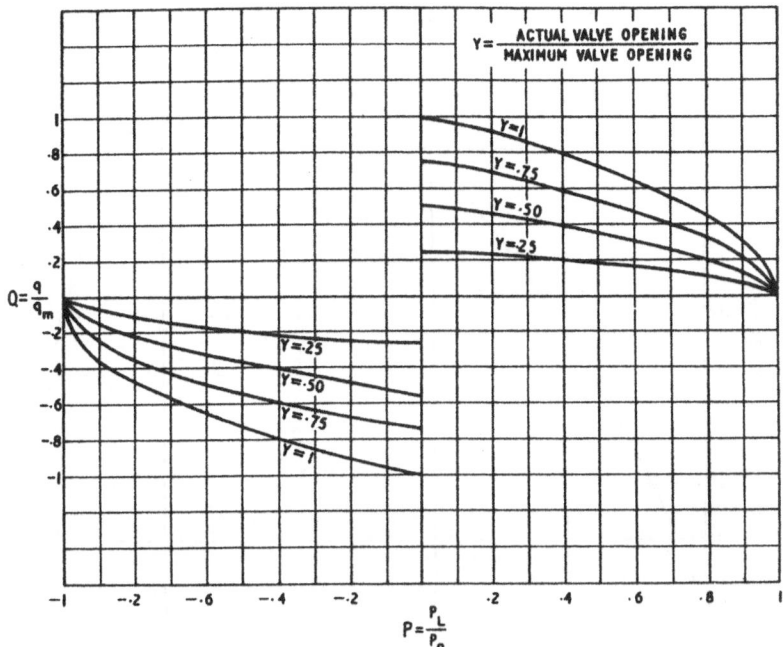

Fig. 9.3 Pressure-flow characteristics of four-way valve.

stiffness or pressure gradient. Valve characteristics, particularly the lap condition, determine the magnitude of these parameters.

Three alternative lap conditions are shown in Fig. 9.5. In an ideal or zero valve the width of the lands equals that of the orifices, in an overlapped valve the lands are wider than the orifices while in an underlapped valve the orifice width exceeds that of the lands.

Flow displacement characteristics of the three types of valve incorporating rectangular orifices are shown in Fig. 9.6. It can be seen that both under- and overlap give rise to discontinuity in the vicinity of the null region. As overlap introduces a dead zone, in which the valve is not responsive to changes of valve displacement, overlap, other than a nominal amount sometimes introduced to reduce quiescent leakage, is not normally suitable for high-performance positional control systems. Overlapped proportional valves are, however, often employed for velocity control applications, since they facilitate arresting the actuator at small signal input.

The slope of the flow displacement curve is called the flow gain or simply the gain of the valve. For an ideal valve the gain is constant over the full

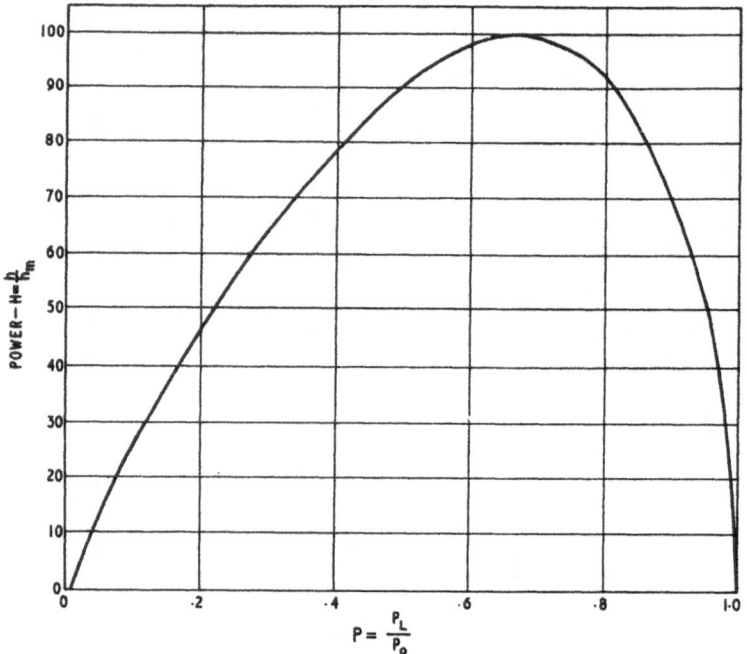

Fig. 9.4 Power-load characteristics of four-way valve.

operating range, while the gain of an underlapped valve within the underlap region is twice that of the rest of the range.

The pressure factor is defined as the slope of the pressure flow curve as plotted in Fig. 9.3.

The third parameter, which can be derived from the valve gain and pressure factor, is the valve output stiffness or pressure gradient. This

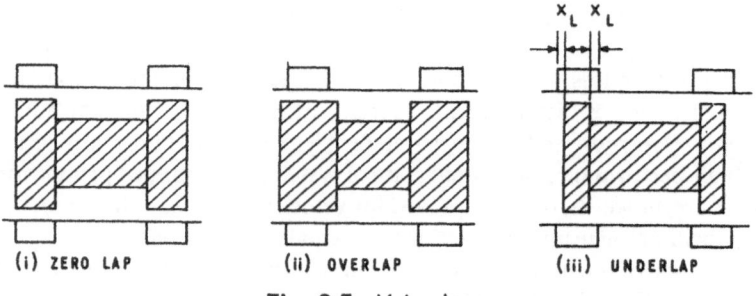

Fig. 9.5 Valve laps.

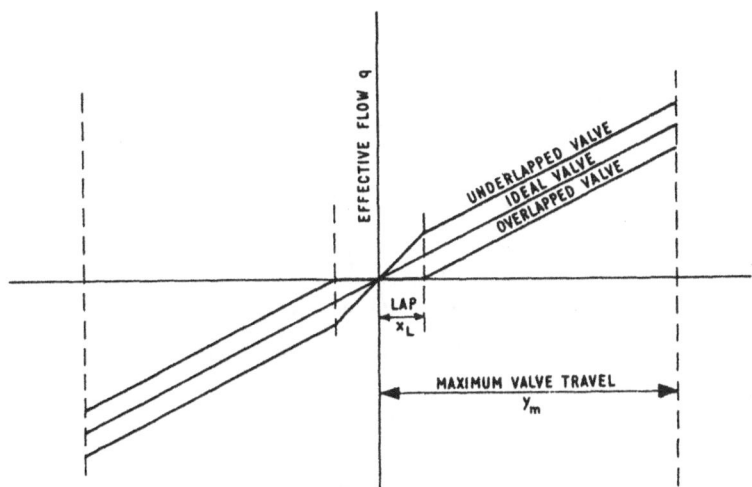

Fig. 9.6 Flow-displacement characteristics.

parameter can also be obtained by plotting load–pressure differential versus valve travel against blocked service ports. The slope of the curve is the valve output stiffness or pressure gradient. Theoretically the pressure gradient of an ideal valve is infinite at zero valve opening, but owing to internal quiescent leakage this condition cannot be achieved in practical valves. Underlap tends to reduce valve output stiffness.

Let the flow gain, pressure factor and valve output stiffness be denoted by K_v, R_v, and S_v respectively. Then

$$K_V = \frac{\partial q}{\partial y}$$

i.e. the flow per unit valve travel at constant load pressure,

$$R_V = \frac{\partial q}{\partial P_L}$$

i.e. the flow per unit load pressure at constant valve travel, and

$$S_V = \frac{\partial P_L}{\partial y}$$

i.e. the load pressure per unit valve travel at constant flow. Also

$$\frac{\partial P_L}{\partial y} = \left(\frac{\partial q}{\partial y}\right)\left(\frac{\partial P_L}{\partial q}\right) = \frac{K_V}{R_V} \tag{9.19}$$

The static output stiffness of the valve is defined as the load pressure per unit valve travel at zero flow.

Both the flow gain K_V and the pressure factor R_V for a zero-lapped valve with rectangular orifices can be obtained by differentiating eqn (9.9)

$$K_V = \frac{\partial q}{\partial y} = kw\sqrt{(P_o - P_L)} \tag{9.20}$$

$$R_V = \frac{\partial q}{\partial P_L} = \frac{-\tfrac{1}{2}kwy}{\sqrt{(P_0 - P_L)}} \tag{9.21}$$

Alternatively the pressure factor can be derived from eqn (9.8), or

$$R_V = \frac{\partial q}{\partial P_L} = \frac{-\tfrac{1}{2}ka}{\sqrt{(P_0 - P_L)}}$$

$$= \frac{-\tfrac{1}{2}q}{(P_0 - P_L)} \tag{9.22}$$

and this equation is applicable to all valve configurations.

The valve output stiffness can now be obtained from eqn (9.19) or

$$S_V = \frac{\partial P_L}{\partial y} = -\frac{2(P_0 - P_L)}{y} \tag{9.23}$$

The flow pattern of an underlapped valve within the underlap region is equivalent to a Wheatstone bridge network as shown in Fig. 9.7. Outside the underlap region, i.e. when the valve travel, y, exceeds the underlap, u, the valve characteristics are identical to those of an ideal valve of similar valve opening.

In considering the characteristics of an underlapped valve it is essential to clearly define the the terms valve travel and valve opening; when dealing with an ideal valve these two terms have identical meanings.

Valve travel is defined as the displacement of the spool from the neutral position, valve opening is the actual opening of an orifice in the direction of the valve travel.

Referring to Fig. 9.7(i), the valve opening, x, when the valve is displaced by an amount y from neutral is $u - y$ at the two metering lands b and d and $u + y$ at the other two lands a and c.

The flow through the actuator is then given by the equation

$$q = q_1 - q_2$$
$$= C_d\sqrt{(2/\sigma)}w[(u + y)\sqrt{(P_0 - P_1)} - (u - y)\sqrt{P_1}] \tag{9.24}$$

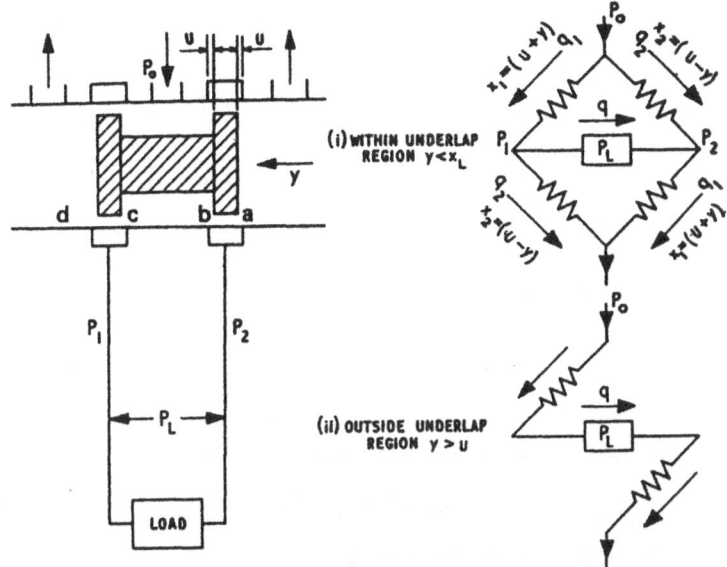

(i) WITHIN UNDERLAP REGION $y < x_L$

(ii) OUTSIDE UNDERLAP REGION $y > u$

Fig. 9.7 Four-way valve control.

or by the equation

$$q = C_d\sqrt{(2/\sigma)}w[(u+y)\sqrt{P_2} - (u-y)\sqrt{(P_0 - P_2)}] \qquad (9.25)$$

Also

$$P_1 = \tfrac{1}{2}(P_0 + P_L)$$

and

$$P_2 = \tfrac{1}{2}(P_0 - P_L)$$

Substituting for P_1 in eqn (9·24) or for P_2 in eqn (9.25), yields the expression

$$q = kw[(u+y)\sqrt{(P_0 - P_L)} - (u-y)\sqrt{(P_0 + P_L)}] \qquad (9.26)$$

where $k = C_d/\sqrt{\sigma}$.

Both the flow gain K_V and the pressure factor R_V can now be obtained by differentiating eqn (9.26).

$$K_V = \frac{\partial q}{\partial y} = kw[\sqrt{(P_0 - P_L)} + \sqrt{(P_0 + P_L)}] \qquad (9.27)$$

$$R_V = \frac{\partial q}{\partial P_L} = -\tfrac{1}{2}kw\left[\frac{u+y}{\sqrt{(P_0 - P_L)}} + \frac{u-y}{\sqrt{(P_0 + P_L)}}\right] \qquad (9.28)$$

The valve output stiffness can be obtained by dividing eqn (9.27) by eqn (9.28), then

$$S_\mathrm{v} = \frac{\partial P_\mathrm{L}}{\partial y} = -2\ \frac{\sqrt{(P_0 - P_\mathrm{L})} + \sqrt{(P_0 + P_\mathrm{L})}}{\dfrac{u + y}{\sqrt{(P_0 - P_\mathrm{L})}} + \dfrac{u - y}{\sqrt{(P_0 + P_\mathrm{L})}}} \qquad (9.29)$$

By comparing eqns (9.22) and (9.23), applicable to an ideal valve, with the corresponding equations for underlapped valves, it can be seen that the pressure factor for an ideal valve at neutral is zero, giving rise to infinite static output stiffness, while the static output stiffness of an underlapped valve is finite.

The above equations can be simplified by expressing them in a non-dimensional form. Since the flow rating of the valve will be at its maximum at zero load and maximum valve opening, from eqn (9.9), maximum flow

$$q_\mathrm{m} = kwx_\mathrm{m}\sqrt{P_0} \qquad (9.30)$$

where x_m is the maximum valve opening.

Let $y/x_\mathrm{m} = Y$, $P_\mathrm{L}/P_0 = P$, $q/q_\mathrm{m} = Q$ and $u/x_\mathrm{m} = U$, then for an ideal valve with rectangular orifices, flow, flow gain, pressure factor and valve output stiffness, given in eqns (9.10), (9.20), (9.21) and (9.23) respectively, can be re-written as follows:

Flow:

$$Q = Y\sqrt{(1 - P)} \qquad (9.31)$$

Flow gain:

$$\frac{\partial Q}{\partial Y} = \sqrt{(1 - P)} \qquad (9.32)$$

Pressure factor:

$$\frac{\partial Q}{\partial P} = \frac{-\tfrac{1}{2}Y}{\sqrt{(1 - P)}} \qquad (9.33)$$

Valve output stiffness:

$$\frac{\partial P}{\partial Y} = \frac{-2(1 - P)}{Y} \qquad (9.34)$$

Similarly the corresponding equations applicable to underlapped valves, (9.26) to (9.29), can be expressed non-dimensionally as:

Flow:

$$Q = (U + Y)\sqrt{(1 - P)} - (U - Y)\sqrt{(1 + P)} \qquad (9.35)$$

Flow gain:

$$\frac{\partial Q}{\partial Y} = \sqrt{(1 - P)} + \sqrt{(1 + P)} \qquad (9.36)$$

Pressure factor:

$$\frac{\partial Q}{\partial P} = -\tfrac{1}{2}\left[\frac{U + Y}{\sqrt{(1 - P)}} + \frac{U - Y}{\sqrt{(1 + P)}}\right] \qquad (9.37)$$

Valve output stiffness:

$$\frac{\partial P}{\partial Y} = -2\left[\frac{\sqrt{(1 - P)} + \sqrt{(1 + P)}}{(U + Y)/\sqrt{(1 - P)} + (U - Y)/\sqrt{(1 + P)}}\right] \qquad (9.38)$$

The expressions derived so far are restricted to systems employing valves with rectangular control orifices. In the following chapter the treatment of valve operated systems will be extended to cover systems controlled by valves having round control orifices.

10

Principles of Flow Control for Valve-Operated Systems: Part 2

The first generations of servo valves, which were developed for aero and industrial applications, incorporated rectangular control orifices in order to provide linear flow characteristics. In later versions of servo valves and particularly in the more recent generation of proportional control valves rectangular control orifices were replaced by round orifices, mainly to reduce manufacturing costs. In this chapter the characteristics of round control orifices and their effect on valve parameters will be examined.

As the exact expression for the area of a segment of a circle is somewhat difficult to handle, it is more convenient to use a close approximation.

Let the area of the segment of a circle of radius, r, be equal to a_0, the angle subtended by the segment $= 2\theta$, and the height of the segment $= x$. Then, when

$$\theta < \frac{\pi}{2} \qquad \frac{a_0}{r^2} = \theta - \sin\theta\cos\theta$$

also

$$\cos\theta = 1 - \frac{x}{r}$$

and

$$\sin\theta = \sqrt{\left[\frac{2x}{r}\left(1 - \frac{\frac{1}{2}x}{r}\right)\right]}$$

By means of the binomial theorem, $\sin\theta$ can be expanded to

$$\sin\theta = \sqrt{2}\left[(x/r)^{1/2} - \frac{(x/r)^{3/2}}{4} - \frac{(x/r)^{5/2}}{32}\right]$$

all higher powers of x/r being omitted.

By means of the expansion

$$\sin^{-1} X = X + \frac{X^3}{6} + \frac{3X^5}{40}$$

and the binomial theorem, it can be shown that

$$\theta = \sqrt{2} \left[(x/r)^{1/2} - \frac{(x/r)^{3/2}}{12} + \frac{3(x/r)^{5/2}}{160} \right]$$

Combining the above equations yields

$$a_0/r^2 = \sqrt{2} \left[\frac{4(x/r)^{3/2}}{3} - \frac{(x/r)^{5/2}}{5} \right]$$

$$= \sqrt{2}(4/3)(x/r)^{3/2} \left[1 - \frac{3(x/r)}{20} \right]$$

The above approximation gives a maximum error of 2% at $x/r = 1$ as shown in Fig. 10.1, curve (i).

By modifying the expression to

$$a_0/r^2 = \sqrt{2}(4/3)(x/r)^{3/2} \left[1 - \frac{3(x/r)}{19} \right]$$

the maximum error at $x/r = 1$ is reduced to 1% while reversing the error at $x/r = 0.4$ from $+\frac{1}{4}\%$ to $-\frac{1}{4}\%$ as shown by curve (ii).

Alternatively the area can be expressed in terms of the orifice diameter, d, i.e. when $x/d < \frac{1}{2}$

$$a_0/d^2 = (4/3)(x/d)^{3/2} \left[1 - \frac{6(x/d)}{19} \right] \tag{10.1}$$

and, when $x/d > \frac{1}{2}$,

$$a_0/d^2 = \pi/4 - (4/3)(1 - x/d)^{3/2} \left[1 - \frac{6(1 - x/d)}{19} \right] \tag{10.2}$$

where x is the valve opening and a_0 the area of one orifice. For a zero-lap valve containing n inlet and n outlet control orifices

$$q = kna_0 \sqrt{(P_0 - P_L)} \tag{10.3}$$

The above equations can again be expressed in a non-dimensional form.

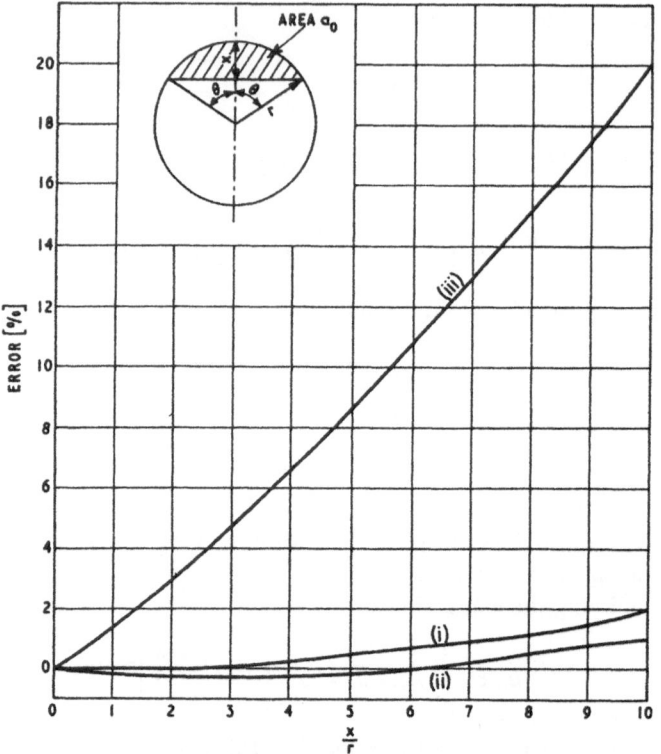

Fig. 10.1 Area of segment of circle.

Let the flow at valve opening d and zero load $= q_0$, $q/q_0 = Q_R$, $x/d = Y_R$, and $P_L/P_0 = P$, then since $q_0 = (k\pi d^2/4)\sqrt{P_0}$, the non-dimensional flow

$$Q_R = \frac{16}{3\pi} Y_R^{1 \cdot 5}\left(1 - \frac{6Y_R}{19}\right)\sqrt{(1 - P)} \qquad (10.4)$$

The flow gain

$$\frac{\partial Q_R}{\partial Y_R} = \frac{8}{\pi}\sqrt{X_R}\left(1 - \frac{10Y_R}{19}\right)\sqrt{(1 - P)} \qquad (10.5)$$

and for small valve openings under no load conditions

$$\frac{\partial Q_R}{\partial Y_R} = \frac{8}{\pi}\sqrt{Y_R} \qquad (10.6)$$

For an underlapped valve at small valve openings, flow

$$q = \frac{4k\sqrt{d}}{3}[\sqrt{(P_0 - P_L)}(u + y)^{1 \cdot 5} - \sqrt{(P_0 + P_L)}(u - y)^{1 \cdot 5}] \quad (10.7)$$

Let $u/d = U_R$ and $y/d = Y_R$ be the non-dimensional underlap and valve travel respectively, then eqn (10.7) becomes

$$Q_R = \frac{16}{3\pi}[\sqrt{(1-P)}(U_R + Y_R)^{1 \cdot 5} - \sqrt{(1 + P)}(U_R - Y_R)^{1 \cdot 5}] \quad (10.8)$$

The corresponding flow gain

$$\frac{\partial Q_R}{\partial Y_R} = \frac{8}{\pi}[\sqrt{(1 - P)}\sqrt{(U_R + Y_R)} + \sqrt{(1 + P)}\sqrt{(U_R - Y_R)}] \quad (10.9)$$

and at zero load

$$\frac{\partial Q_R}{\partial Y_R} = \frac{8}{\pi}[\sqrt{(U_R + Y_R)} + \sqrt{(U_R - Y_R)}] \quad (10.10)$$

From eqns (10.6) and (10.10) a comparison between the zero load flow gains of an ideal and an underlapped valve at equal valve openings can now be obtained. In its neutral position, an underlapped valve has a valve opening equal to the underlap, while for an ideal valve, valve opening and travel have identical meanings. It can be seen from eqn (10.6) that the flow gain of an ideal valve at neutral is zero, while the non-dimensional flow gain of an underlapped valve, obtained from eqn (10.10), at neutral

$$\frac{\partial Q_R \partial Y_R}{\partial Y_R} = \frac{16}{\pi}\sqrt{U_R} \quad (10.11)$$

At valve travel equal to the underlap, i.e. when $Y_R = U_R$, the gain of an ideal valve

$$\frac{\partial Q_R}{\partial Y_R} = \frac{8}{\pi}\sqrt{U_R} \quad (10.12)$$

while the corresponding gain of an underlapped valve

$$\frac{\partial Q_R}{\partial Y_R} = \frac{8\sqrt{2}}{\pi}\sqrt{U_R} \quad (10.13)$$

i.e. the ratio of underlapped to ideal flow gains is $\sqrt{2}$.

Non-dimensional flow-travel characteristics for an ideal and under-lapped valve with rectangular control orifices is shown in Fig. 10.4, and

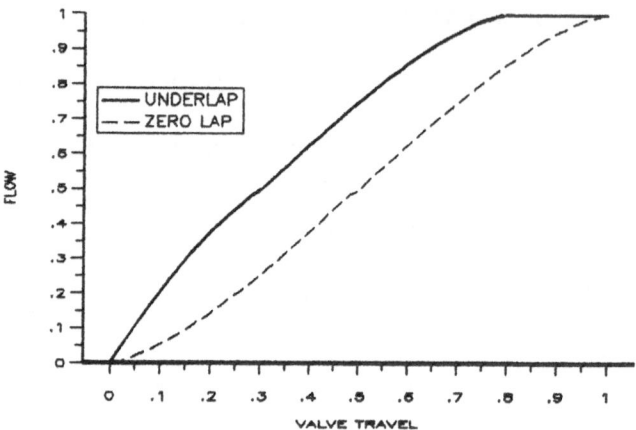

Fig. 10.2 Flow-displacement characteristics, round orifices.

corresponding curves for a valve incorporating round control orifices in Fig. 10.2.

Gain–travel characteristics for ideal and underlapped valves are shown in Figs 10.5 and 10.3 for rectangular and round control orifices respectively.

Since an ideal valve has infinite output stiffness at its null position (see eqn (9.34)), it is more convenient to plot valve output compliance, i.e. the reciprocal of output stiffness, versus valve travel. Output compliance–travel characteristics for a valve with rectangular orifices are plotted non-dimensionally in Fig. 10.6 and for round orifices in Fig. 10.7.

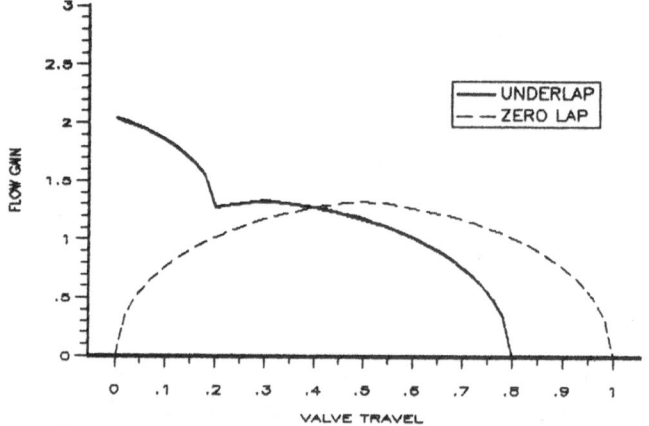

Fig. 10.3 Gain-displacement characteristics, round orifices.

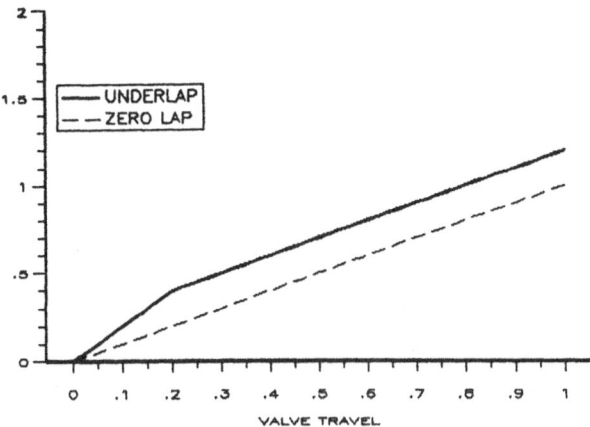

Fig. 10.4 Flow-displacement characteristics, rectangular orifices.

A summary of valve parameters applicable to rectangular control orifices is tabulated in Table 10.1; corresponding parameters for round control orifices are given in Table 10.2.

10.1 EFFECT OF QUIESCENT LEAKAGE ON LINEARITY

As flow gain is proportional to orifice width, as shown by Fig. 10.3, the linearity of a valve with round control orifices can be improved by allowing

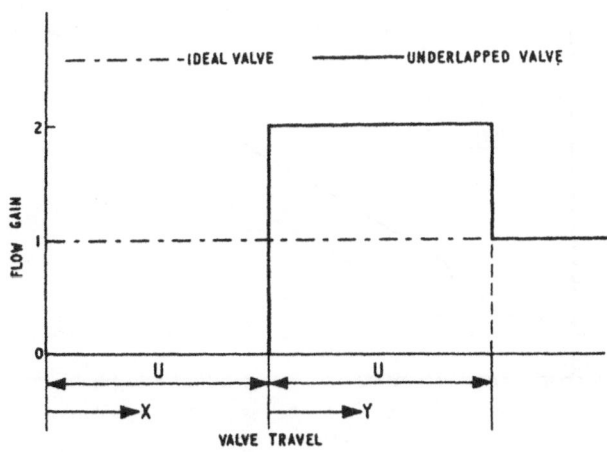

Fig. 10.5 Gain-displacement characteristics, rectangular orifices.

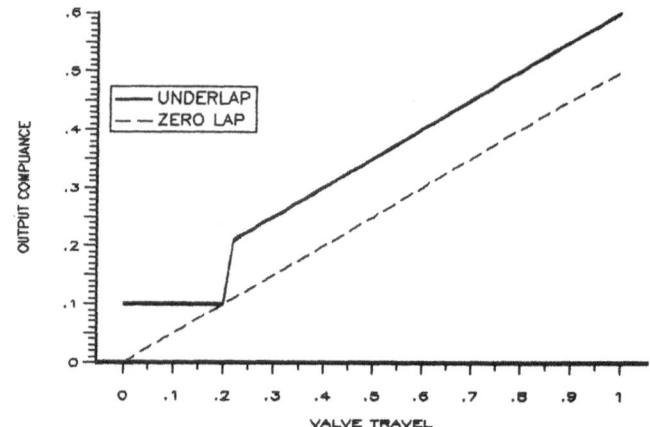

Fig. 10.6 Valve output compliance, rectangular orifices.

some underlap. Underlap introduces leakage flow and reduces output stiffness (see Table 10.2), and therefore improved linearity is gained at the expense of additional leakage and reduced output stiffness.

Before the relationship between orifice size, valve travel, underlap and leakage and their effect on linearity can be established, the terms linearity and leakage have to be clearly defined. Quiescent leakage is defined as the internal flow through the valve due to underlap, with the valve at neutral and under no-load condition. This flow is of the orifice resistance type, i.e. it

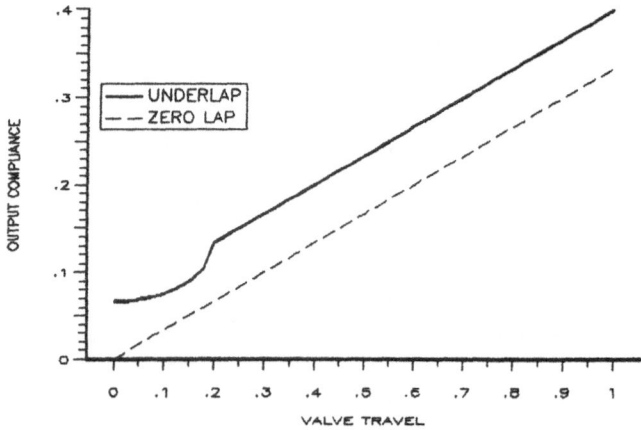

Fig. 10.7 Valve output compliance, round orifices.

Table 10.1 Valve parameters, rectangular orifices

	Rectangular Orifices	
	Zero Lap	*Underlap*
Flow	$Q = X\sqrt{(1-P)}$	$Q = \sqrt{(1-P)(U+Y)} - \sqrt{(1+P)(U-Y)}$
Valve Gain	$\dfrac{\partial Q}{\partial X} = \sqrt{(1-P)}$	$\dfrac{\partial Q}{\partial Y}\sqrt{(1-P)} + \sqrt{(1+P)}$
Pressure Factor	$\dfrac{\partial Q}{\partial P} = -\dfrac{X}{2\sqrt{(1-P)}}$	$\dfrac{\partial Q}{\partial P} = -\dfrac{1}{2}\left[\dfrac{U+Y}{\sqrt{(1-P)}} + \dfrac{U-Y}{\sqrt{(1+P)}}\right]$
Valve Output Stiffness	$\dfrac{\partial P}{\partial X} = -\dfrac{2(1-P)}{X}$	$\dfrac{\partial P}{\partial Y} = -\dfrac{2[\sqrt{(1-P)}+\sqrt{(1+P)}]}{\dfrac{U+Y}{\sqrt{(1-P)}}+\dfrac{U-Y}{\sqrt{(1+P)}}}$
Valve Output Stiffness when $P \ll 1$	$\dfrac{\partial P}{\partial X} = -\dfrac{2}{X}$	$\dfrac{\partial P}{\partial Y} = -\dfrac{2}{U}$
Valve Output Stiffness when $P \ll 1$	∞	$\dfrac{\partial P}{\partial Y} = -\dfrac{2}{U}$

Table 10.2 Valve parameters, round orifices

Round Orifices

	Zero Lap	*Underlap*
Flow	$Q_R = \dfrac{16}{3\pi} X_R^{\frac{3}{2}} \sqrt{(1-P)}$	$Q_R = \dfrac{16}{3\pi}[\sqrt{(1-P)}(U_R+Y_R)^{\frac{3}{2}} - \sqrt{(1+P)}(U_R-Y_R)^{\frac{3}{2}}]$
Valve Gain	$\dfrac{\partial Q_R}{\partial X_R} = \dfrac{8}{\pi}\sqrt{[X_R(1-P)]}$	$\dfrac{\partial Q_R}{\partial Y_R} = \dfrac{8}{\pi}\{\sqrt{[(1-P)(U_R+Y_R)]} + \sqrt{[(1+P)(U_R-Y_R)]}\}$
	when $X_R = 0,\ \dfrac{\partial Q_R}{\partial X_R} = 0$	when $Y_R = 0,\ \dfrac{\partial Q_R}{\partial Y_R} = \dfrac{8}{\pi}\sqrt{U_R}[\sqrt{(1-P)} + \sqrt{(1+P)}]$
Pressure Factor	$\dfrac{\partial Q_R}{\partial P} = -\dfrac{8}{3\pi} X_R^{\frac{3}{2}} \dfrac{1}{\sqrt{(1-P)}}$	$\dfrac{\partial Q_R}{\partial P} = -\dfrac{8}{3\pi}\left[\dfrac{(U_R+Y_R)^{\frac{3}{2}}}{\sqrt{(1-P)}} + \dfrac{(U_R-Y_R)^{\frac{3}{2}}}{\sqrt{(1+P)}}\right]$
Valve Output Stiffness	$\dfrac{\partial P}{\partial X_R} = -\dfrac{3(1-P)}{X_R}$	$\dfrac{\partial P}{\partial Y_R} = -\dfrac{3\{\sqrt{[(1-P)(U_R+Y_R)]} + \sqrt{[(1+P)(U_R-Y_R)]}\}}{\dfrac{(U_R+Y_R)^{\frac{3}{2}}}{\sqrt{(1-P)}} - \dfrac{(U_R-Y_R)^{\frac{3}{2}}}{\sqrt{(1+P)}}}$
Valve Output Stiffness when $P \ll 1$	$\dfrac{\partial P}{\partial X_R} = -\dfrac{3}{X_R}$	$\dfrac{\partial P}{\partial Y_R} = -\dfrac{3[\sqrt{(U_R+Y_R)} + \sqrt{(U_R-Y_R)}]}{(U_R+Y_R)^{\frac{3}{2}} + (U_R-Y_R)^{\frac{3}{2}}}$
Valve Output Stiffness when $P \ll 1$	∞	$\dfrac{\partial P}{\partial Y_R} = -\dfrac{3}{U_R}$

Note: Expressions applicable to small valve openings. For simplification, correction factor has been omitted.

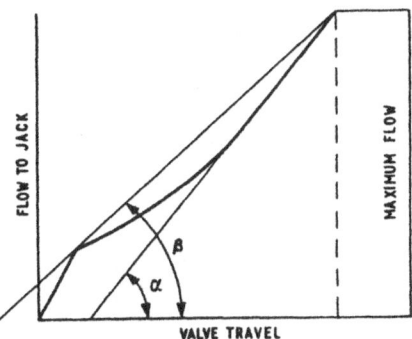

Fig. 10.8 Linearity of underlapped valve.

is proportional to the orifice area and the square root of the pressure drop and is substantially independent of fluid temperature; total leakage flow does, of course, include viscous flow which is a function of diametral clearances and fluid temperature.

At first sight, an obvious way of defining linearity would be to draw two radial lines from a point corresponding to valve neutral and zero flow on a flow–travel curve, in such a manner as to embrace all points to maximum travel and flow and to define linearity as the ratio of the difference of the slopes to mean slope.

If this definition is applied to a valve with rectangular orifices, it can be seen from Fig. 10.4 that deviation from linearity will increase with a reduction of underlap. As deviation from linearity must approach zero as the underlap approaches zero, the above definition is obviously not valid for rectangular ports unless threshold is taken into account. This limitation can, however, be overcome by shifting the origin of the boundary lines to a point corresponding to maximum flow and travel. Deviation from linearity, or simply linearity, can then be defined as the ratio of the difference of the slopes to mean slope of the two radial boundary lines drawn from a point corresponding to maximum valve travel and flow and embracing all points to valve neutral and zero flow, or, referring to Fig. 10.8,

$$\text{Linearity} = \pm \frac{\tan \alpha - \tan \beta}{\tan \alpha + \tan \beta} \qquad (10.14)$$

The relationship between quiescent leakage and linearity for valves with round orifices can now be established.

$$q_L = 2kna_0\sqrt{P_0} \qquad (10.15)$$

where q_L is the quiescent leakage, and a_0 the area of one orifice due to underlap u. Expressing quiescent leakage and maximum flow rate in a non-dimensional form yields the following equations.

$$q_L/q_0 = \frac{32}{3\pi} U_R^{1.5} \left(1 - \frac{6U_R}{19}\right) \qquad (10.16)$$

and

$$q_m/q_0 = \frac{16}{3\pi} X_m^{1.5} \left(1 - \frac{6X_m}{19}\right) \qquad (10.17)$$

where q_m is the flow rate at maximum valve opening x_m and the non-dimensional maximum valve opening $X_m = x_m/d$.

The leakage ratio, defined as the ratio of quiescent to maximum flow, or $Q_L = q_L/q_m$ can now be obtained by dividing eqn (10.16) by eqn (10.17).

The above equations were used to show graphically the effect of underlap on leakage and linearity and specifically the relationship between linearity and leakage ratio for a given maximum valve opening.

It can be seen from Fig. 10.9, in which linearity is plotted against valve

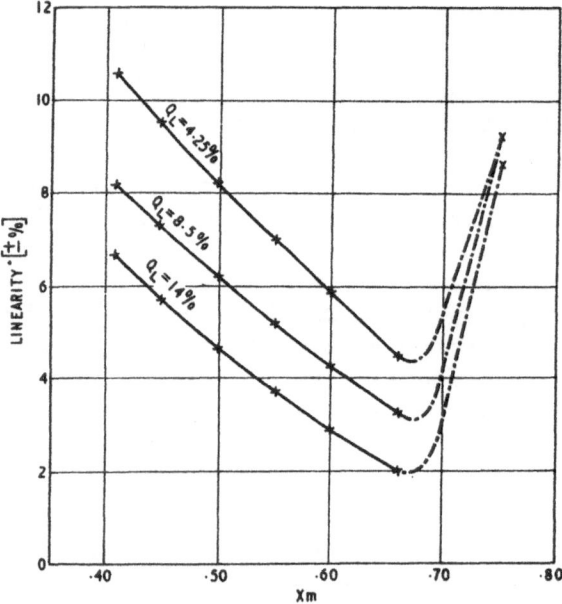

Fig. 10.9 Effect of valve opening on linearity.

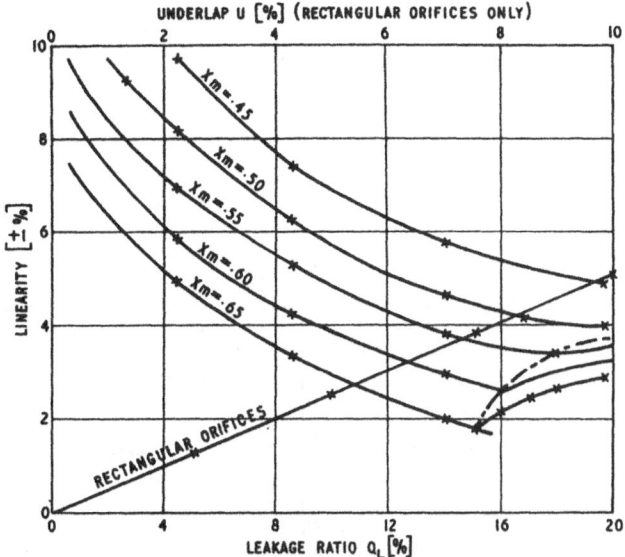

Fig. 10.10 Linearity versus leakage.

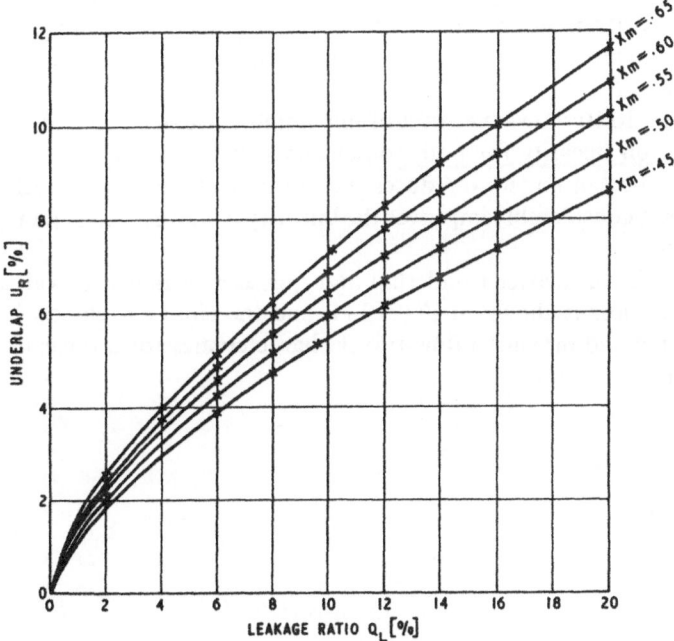

Fig. 10.11 Underlap versus leakage.

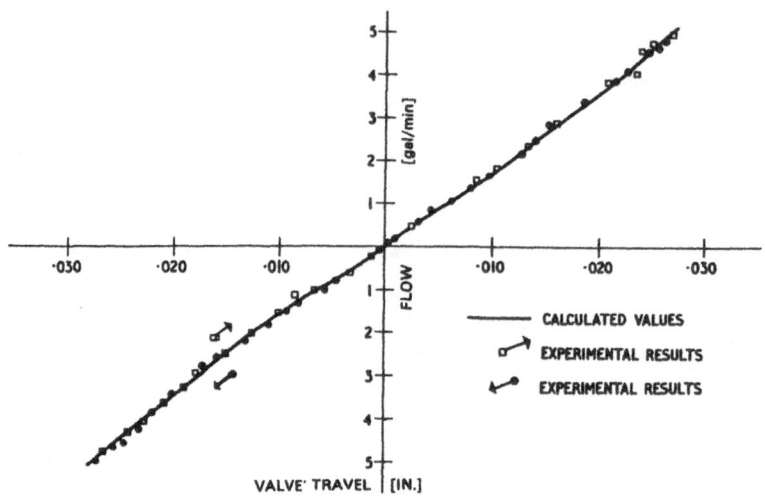

Fig. 10.12   Typical flow characteristics, round orifices.

opening ratio, X_m, at constant leakage ratios Q_L of $4\frac{1}{4}\%$, $8\frac{1}{2}\%$ and 14%, that linearity is improved by increasing X_m up to a limiting value of approximately 0·67. Beyond this value deviation from linearity increases rapidly.

Figure 10.10 shows the effect of maximum valve opening and quiescent leakage on linearity for both round and rectangular orifices. While the introduction of underlap reduces the linearity of valves with rectangular orifices, it considerably improves the linearity of valves with round control orifices.

The relation between underlap and quiescent leakage for valves with round orifices is shown in Fig. 10.11, and the close correlation between calculated and measured flow-travel characteristics for a typical valve is plotted in Fig. 10.12.

11

Introduction to System Analysis

This chapter is intended as a brief introduction to the basic concepts used in analysing hydraulic control systems. To simplify the analysis, we shall consider a mechanical–hydraulic valve-operated cylinder system, shown diagrammatically in Fig. 11.1, and represented by the block diagram Fig. 11.2. In the position control system shown in Fig. 11.1, the cylinder piston rod is anchored, and the input acts directly on the valve spool, so that the displacement of the valve spool is a direct measure of system error, or

$$\theta_i - \theta_o = y \qquad (11.1)$$

As the compressibility of the fluid has a destabilizing effect on the system, the compliance (reciprocal of stiffness) of the column of oil in the cylinder has to be taken into account. The compliance of the cylinder mounting is

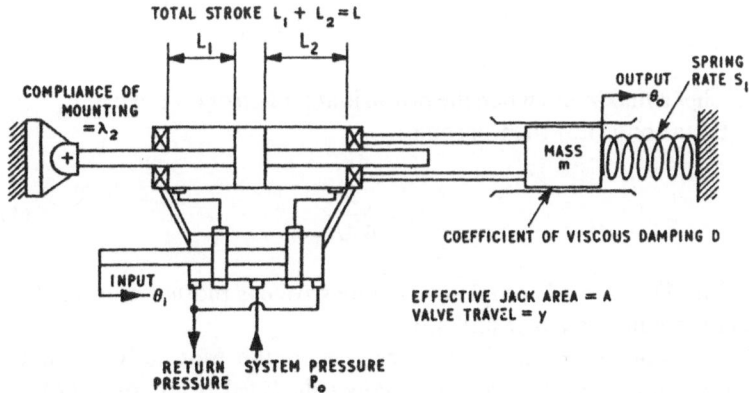

Fig. 11.1   Diagram of mechanical–hydraulic position control.

added to the oil compliance to arrive at the total effective actuator compliance.

Let λ_1 denote the oil, λ_2 the mounting, and λ the total actuator compliance, then

$$\lambda_1 + \lambda_2 = \lambda \tag{11.2}$$

When the mounting compliance is small in relation to the oil compliance, the cylinder can be regarded as rigidly mounted, or $\lambda = \lambda_1$. Oil compliance can be expressed in terms of the effective piston area, A, the total stroke of the actuator, L, and the bulk modulus of the fluid, N.

When ∂P, ∂F, ∂V and ∂L are small increments of pressure, force, volume and stroke respectively, and V denotes the volume of fluid under compression,

$$N = \partial P \frac{V}{\partial V} = \frac{\partial F}{\partial L} \frac{V}{A^2}$$

Referring to Fig. 11.1, the compliance of the columns of oil on either side of the piston is given by the expressions

$$\frac{L_1}{AN} \quad \text{and} \quad \frac{L_2}{AN}$$

giving a total compliance

$$\lambda_1 = \frac{1}{AN[(1/L_1) + (1/L_2)]} \tag{11.3}$$

λ_1 reaches a maximum when the piston is at mid-stroke, i.e. when $L_1 = L_2 = L/2$. Equation (11.3) then becomes

$$\lambda_1 = \frac{L}{4AN} \tag{11.4}$$

In Fig. 11.3, the oil compliance at any stroke is plotted as a function of maximum compliance at mid-stroke.

As the mid-stroke position gives rise to the most adverse stability condition, fluid compliance is normally calculated from eqn (11.4).

If the oil compliance of the pipes connecting the valve to the actuator is

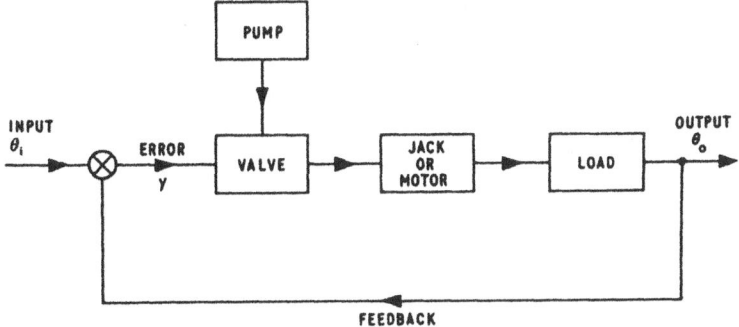

Fig. 11.2 Block diagram of mechanical–hydraulic position control.

taken into account, the total oil compliance then becomes

$$\lambda_1 = \frac{L}{4AN} + \frac{V_P}{4A^2N} \qquad (11 \cdot 5)$$

where V_p is the total oil volume contained in the pipes.

In a position control system, as shown in Fig. 11.1, the actuator acts as an integrator and the output position can then be expressed in terms of the flow, q, the external load, F, the effective piston area, A, and the compliance, λ, by the equation

$$\theta_0 = \frac{1}{A}\int q\,\mathrm{d}t - \lambda F$$

$$= \frac{1}{A}\frac{q}{s} \lambda F \qquad (11.6)$$

It can be seen from eqn (9.9) and Fig. 9.3 that hydraulic systems employing flow control are essentially non-linear. By applying a small perturbation technique to the solution of the equation of motion of the hydraulic transmission, the resulting differential equations can be linearized, thus simplifying the derivation of transfer functions.

Treating the supply pressure as a constant quantity, the controlled flow is a function of two variables, the valve travel and the load pressure, or $q = f(y, P_L)$.

Considering small perturbations about a fixed steady-state operating point,

$$q = \frac{\partial q}{\partial y}y + \frac{\partial q}{\partial P_L}P_L \qquad (11.7)$$

and combining eqns (11.6) and (11.7),

$$\theta_0 = \frac{1}{As}\left(\frac{\partial q}{\partial y}y + \frac{\partial q}{\partial P_L}P_L\right) - \lambda F$$

$$= \frac{K_V}{As}\left(y + \frac{R_V}{K_V}\frac{F}{A}\right) - \lambda F \tag{11.8}$$

Alternatively, the output can be expressed in terms of the valve output stiffness, S_V, or

$$\theta_0 = \frac{K_V}{As}\left(y + \frac{F}{S_V A}\right) - \lambda F \tag{11·9}$$

For the simple mechanical–hydraulic system of Fig. 11.1, the overall system parameters loop gain and output stiffness are given by the expressions

$$K = \frac{K_V}{A} \tag{11.10}$$

and

$$S = \frac{\partial F}{\partial y} = -S_V A \tag{11.11}$$

As the valve output stiffness, previously derived in eqns (9.23) and (9.29), is negative, a negative sign is introduced in eqn (11.11) in order to keep system output stiffness positive.

From eqns (11·9), (11.10) and (11.11),

$$\theta_0 = \frac{K}{s}\left(y - \frac{F}{S}\right) - \lambda F \tag{11.12}$$

In the system shown in Fig. 11.1, the actuator is subjected to an exernal force comprising inertia loading, viscous damping and a spring force. The differential equation of motion is then

$$F = (ms^2 + Ds + S_1)\theta_0 \tag{11.13}$$

The open loop system transfer function can now be derived by combining eqns (11.12) and (11.13).

$$KG(s) = \frac{\theta_0}{y} = \frac{K}{\lambda ms^3 + (\lambda D + Km/S)s^2 + (\lambda S_1 + KD/S + 1)s + (KS_1/S)}$$

$$\tag{11.14}$$

The most adverse stability condition arises when the actuator is controlling a pure inertia load. Equation (11.14) then simplifies to

$$KG(s) = \frac{\theta_0}{y} = \frac{K}{s[\lambda m s^2 + (K m s/S) + 1]} \quad (11.15)$$

Alternatively, the above transfer function can be expressed in terms of the hydraulic actuator natural frequency and damping factor, w_0 and ζ.

$$KG(s) = \frac{K}{s[(s^2/\omega_0^2) + (2\zeta s/\omega_0) + 1]} \quad (11.16)$$

where

$$w_0 = \frac{1}{\sqrt{(\lambda m)}} \quad (11.17)$$

and

$$\zeta = \frac{K m \omega_0}{2S} \quad (11.18)$$

Equation (11.16) is identical to the transfer function of the position servo given as eqn (8.17) in Table 8.1(iv), the corresponding closed loop transfer function is given by eqn (8.18). Nyquist and Bode diagrams applicable to a system of this type are shown in Figs 8.8 and 8.9.

The block diagram, Fig. 11.2 can be reduced to a single block diagram, Fig. 11.4. By applying the stability criteria discussed in Section 8.1.6, the limiting loop gain for a given range of operating conditions can now be obtained by plotting a family of curves on a Nichols chart. Although this

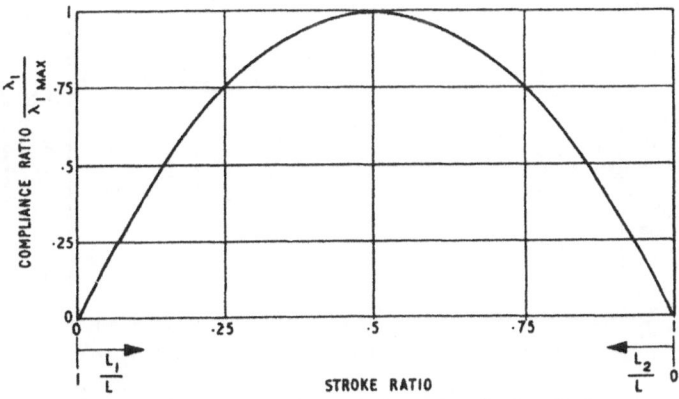

Fig. 11.3 Oil compliance of symmetrical cylinder.

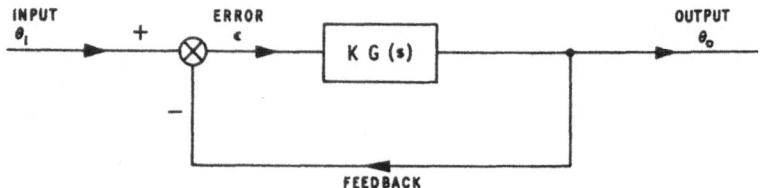

Fig. 11.4 System block diagram.

approach could be used for the simple control system under descussion, it would not be feasible to analyse practical electro-hydraulic control systems, which are almost invariably represented by fifth or higher order transfer functions, by a manual method. The computer program used to obtain the performance curves Figs 11.5 to 11.9 will be dealt with in some depth in subsequent chapters.

Equation (11.16) can be re-written as a function of a complex variable, as previously explained in Section 8.1.5; then

$$KG(j\omega) = \frac{K}{j\omega[1-(\omega/\omega_0)^2 + j2\zeta(\omega/\omega_0)]} \qquad (11.19)$$

By making the substitution $K/\omega_0 = K'$ and $\omega/\omega_0 = \omega'$, eqn (11.19) can be expressed in a non-dimensional form.

$$KG(j\omega') = \frac{K'}{j\omega'(1 - \omega'^2 + j2\zeta\omega')} \qquad (11.20)$$

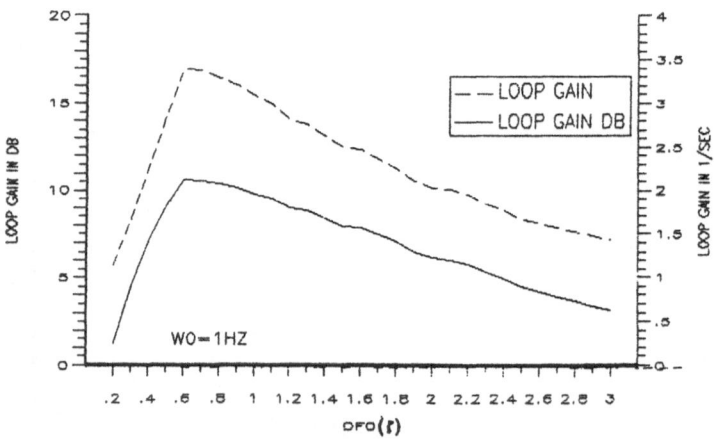

Fig. 11.5 Stability boundary.

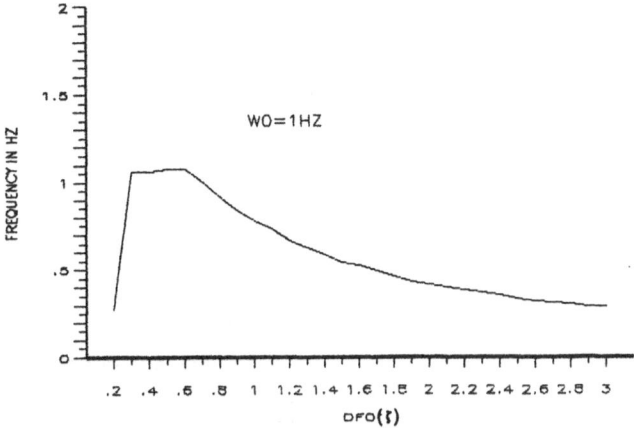

Fig. 11.6 Frequency bandwidth.

In Fig. 11.5 the limiting loop gain is plotted as a function of the damping factor. It can be seen that the loop gain reaches a maximum at 0·6 damping; the value of the non-dimensional loop gain K' being $3·4\,\text{s}^{-1}$ or 10·6 dB. Corresponding frequency bandwidth at 4 dB attenuation is plotted in Fig. 11.6. The non-dimensional bandwidth, i.e. when $\omega_0 = 1$ Hz reaches a maximum value of 1·1 between 0·3 and 0·6 damping. The frequency bandwidth, sometimes referred to as the cut-off frequency, is a measure of the speed of response to transients; i.e. the system will respond to a

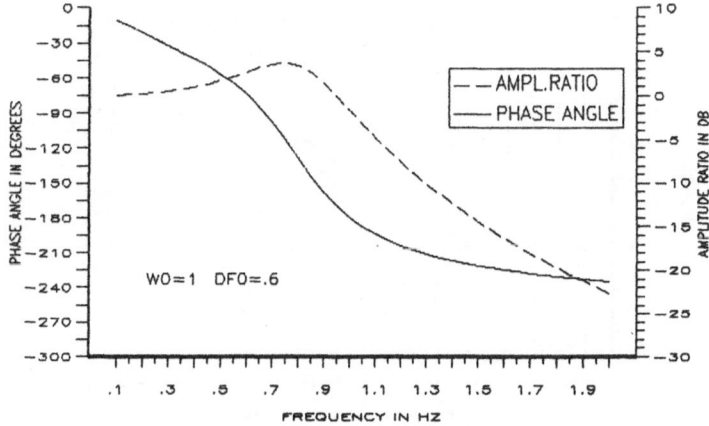

Fig. 11.7 Closed loop Bode diagram.

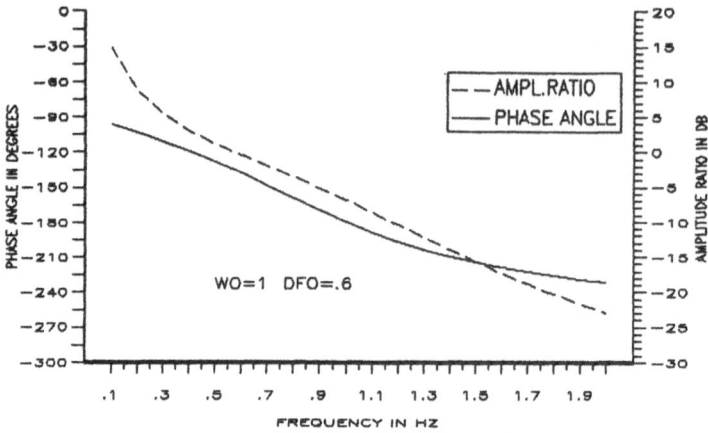

Fig. 11.8   Open loop Bode diagram.

sinusoidal input up to the cut-off frequency, but the output will attenuate rapidly beyond the frequency bandwidth. This is clearly shown by the closed loop Bode diagram, Fig. 11.7, where the amplitude ratio is -2 dB at a frequency of 1 Hz, -4 dB at 1·1 Hz, -14 dB at 1·5 Hz and -23 dB at 2 Hz. Corresponding output attenuation is 20·5%, 37%, 80% and 93%. The closed loop Bode diagram, Fig. 11.7, shows system frequency response at the optimum operating condition indicated by the stability boundary

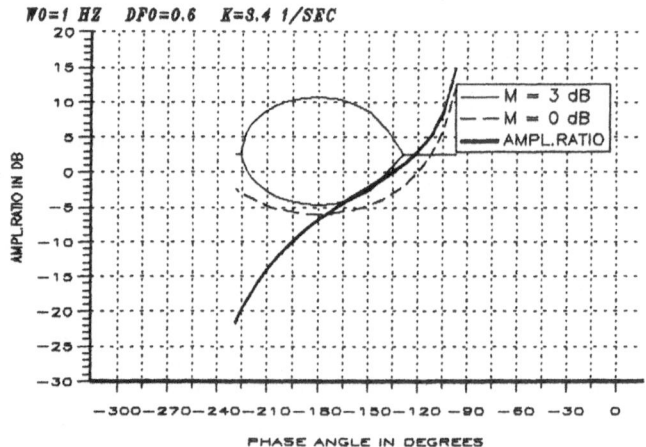

Fig. 11.9   Nichols chart.

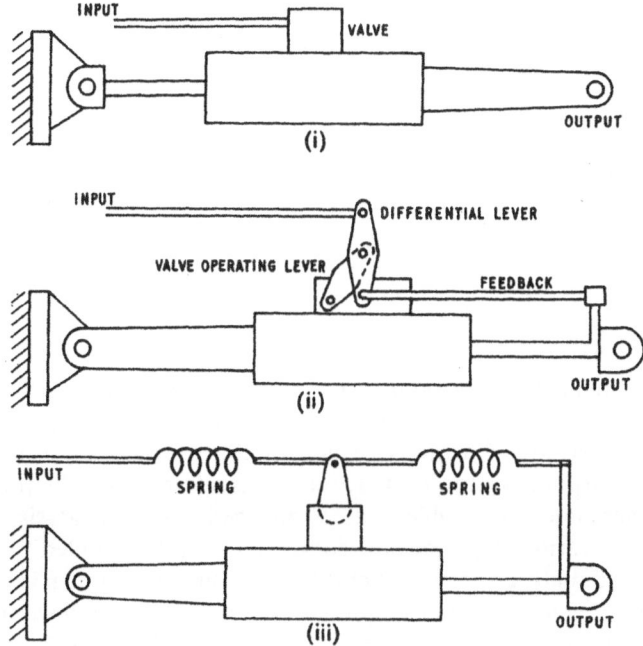

Fig. 11.10 Mechanical–hydraulic servo actuator arrangements.

plot, Fig. 11.5. An open loop Bode diagram and Nichols chart for the optimum operating condition are plotted in Figs 11.8 and 11.9. Alternative arrangements of mechanical hudraulic servo actuators are shown in Fig. 11.10.

12

System Analysis of Electro-hydraulic Control Systems

The system analysis introduced in the previous chapter will now be expanded to cover several types of electro-hydraulic control systems. The method described in Chapter 11 is not suitable for the analysis of control systems incorporating flow feedback elements referred to in Chapters 5 and 6. To overcome this problem, a rationalized system analysis method, suitable for various types of flow controls, has to be adopted.

A diagrammatic arrangement of the valve controlled system is shown in Fig. 12.1, and block diagrams representing systems with and without flow feedback shown in Figs 12.3 and 12.2 respectively. Q_0 denotes the valve-

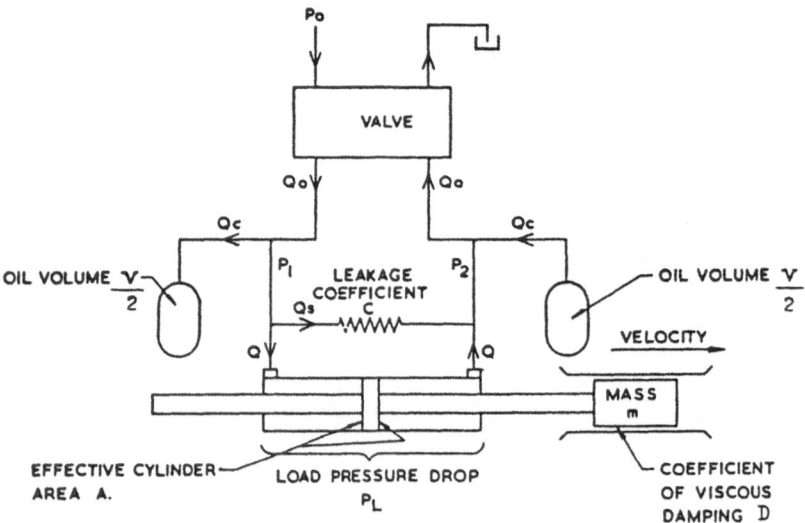

Fig. 12.1 Valve system diagram.

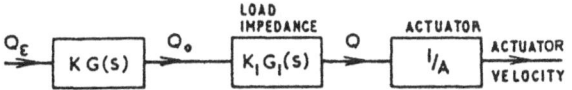

Fig. 12.2 System block diagram (no flow feedback).

controlled flow, Q the effective flow converted into motion by the actuator, Q_i the demanded (or ideal) flow and Q_ε the flow error. For non-flow feedback systems $Q_\varepsilon = Q_i$.

The following equations are then generally applicable. For small perturbations, since $Q_0 = f(Q_\varepsilon, P_L)$,

$$Q_0 = \frac{\partial Q_0}{\partial Q_\varepsilon} Q_\varepsilon + \frac{\partial Q_0}{\partial P_L} P_L \tag{12.1}$$

$$= KQ_\varepsilon + R_V P_L$$

where

$$K = \frac{\partial Q_0}{\partial Q_\varepsilon} \quad \text{and} \quad R_V = \frac{\partial Q_0}{\partial P_L}$$

Since

$$Q_0 = k\sqrt{(P_0 - P_L)}, \quad R_V = \frac{-\frac{1}{2}Q_0}{P_0 - P_L} \tag{12.2}$$

$$Q_s + Q = Q_0 - Q_c \tag{12.3}$$

$Q_c = \frac{1}{2}sP_1(V/N)$, and since $P_1 = \frac{1}{2}(P_0 + P_L)$,

$$Q_c = \frac{1}{4}sP_L \frac{V}{N} \tag{12.4}$$

$$P_L = \frac{Q}{A^2}(sm + D) \tag{12.5}$$

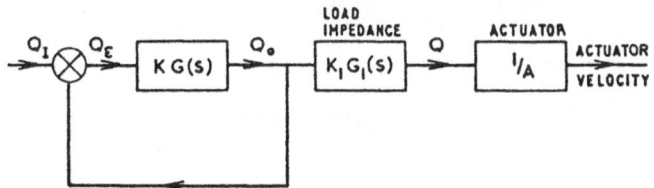

Fig. 12.3 System block diagram (flow feedback).

Combining the above equations yields the open loop transfer functions:

$$\frac{Q_0}{Q_\varepsilon} = \frac{K_0\left(\dfrac{s^2}{\omega_3^2} + 2\zeta_3\dfrac{s}{\omega_3} + 1\right)}{\dfrac{s^2}{\omega_0^2} + 2\zeta_0\dfrac{s}{\omega_0} + 1} \tag{12·6}$$

$$\frac{Q}{Q_0} = \frac{K_1}{\dfrac{s^2}{\omega_3^2} + 2\zeta_3\dfrac{s}{\omega_3} + 1} \tag{12·7}$$

where

$$\omega_3 = 2\sqrt{\left[\frac{N}{V}\frac{A^2}{m}\left(\frac{cD}{A^2} + 1\right)\right]} \tag{12.8}$$

$$\omega_0 = 2\sqrt{\left\{\frac{N}{V}\frac{A^2}{m}\left[\frac{D}{A^2}(c + R_V) + 1\right]\right\}} \tag{12.9}$$

$$\zeta_3 = \frac{\dfrac{cm}{A^2} + \dfrac{D}{A^2}\dfrac{V}{4N}}{\left(\dfrac{cD}{A^2}\right) + 1}\left(\frac{\omega_3}{2}\right) \tag{12.10}$$

$$\zeta_0 = \frac{\dfrac{m}{A^2}(c + R_V) + \dfrac{D}{A^2}\dfrac{V}{4N}}{\dfrac{D}{A^2}(c + R_V) + 1}\left(\frac{\omega_0}{2}\right) \tag{12.11}$$

For systems incorporating flow feedback,

$$\frac{K_0}{K} = \frac{1}{1 + \dfrac{R_V}{c + (A^2/D)}} \tag{12.12}$$

For systems without flow feedback,

$$\frac{K_0}{K} = \frac{1}{\dfrac{D}{A^2}(c + R_V) + 1} \tag{12.13}$$

$$K_1 = \frac{1}{cD/A^2 + 1} \tag{12.14}$$

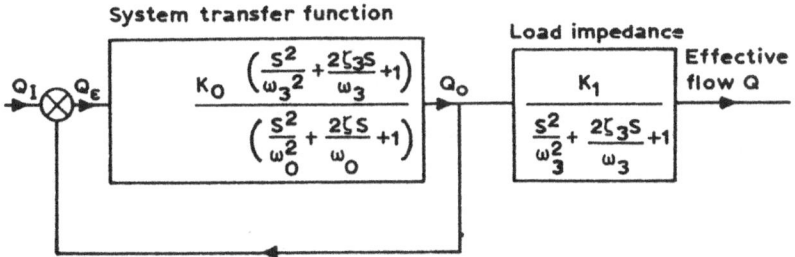

Fig. 12.4 System transfer function (flow feedback).

A system without flow feedback then simplifies to the transfer function given in Fig. 12.5, while the system with flow feedback is represented by the transfer function Fig. 12.4.

The method applied to analyse the valve-controlled cylinder system can equally be used to analyse a hydrostatic transmission described in Section 5.2 and represented by block diagram Fig. 5.5. A system comprising a variable displacement pump and a fixed displacement motor is shown diagrammatically in Fig. 12.6 and in block diagram form in Fig. 12.7.

Equations (12.1), (12.3) and (12.4) are applicable. Since

$$Q_0 = C_p\omega - c_p P_L \qquad R_v = -c_p \tag{12.15}$$

The equation of motion for a motor-driven system is:

$$P_L = \frac{Q}{C^2}(sI + D) \tag{12.16}$$

Combining the above equations yields the open loop transfer functions expressed by eqns (12.6) and (12.7). For eqns (12.8) to (12.14), $R_v = c_p$, the motor displacement, C, replaces the piston area, A, and the moment of inertia, I, replaces mass, m.

For a valve-controlled, motor-driven system, eqns (12.1) to (12.14) are applicable, provided A is replaced by C and m replaced by I.

System transfer function

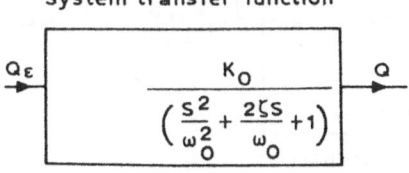

Fig. 12.5 System transfer function (no flow feedback).

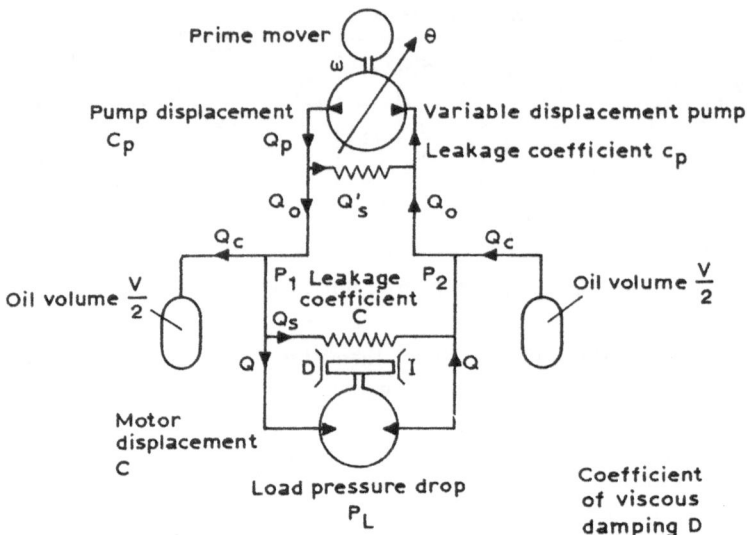

Fig. 12.6 Hydrostatic transmission diagram.

So far we have established the transfer functions for the hydraulic transmission of flow-controlled systems, but since we are dealing with electro-hydraulic control systems, we have to include the dynamic characteristics of additional control and data transmission elements, e.g. servo or proportional control valves and feedback transducers, as well as electronic circuitry such as three-term controllers, passive networks and signal conditioning. Electro-hydraulic control valves and feedback transducers can usually be adequately represented by second order and electronic circuits by first order transfer functions.

These additional transfer functions can be cascaded with the transfer function describing the hydraulic transmission, to give the overall system

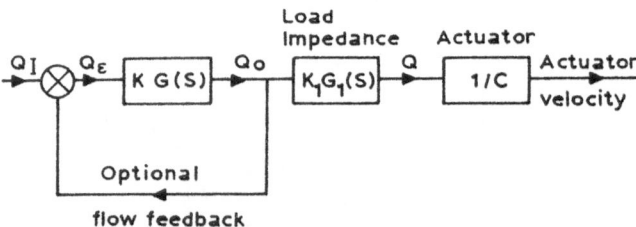

Fig. 12.7 Hydrostatic transmission block diagram.

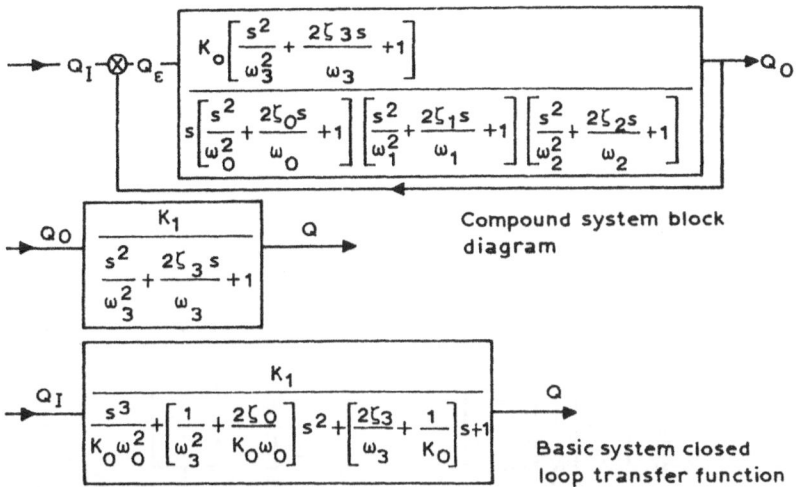

Compound system block diagram

Basic system closed loop transfer function

Frequency response subroutine

Inputs: K_0 K_1 ω_1 ω_2 ω_0 ω_3 ζ_0 ζ_1 ζ_2 ζ_3

Table 1:

	Rectangular Co-ordinates		Polar Co-ordinates	
Imaginary	Real		Angle	Amplitude
$J_1 = c\omega - a\omega^3$	$R_1 = 1 - b\omega^2$		ϕ_1	A_1
$J_2 = 2\zeta_1\omega/\omega_1$	$R_2 = 1 - (\omega/\omega_1)^2$		ϕ_2	A_2
$J_3 = 2\zeta_2\omega/\omega_2$	$R_3 = 1 - (\omega/\omega_2)^2$		ϕ_3	A_3

$$\phi = \tan^{-1} J/R \quad \text{and} \quad A = \sqrt{J^2 + R^2}$$

$$a = \frac{1}{K_0\omega_0^2} \quad b = \frac{1}{\omega_3^2} + \frac{2\zeta_0}{K_0\omega_0} \quad c = \frac{2\zeta_3}{\omega_3} + \frac{1}{K_0}$$

Outputs: Phase shift ϕ (degrees)
            Amplitude ratio $A(d\text{B})$

Variants: Open loop frequency response—Compound system
            Open loop frequency response—Basic system
            Closed loop frequency response—Compound system
            Closed loop frequency response—Basic system

Fig. 12.8   Mathematical model, ('a' configuration).

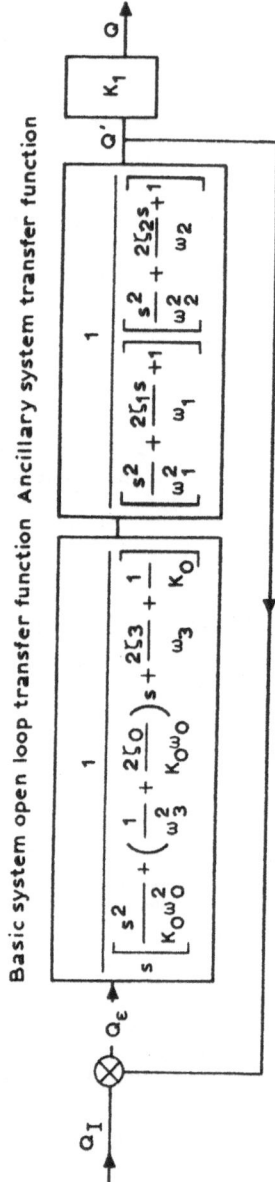

Basic system open loop transfer function   Ancillary system transfer function

$$\left[\dfrac{s^2}{\dfrac{s^2}{\omega_1^2} + \dfrac{2\zeta_1 s}{\omega_1} + 1} \right]\left[\dfrac{1}{\dfrac{s^2}{\omega_2^2} + \dfrac{2\zeta_2 s}{\omega_2} + 1} \right]$$

$$s\left[\dfrac{s^2}{K_O\omega_O^2} + \left(\dfrac{1}{\omega_3^2} + \dfrac{2\zeta_O}{K_O\omega_O} \right)s + \dfrac{2\zeta_3}{\omega_3} + \dfrac{1}{K_O} \right]$$

Equivalent compound system transfer function

Fig. 12.9   Mathematical model. ('b' configuration).

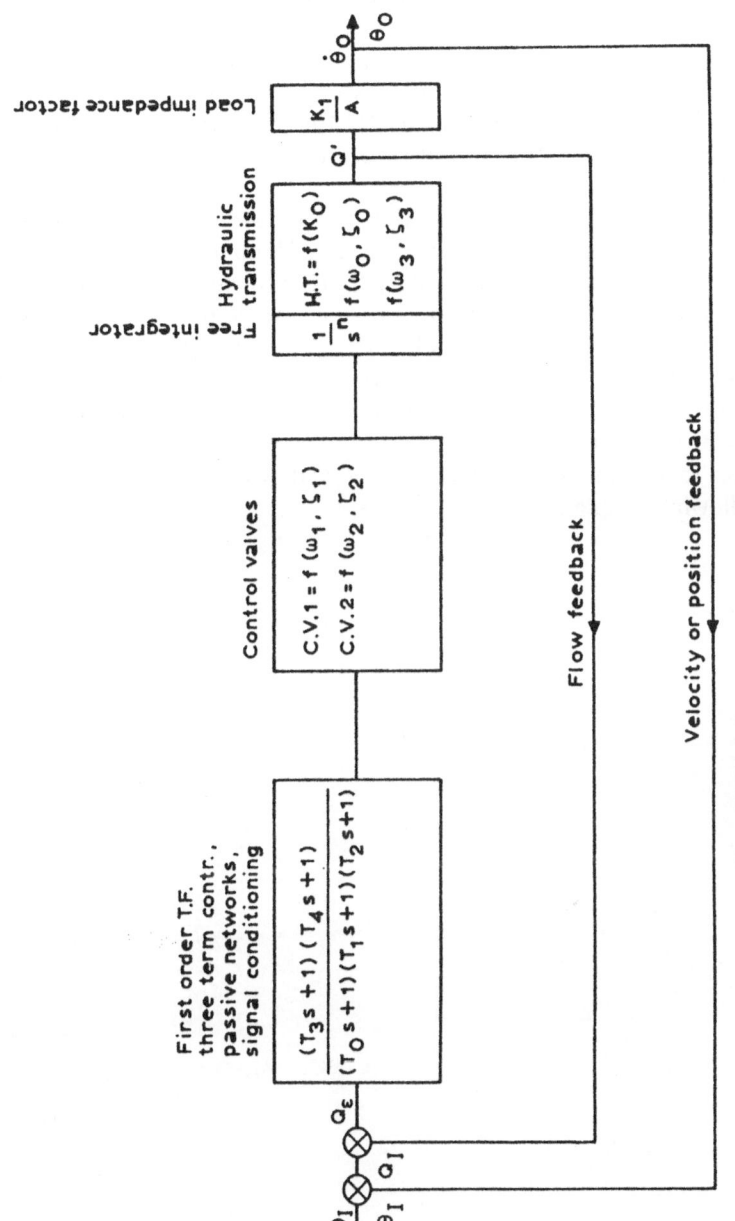

Fig. 12.10 Compound system block diagram.

transfer function. Electro-hydraulic control systems can then be represented by a compound system block diagram, as shown in Fig. 12.10. A free integrator $1/s^n$ is included; in the position control system analysed in the previous chapter, the actuator acts as an integrator depicted by the free integrator $1/s$ in eqns (11.15) and (11.16). Free integrators are also introduced by pilot stages of multi-stage valves, and can be inserted by means of electronic circuits.

The more complex transfer function of the flow feedback system, Fig. 12.4 can be simplified to the basic system closed loop transfer function, Fig. 12.8 and then combined with ancillary transfer functions in the form of a basic system open loop transfer function, as shown in Fig. 12.9.

Since we are now dealing with higher order transfer functions and, in relation to loop gain optimization, with differential equations containing non-constant coefficients, as shown by Figs 12.8 and 12.9, a manual method of system analysis and optimization is no longer feasible. We therefore have to resort to a computerized analytical method, which will be described in the following chapter.

13

Modular Optimized System Simulation

The need for a rapid and easy means of analysing hydraulic control systems and predicting system performance became apparent many years ago. For any such program to be of practical use, it has to satisfy several basic requirements, which can be summarized as:

(1) The ability to accommodate high-order system transfer functions, e.g. up to 15th order.
(2) An automatic analytical optimization procedure to any given stability criteria.
(3) Transient response to any given composite duty cycle.
(4) Comprehensive graphics display and print-out facilities.
(5) A database containing system data, component data and graphics display files.
(6) Has been demonstrated to provide close correlation between predicted and actual performance.
(7) Is structured to be user-friendly, i.e. is written in a conversational mode, does not assume any specialized knowledge of control theory, all input and output values are in practical units, incorporates safety warnings in case limiting operating conditions are exceeded.

The modular structure, as shown in flow chart, Fig. 13.1, was chosen to facilitate extension to other types of control system, e.g. electrical, electro-pneumatic, since all modules other than system identification modules are common to all configurations. The program structure block diagram, Fig. 13.2 gives an overview of the layout of the software package 'HYDRASOFT'. S, F, T, P and L are functional modules; C and D provide access to the component and system databases.

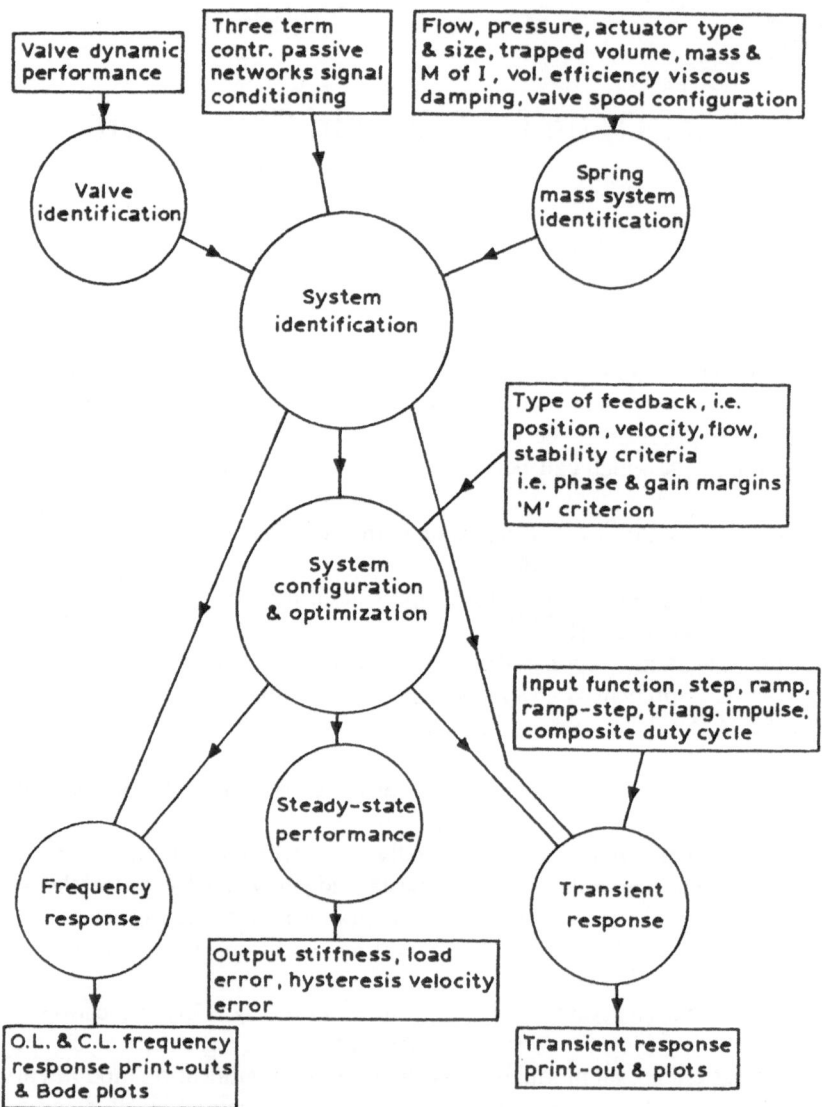

Fig. 13.1 Computer model flow chart.

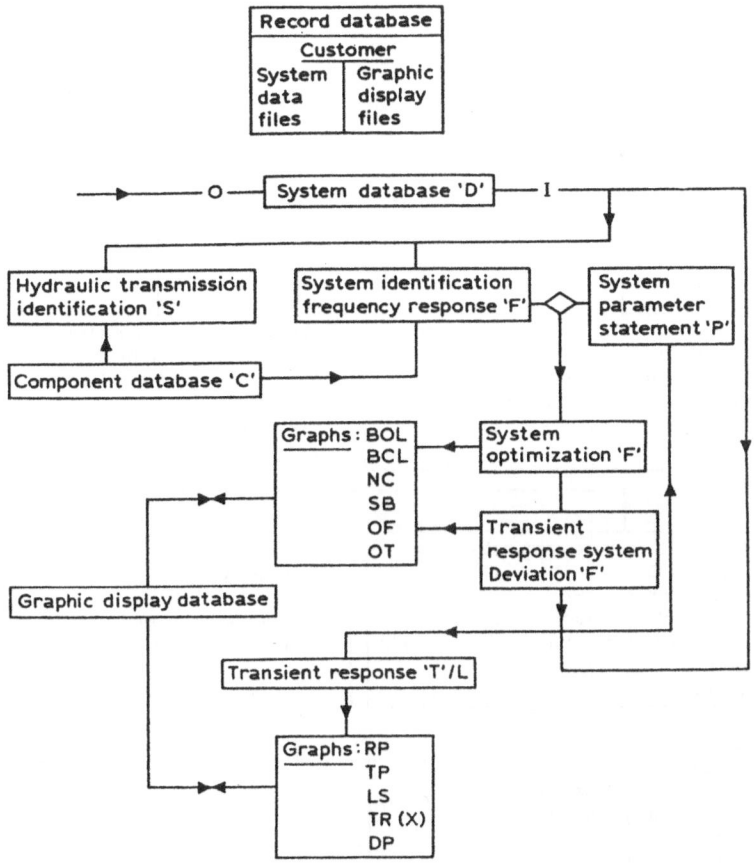

Fig. 13.2 Program structure block diagram.

For the benefit of any readers willing to write their own programs, some of the guidelines adopted in the preparation of HYDRASOFT will now be reiterated.

All calculations are performed using SI units, all input and output values are in practical units. A typical hydraulic transmission identification algorithm is shown in Fig. 13.3. This is basically a number-crunching procedure, evaluating the equations derived in previous chapters. The algorithm can of course be expanded to include additional system configurations.

A frequency response algorithm applicable to the system transfer

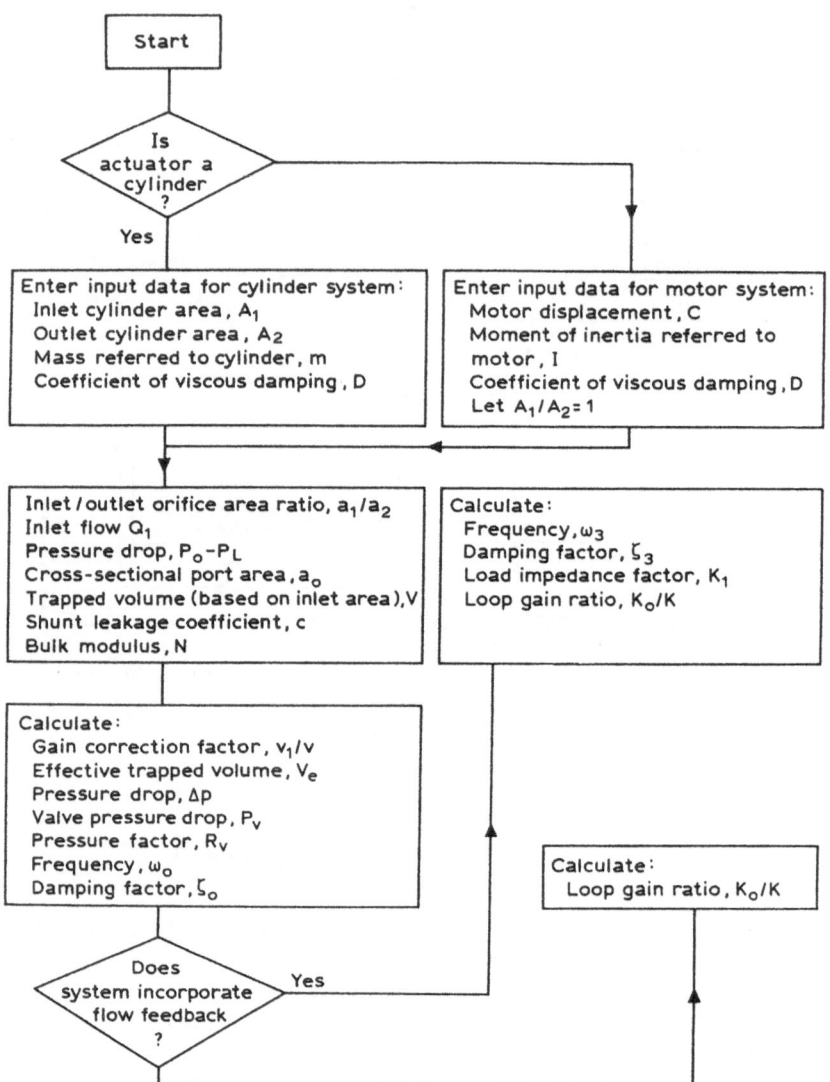

Fig. 13.3 System identification algorithm.

function, Fig. 13.4 is shown in Fig. 13.5, which merely involves the cascading of several functions of complex variables. A manual method to establish the loop gain compatible with a number of given stability criteria was described in Section 8.1.6. Referring to Fig. 8.10, this involves plotting contours, each contour representing a specific operating condition. This is obviously a laborious procedure, even for a third order transfer function as given by eqn (11.19), and therefore not suitable for the analysis of higher order systems. In order to overcome this problem, we have to devise an analytical computerized method which replaces the manual procedure of

Open loop system transfer function

$$\frac{\theta_0}{\varepsilon} = \frac{K(\tau_2 s + 1)(\tau_4 s + 1)}{s^n \left(\frac{s^2}{\omega_1^2} + \frac{2\zeta_1 s}{\omega_1} + 1\right)\left(\frac{s^2}{\omega_0^2} + \frac{2\zeta s}{\omega_0} + 1\right)(\tau_1 s + 1)(\tau_3 s + 1)(\tau S + 1)}$$

Frequency response subroutine

Inputs: K n ω τ_1 τ_2 τ_3 τ_4 τ ω_1 ζ_1 ω_0 ζ

Table 1:

	Rectangular co-ordinates		Polar co-ordinates	
Imaginary	Real		Angle	Amplitude
$J_1 = \tau_1\omega$	$R_1 = 1$		ϕ_1	A_1
$J_2 = \tau_2\omega$	$R_2 = 1$		ϕ_2	A_2
$J_3 = \tau_3\omega$	$R_3 = 1$		ϕ_3	A_3
$J_4 = \tau_4\omega$	$R_4 = 1$		ϕ_4	A_4
$J_5 = \tau\omega$	$R_5 = 1$		ϕ_5	A_5
$J_6 = 2\zeta\omega/\omega_0$	$R_6 = 1 - (\omega/\omega_0)^2$		ϕ_6	A_6
$J_7 = 2\zeta_1\omega/\omega_1$	$R_7 = 1 - (\omega/\omega_1)^2$		ϕ_7	A_7

$$\phi = \tan^{-1} J/R \quad \text{and} \quad A = \sqrt{J^2 + R^2}$$

Outputs: Phase shift ϕ (degrees)
             Amplitude ratio A (dB)

Variants: No free integrator ($n = 0$)
             Free single integrator ($n = 1$)
             Free double integrator ($n = 2$)
             Open loop frequency response
             Closed loop frequency response

Fig. 13.4 Compound system transfer function.

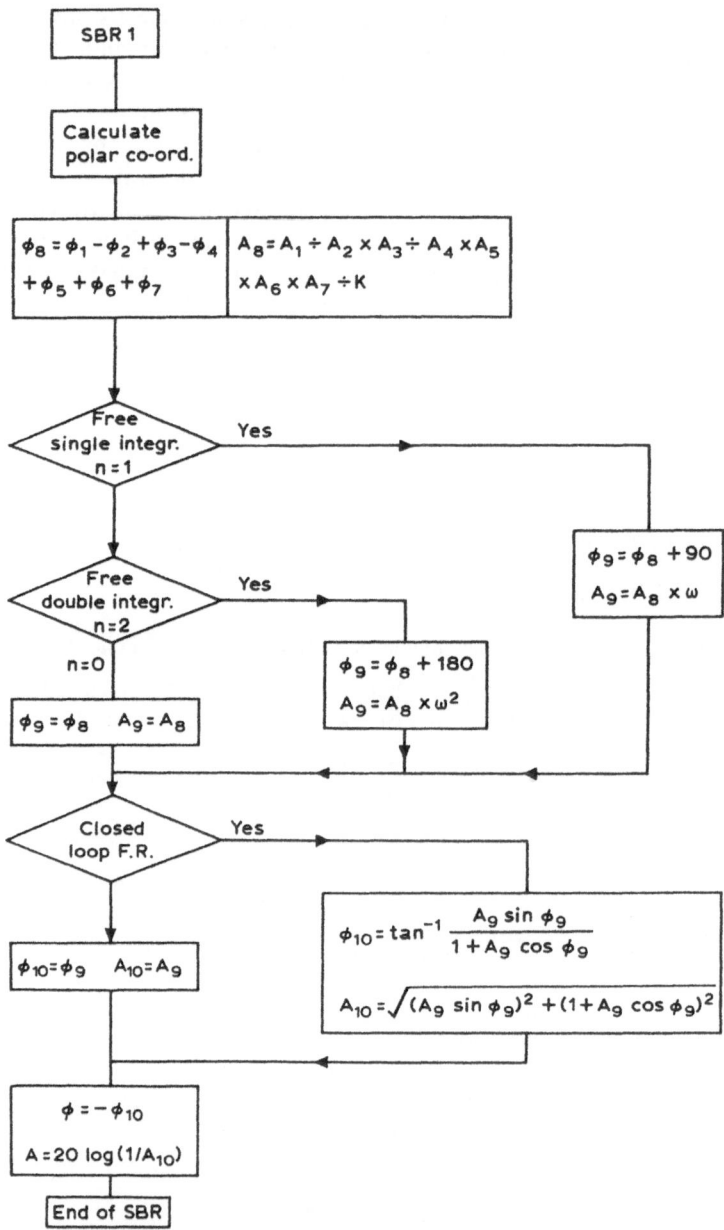

Fig. 13.5 Frequency response algorithm.

plotting system contours on a template which is moved vertically up or down a Nichols chart until either the set phase or gain margin is reached, the vertical displacement being a measure of the maximum permissible loop gain.

A loop gain optimization algorithm is shown in a qualitative form in Fig. 13.6, and quantified in Fig. 13.7. The frequency response algorithm, Fig. 13.5 constitutes a subroutine, denoted as SBR1. The various stages of the optimization procedure will now be described with reference to the sequence numbers 1 to 14 in Fig. 13.6. Since it is not known in advance whether the phase or gain margin is going to be the critical stability criterion, the algorithm has to be written on the premise that one of the two criteria is the initial stability criterion to be investigated. The algorithm shown in Figs 13.6 and 13.7 assumes the phase margin to be the initial critical criterion.

(1) Having set an initial value for the frequency ω and the loop gain K, the program homes in on the 0 dB open loop amplitude ratio by means of an iterative loop '8' $= f(\omega)$ until the amplitude ratio is within the set limits of 0·1 dB.

(2) In accordance with our initial premise, the phase margin is assumed to be the critical stability criterion.

(3) Homing in on selected phase margin by means of iterative loop '7' $=$ $f(K)$ and loop '8' until phase margin is within set limits of 1°, thus establishing optimum loop gain compatible with phase margin criterion.

(4) Homing in on $-180°$ open loop phase angle via iterative loop '10' $=$ $f(\omega)$ until phase angle is within set limits of 2°.

(5) Phase margin is still assumed to be the critical criterion.

(6) Actual gain margin is greater than gain margin criterion, hence initial assumption that phase margin is critical criterion is true and calculated loop gain is identical with optimum loop gain.

(7) See stage (3).

(8) See stages (1) and (3).

(9) Convergence check on iteration '7'.

(10) See stage (4).

(11 & 12) Gain margin is critical stability criterion; optimum loop gain compatible with gain margin criterion is calculated.

(13) In some cases where gain margin is the predominantly critical criterion, iteration '7' does not converge and is then aborted and the flow path re-routed to stage (3).

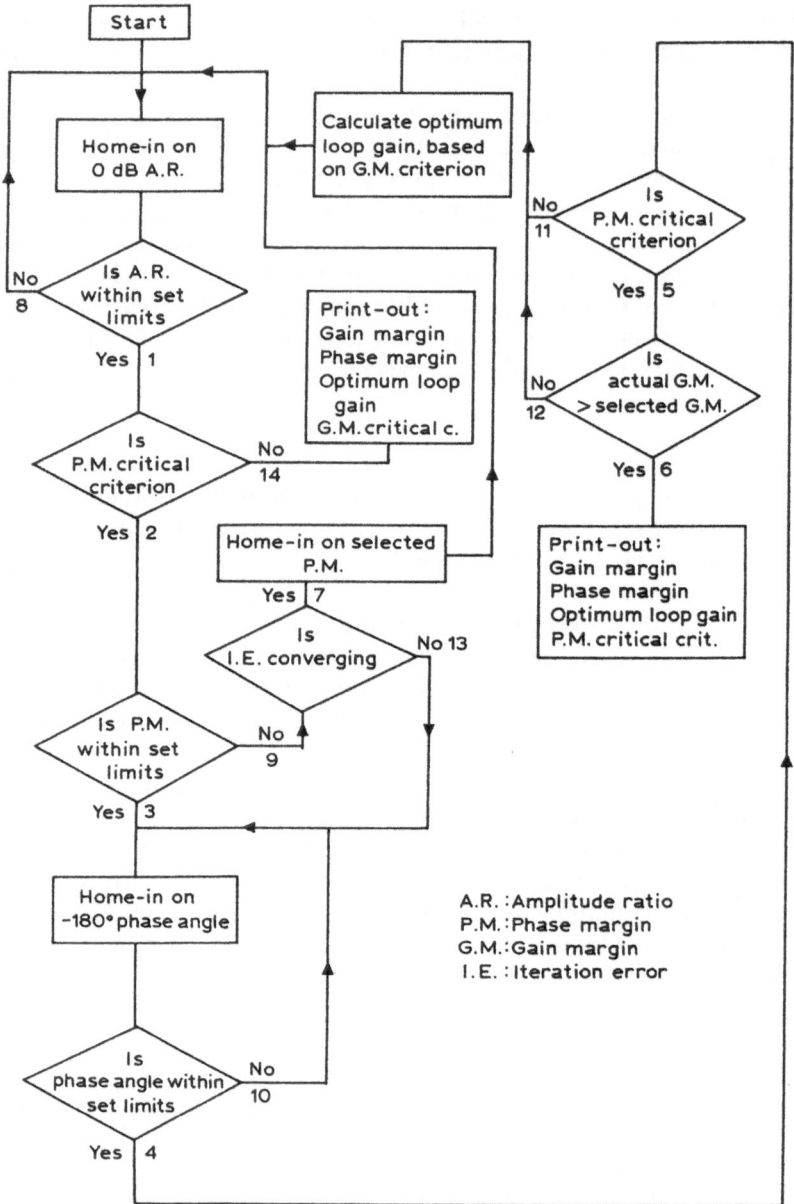

Fig. 13.6 Qualitative optimization algorithm—'A'.

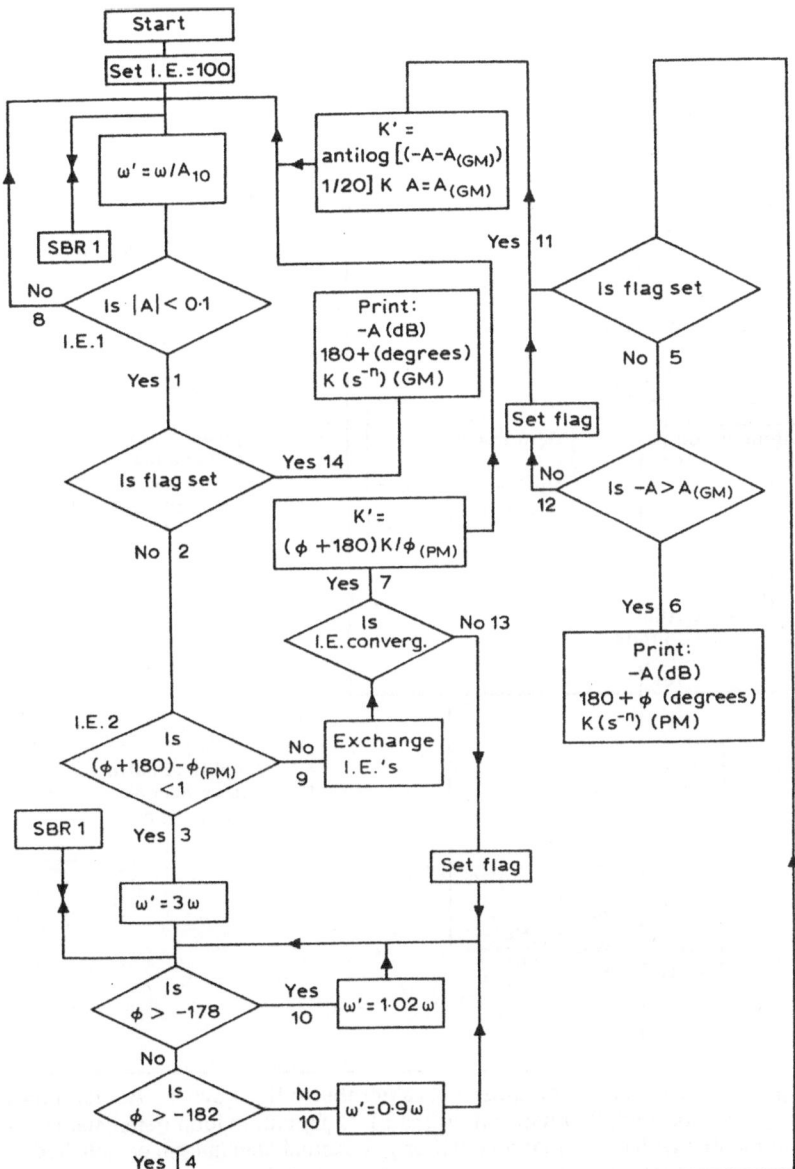

Fig. 13.7 Quantitative optimization algorithm—'A'.

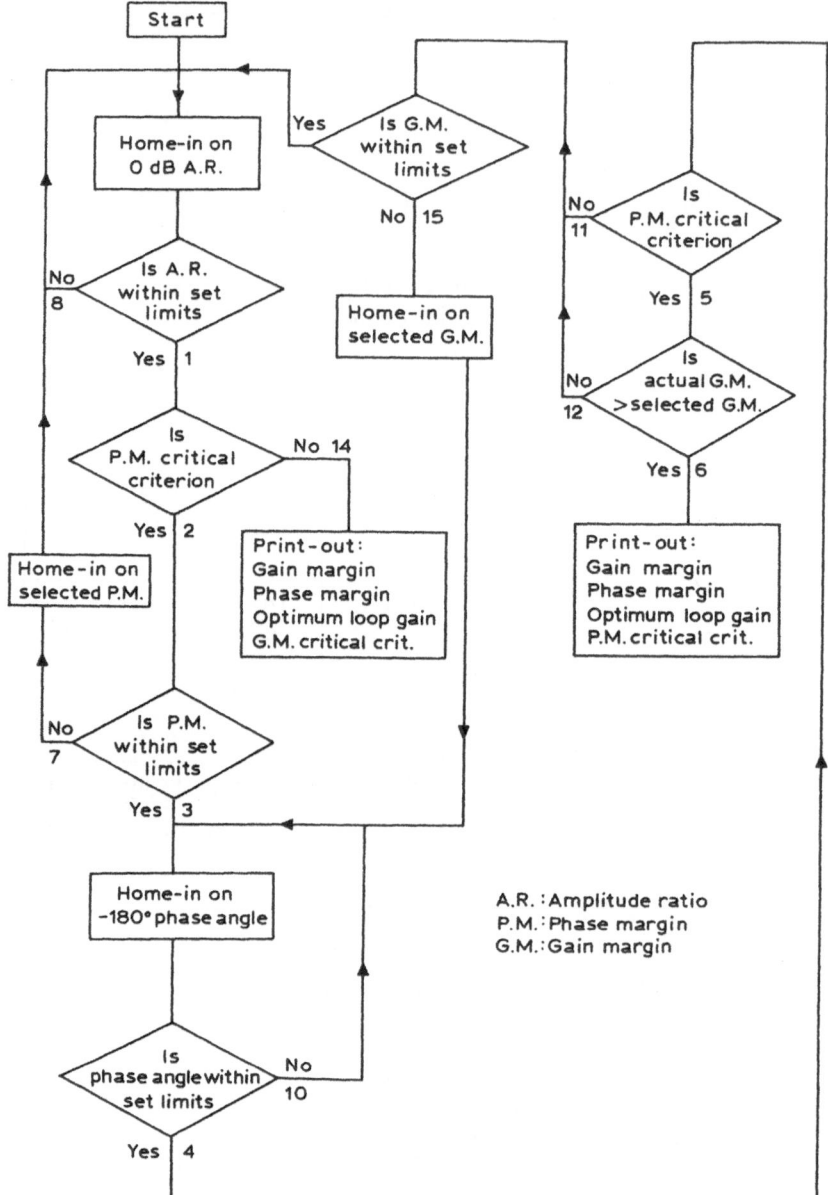

Fig. 13.8 Qualitative optimization algorithm—'B'. Key: 1, Actual phase margin established; 2, Initial setting (as 5); 3, Optimum loop gain established (compatible with phase margin criterion); 4, Actual gain margin established; 5, Initial setting (as 2); 6, Print-out; 7, Iteration f (K); 8, Iteration f (ω); 9, Check whether iteration converging; 10, Iteration f (ω); 11, Triggered by 12 or 13; 12, Triggers 11 & 14; 13, By-passes P.M. criterion; 14, Triggered by 12 or 13; Print-out.

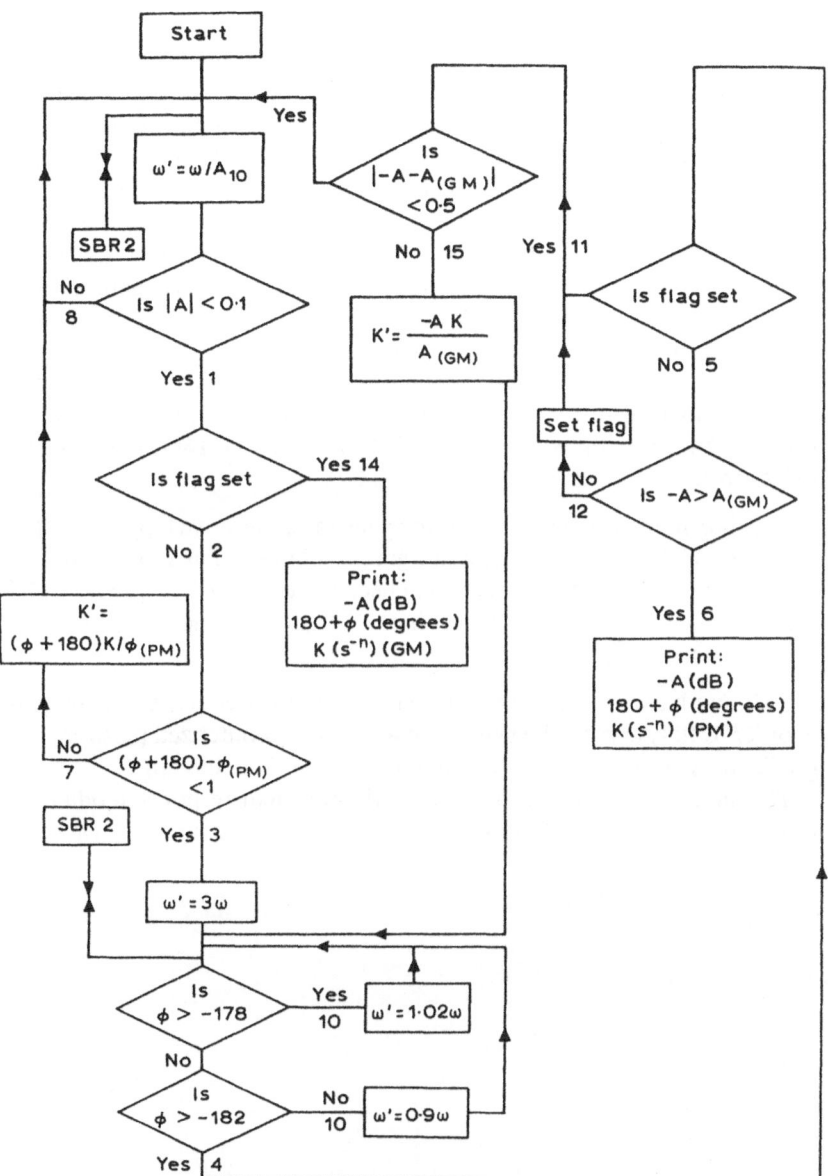

Fig. 13.9 Quantitative optimization algorithm—'B'.

(14)   When gain margin is the critical criterion, the actual phase margin has to be re-established.

The above algorithm is suitable for system transfer functions containing differential equations with constant coefficients, but has to be modified if it is to be applied to transfer functions of the type shown in Figs 12.8 and 12.9. It can be seen clearly from Fig. 12.9 that the loop gain, K_0 forms part of the coefficients of the basic system open loop transfer function, resulting in a continuously changing contour on a Nyquist diagram or Nichols chart as the loop gain is varied. As the calculation performed in stages (11 & 12) to obtain the optimum loop gain compatible with the given gain margin criterion is equivalent to the vertical movement of a given contour on a Nichols chart, it can only be applied to constant contour lines. To be able to cope with transfer functions containing non-constant coefficients, the direct calculation in stages (11 & 12) is replaced by an iterative loop, as shown in the algorithms of Figs 13.8 and 13.9. Stages (11 & 12) can then be redefined as:

Homing in on selected gain margin by means of iterative loop '15' $= f(K)$ and loop '10' $= f(\omega)$ until gain margin is within set limits of 0·2 dB. For the sake of clarity, the convergence check on iteration '7', represented by stages (9) and (13) in the algorithms of Figs 13.6 and 13.7, has been omitted from Figs 13.8 and 13.9.

So far we have analysed the control system in the frequency domain, but in order to fully exploit the benefits of a digital computerized performance prediction package we have to extend it to the time domain.

The methods adopted in developing the transient response module 'T' will be discussed in the following chapter.

14

System Analysis in the Time Domain

The algorithms discussed in the preceding chapter enable us to manipulate high-order system transfer functions in the frequency domain. To obtain the corresponding time response involves the solution of high-order differential equations, using the inverse Laplace transformation, which is a laborious method even for simple input functions such as step and ramp-step demands.

A realistic assessment of system performance in the time domain requires the capability to simulate the actual duty cycle which frequently comprises a compound input function passing through several transitional stages. The procedure adopted in the transient response module of the computer software package Hydrosoft will now be described in detail.

The transient response procedure has two distinct phases:

(1) Derivation of equivalent second order system.
(2) Calculation of transient response to given duty cycle by super-position method.

The equivalent system is defined as a second order system having identical frequency response characteristics to the high-order system at 90° phase lag. Closed loop Bode diagrams for fifth and ninth order systems together with their equivalent system plots are shown in Figs 14.1 and 14.2 respectively. It can be seen that, in accordance with the definition, the frequency and amplitude ratio of the high-order and equivalent systems are identical at 90° phase lag. Although there is some deviation at frequencies below and above the natural frequency of the equivalent system, this does not significantly affect transient response characteristics, and evaluation of high-order systems has produced close correlation between predicted and actual performance. We shall now consider the transient response of a second order system to a compound duty cycle.

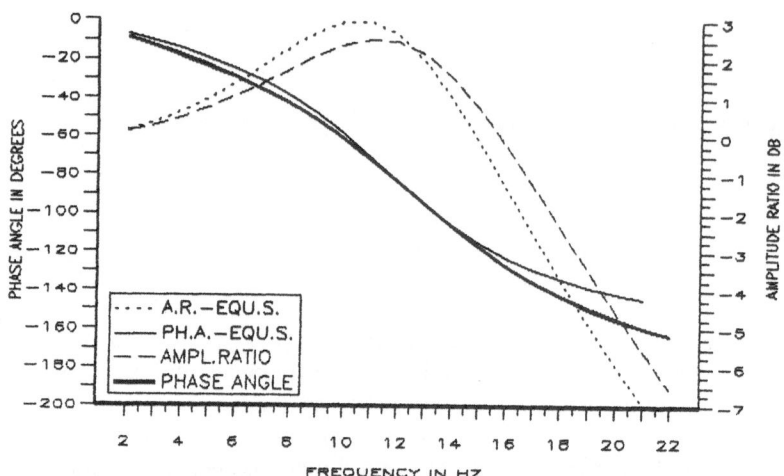

Fig. 14.1   Equivalent system Bode diagram, fifth order system.

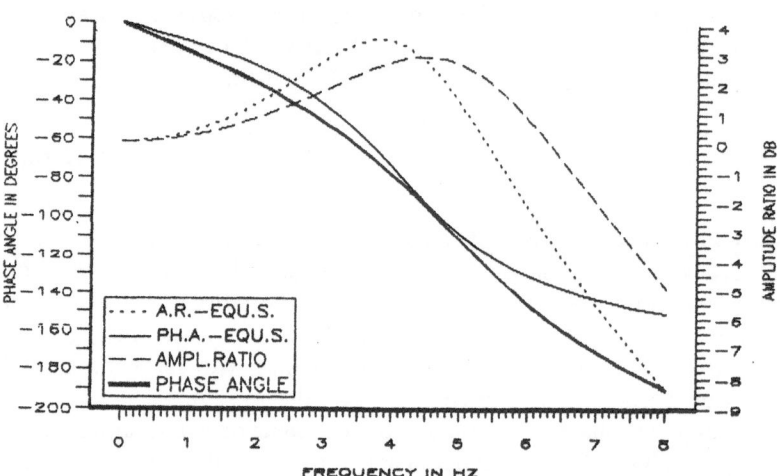

Fig. 14.2   Equivalent system Bode diagram, ninth order system.

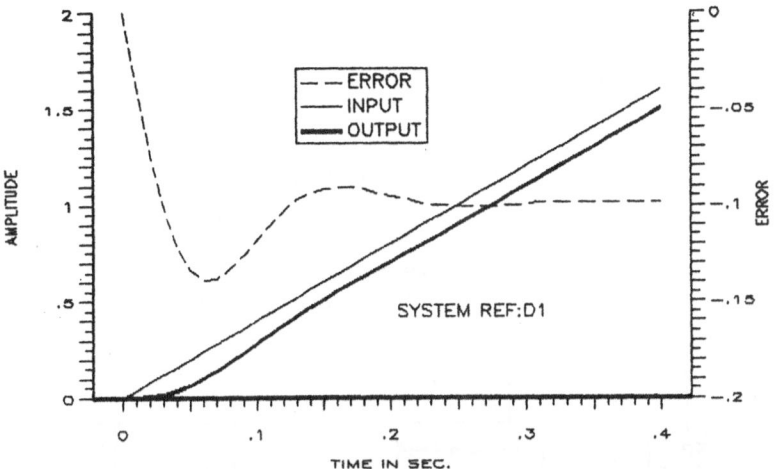

Fig. 14.3 Transient response to ramp input.

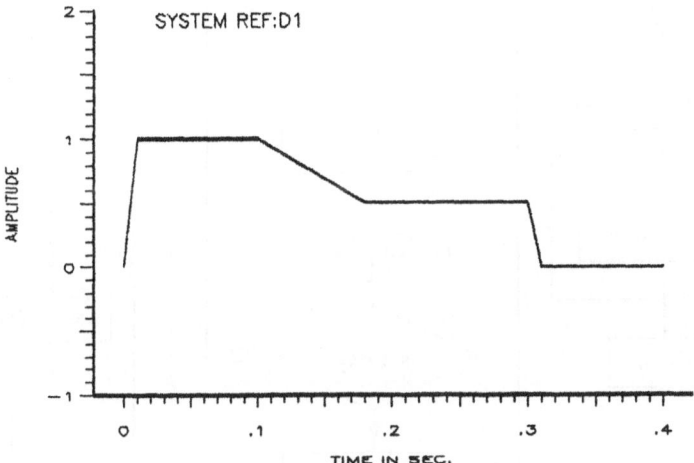

Fig. 14.4 Compound duty cycle.

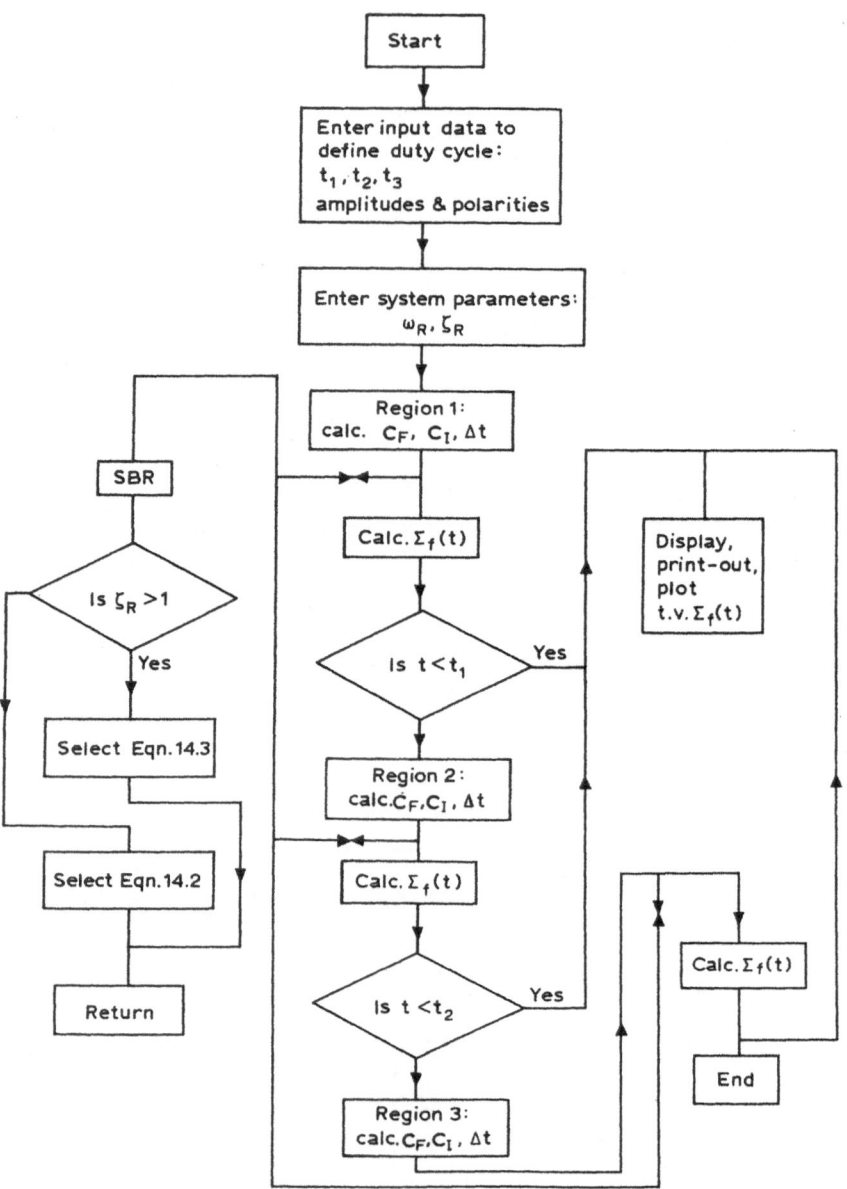

Fig. 14.5 Transient response algorithm.

Let the transfer function of the equivalent system,

$$f(s) = \frac{1}{(s^2/\omega_n^2) + 2\zeta_n(s/\omega_n) + 1} \tag{14.1}$$

The inverse Laplace transform for a unit ramp input is then given by the expression,

$$f(t) = 1/(\omega_n t_1)\{\omega_n t - 2\zeta_n + e^{-\zeta_n \omega_n t}[2\zeta_n \cos(Bt) - D\sin(Bt)]\} \tag{14.2}$$

where the unit ramp time $= t_1$, $B = \omega_n\sqrt{(1 - \zeta_n^2)}$, and $D = (1 - 2\zeta_n^2)/\sqrt{(1 - \zeta_n^2)}$, when $\zeta_n < 1$.

When the damping factor $\zeta_n > 1$, the trigonometric functions $\sin(Bt)$ and $\cos(Bt)$ become hyperbolic functions $\sinh(Bt)$ and $\cosh(Bt)$, where $\sinh(Bt) = \frac{1}{2}(e^{Bt} - e^{-Bt})$, $\cosh(Bt) = \frac{1}{2}(e^{Bt} + e^{-Bt})$, $B = \omega_n\sqrt{(\zeta_n^2 - 1)}$, and

$$D = \frac{1 - 2\zeta_n^2}{\sqrt{(\zeta_n^2 - 1)}} \tag{14.3}$$

The transient response of a typical position control system to a 0·25 s unit ramp is shown in Fig. 14.3. The system reaches a maximum transient error of 0·14 units after 0·06 s and settles down to a steady-state following error of 0·1 units after approximately 0·23 s.

By making use of the superposition method we shall now formulate an algorithm for the calculation of system transient response to a compound duty cycle. A duty cycle comprising six transitional stages is shown in Fig.

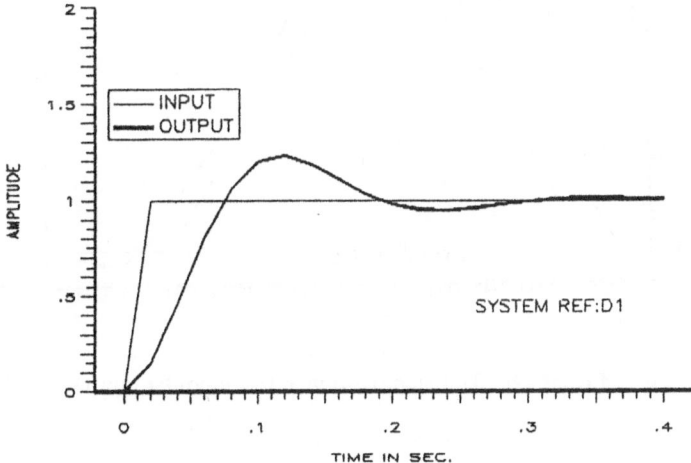

Fig. 14.6 Transient response to step input.

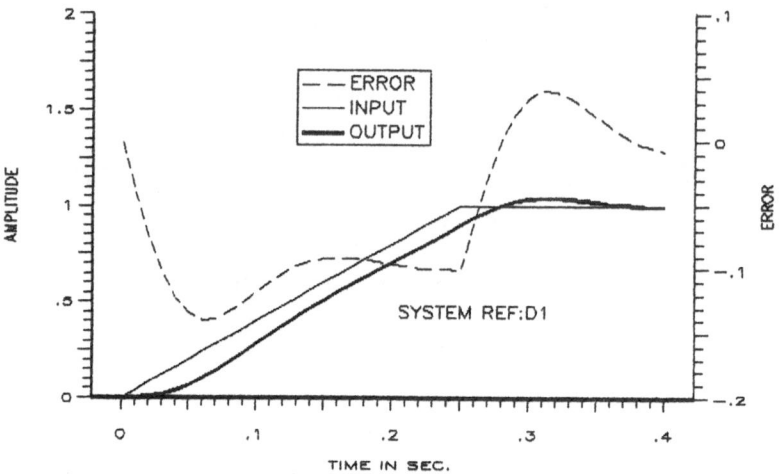

Fig. 14.7 Transient response to ramp-step input.

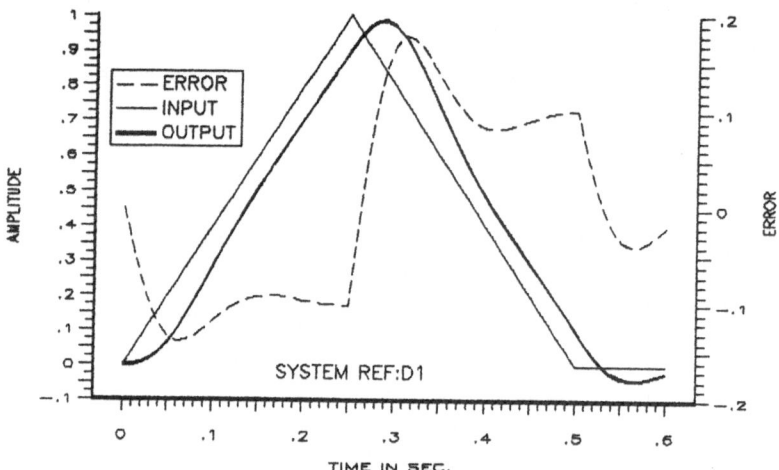

Fig. 14.8 Transient response to triangular input.

14.4, each stage being identified by a time period, t_n, and a polarized amplitude variation, Y_n. Before we can use the ramp equation (14.2) as a subroutine of a transient response algorithm, we have to adapt it for a compound duty cycle. Equation (14.2) can be re-written as:

$$f(t) = \frac{(C_F - C_I)}{\omega_n} \{\omega_n \partial t - 2\zeta_n + e^{-\zeta_n \omega_n \partial t}[2\zeta_n \cos(B\delta t) - D\sin(B\delta t)]\} \quad (14.4)$$

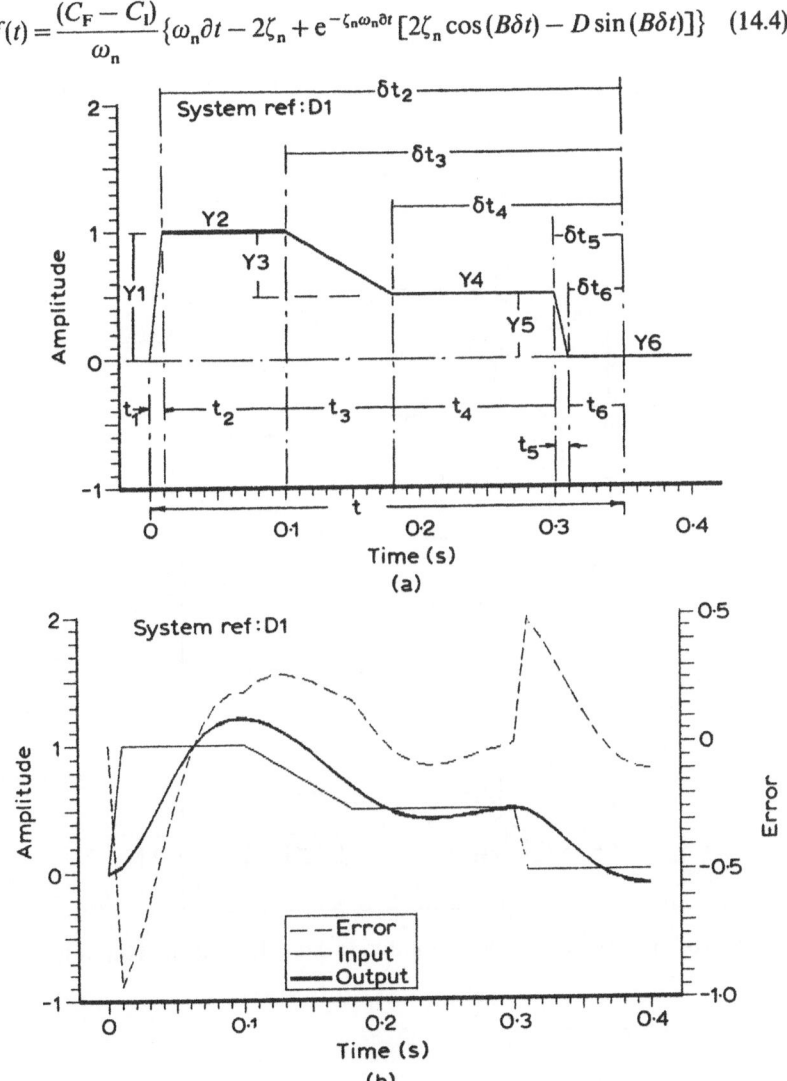

Fig. 14.9 Transient response to compound duty cycle.

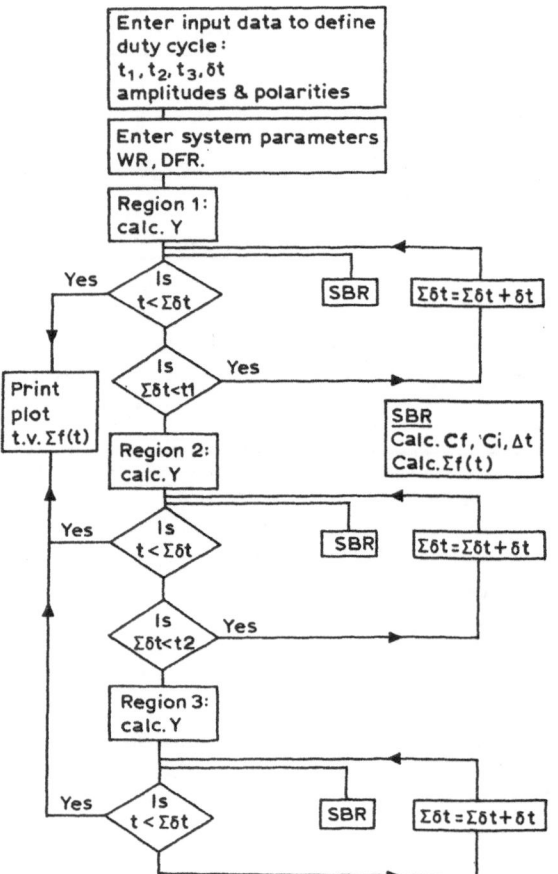

Fig. 14.10 Transient response algorithm to digitally generated input function.

where initial slope $C_1 = Y_1/t_1$, final slope $C_F = Y_F/t_F$ and $\partial t = t - \Sigma t_n$. Referring to Fig. 14.4,

$Y_1 = 1$ unit	$t_1 = 0.0001$ s (for step input)	$\delta t_1 = t$
$Y_2 = 0$	$t_2 = 0.1$	$\delta t_2 = t - t_1$
$Y_3 = -0.5$	$t_3 = 0.8$	$\delta t_3 = t - t_1 - t_2$
$Y_4 = 0$	$t_4 = 1.2$	$\delta t_4 = t - t_1 - t_2 - t_3$
$Y_5 = -0.5$	$t_5 = 0.0001$	$\delta t_5 = t - t_1 - t_2 - t_3 - t_4$
$Y_6 = 0$	$t_6 = > t_{max}$	$\delta t_6 = t - t_1 - t_2 - t_3 - t_4 - t_5$

In calculating the transient response of the system, the parameters appropriate to the initial and final regions have to be allocated to eqn (14.4) and the output amplitude obtained by a summation of all regions within the selected time envelope. A transient response algorithm covering three regions is shown in Fig. 14.5.

Four additional transient response plots of the positional control system, ref. D1, are shown in Figs 14.6 to 14.9. The response to a step input, Fig. 14.6, shows one over- and undershoot with the system settling down after approximately 0.3 s. By applying a 0.25 s ramp-step input, the overshoot can be reduced from 25% to under 5% as shown in graph Fig. 14.7. During the initial 0.25 s period of the ramp, ramp-step and triangular input as shown in Fig. 14.8 transient response is, of course, identical. The demand,

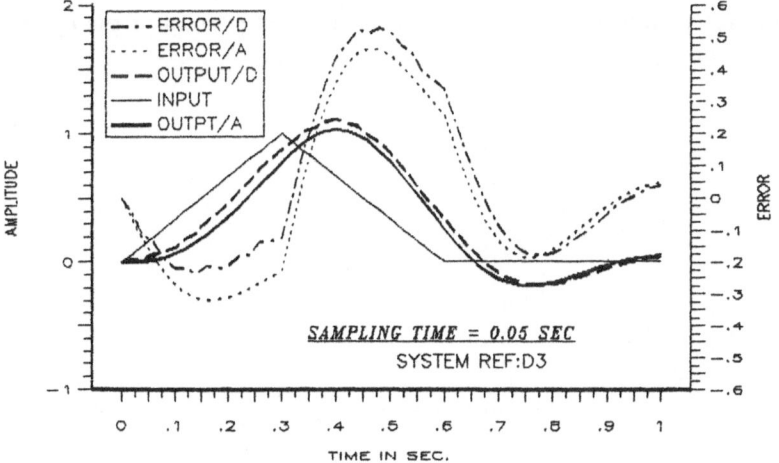

Fig. 14.11 Transient response to digitally generated input.

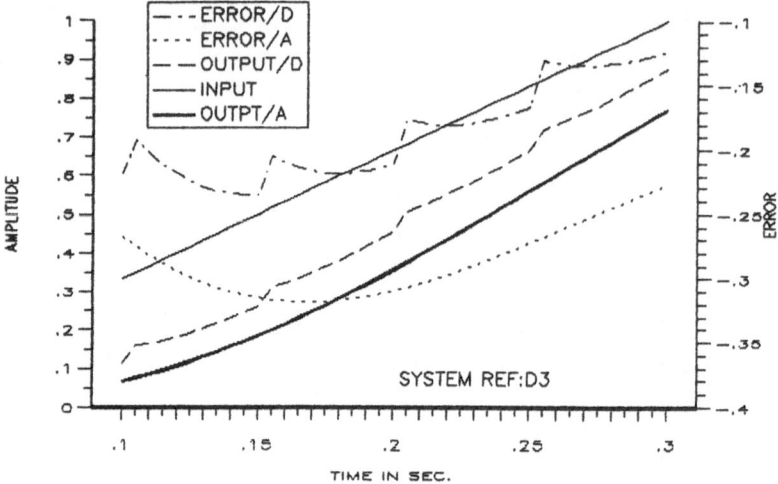

Fig. 14.12 Transient response to digitally generated input expanded portion.

output and error plots applicable to the compound duty cycle of Fig. 14.4 are shown in Fig. 14.9(a) and (b).

By introducing additional loops, the algorithm, Fig. 14.5 can be extended to calculate the transient response to a digitially generated input function. Such an algorithm is shown in Fig. 14.10. The output and error to a digitally generated triangular input is compared to a corresponding analogue demand in Fig. 14.11. The different behaviour of the two types of system can best be illustrated by plotting an expanded portion of the cycle, as shown in Fig. 14.12.

15

Transient Response Characteristics

By using a computer program of the type described in the previous two chapters, we are in a position to analyse complex systems in both the frequency and time domains. Nevertheless it is instructive to examine some basic trends, particularly in the time domain, to enhance our system design capability. As an example we shall extend the treatment of the simple third order system, represented by the transfer functions eqns (11.16), (11.19) and (11.20), into the time domain. By means of the optimization algorithms described in Chapter 13, we can establish the equivalent system parameters corresponding to the limiting loop gain, plotted in Fig. 11.5 as a function of the hydraulic transmission damping factor, ζ. It can be seen from the graph of Fig. 15.1 that the natural frequency of the equivalent system ω_n is 67% of

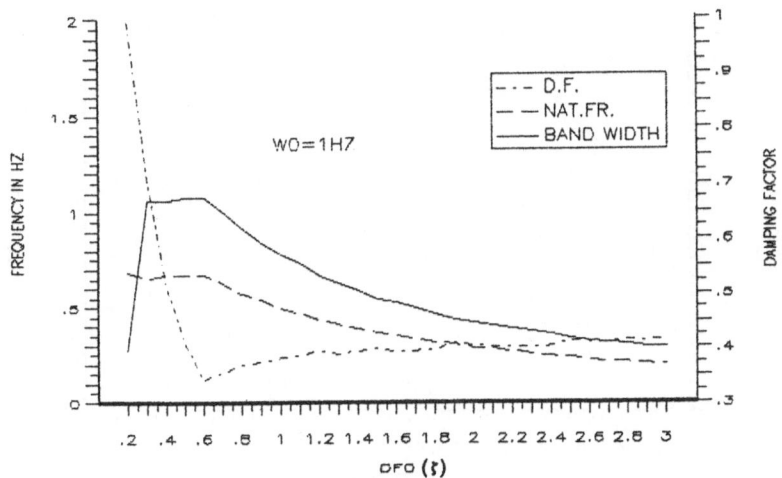

Fig. 15.1 Optimized frequency response parameters.

119

the hydraulic transmission natural frequency ω_0 up to a value of 0·6 hydraulic transmission damping factor ζ. At larger values of ζ, the system to hydraulic transmission natural frequency ratio diminishes. The system damping factor ζ_n reaches a minimum value of 0·34 at the optimum loop gain operating condition of $\zeta = 0·6$. It is of interest to note that up to the optimum loop gain operating condition, low hydraulic transmission damping results in high system damping, whereas beyond the optimum condition system damping remains virtually constant.

Transient response to a step input is frequently defined by two parameters, i.e. the time taken to initially reach 90% of the steady-state value and the maximum overshoot. These two parameters are plotted as a function of hydraulic transmission damping at the maximum permissible loop gain in Fig. 15.2. The optimized transient response parameters reach a minimum response time of 0·43 s with a corresponding 32% maximum overshoot. Overshoot diminishes rapidly at lower values of hydraulic transmission damping, while remaining substantially constant at increased values, levelling off at approximately 24%. A step response plot at the optimum operating condition is shown in Fig. 15.3. This shows that after one over- and undershoot, the system settles down to its steady-state level after approximately 3 s.

For applications where a 32% overshoot would be undesirable, we can either change the input function from a step to a ramp-step demand, or reduce the loop gain. Let us consider an application where the maximum

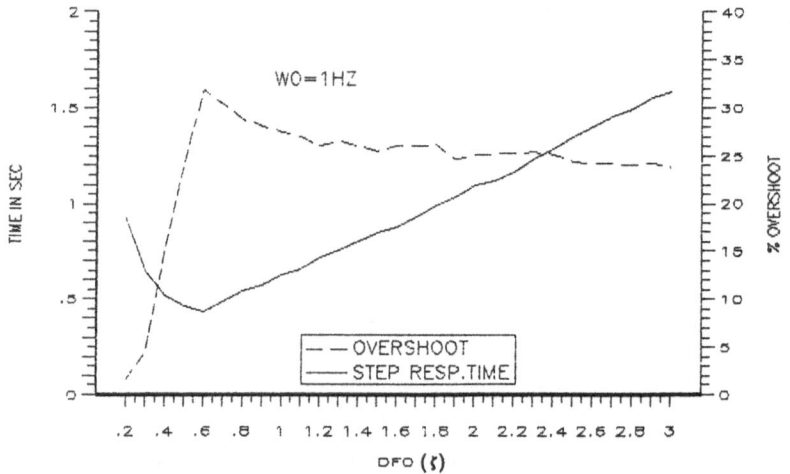

Fig. 15.2 Optimized transient response parameters.

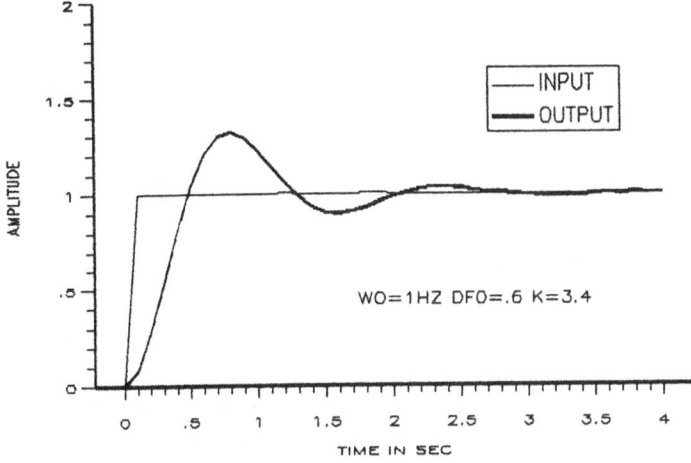

Fig. 15.3 Transient response to step input.

permissible overshoot is 10%. It can be seen from Fig. 15.4 that a 1·2 s ramp-step demand will reduce the maximum overshoot to 10% while increasing the 90% response time from 0·43 to 1·2 s. The settling times for the step and ramp-step demands are almost identical.

For the alternative approach, we shall investigate the effect of reducing the loop gain by 40% to 2·04 s^{-1}. The Nichols chart for the downgraded

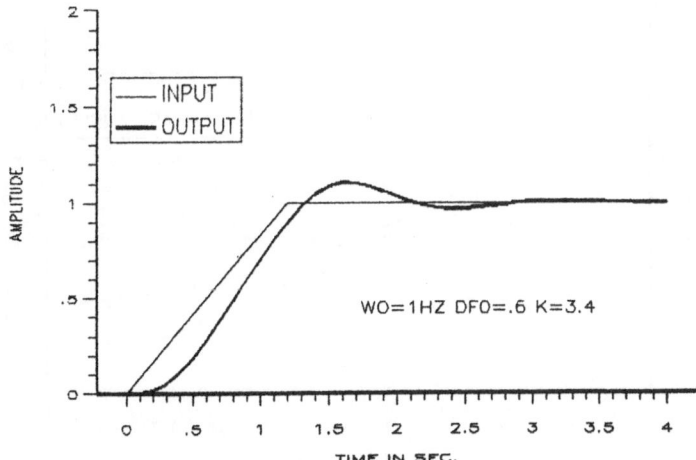

Fig. 15.4 Transient response to ramp-step input.

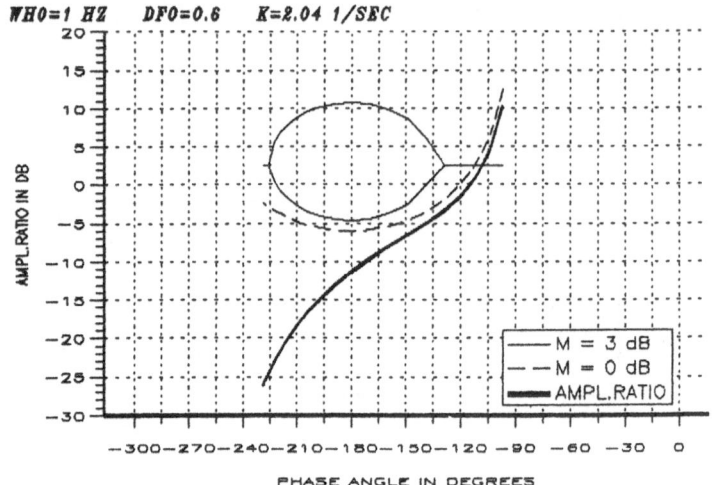

Fig. 15.5 Nichols chart.

system, Fig. 15.5, shows increased phase and gain margins of 67° and 11·5 dB respectively. Corresponding open and closed loop Bode diagrams are shown in Figs 15.6 and 15.7, and the transient response plot to a step input in Fig. 15.8.

As for the ramp-step demand, the maximum overshoot has been reduced to 10% with an improved response time of 0·7 s and settling time of

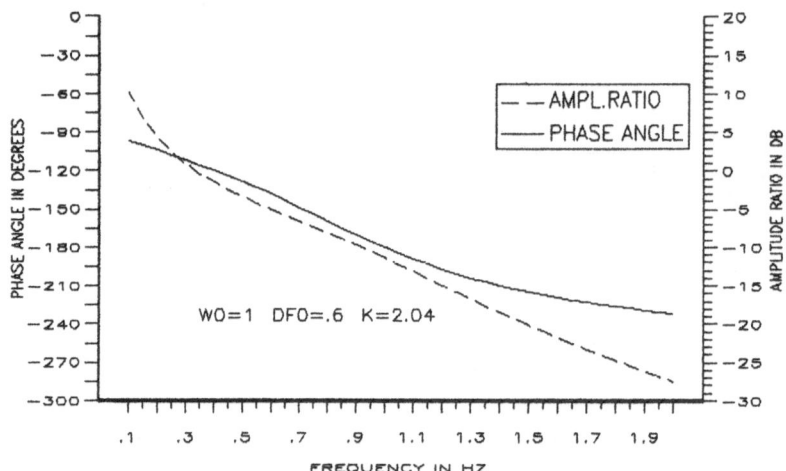

Fig. 15.6 Open loop Bode diagram.

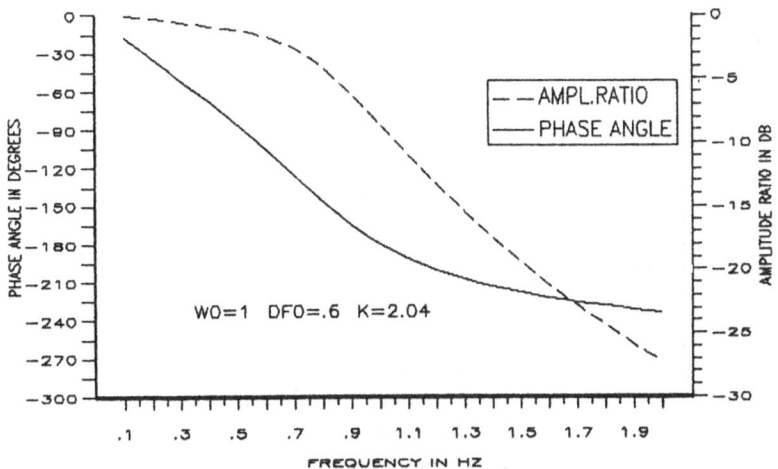

Fig. 15.7   Closed loop Bode diagram.

approximately 2 s. At first glance the second solution appears to be the preferred alternative for all applications, but we must bear in mind that reducing the loop gain downgrades both steady-state and transient response over the entire operating spectrum, and for that reason the first approach might well offer the more acceptable solution in some cases.

Transient response curves to various triangular input functions at the

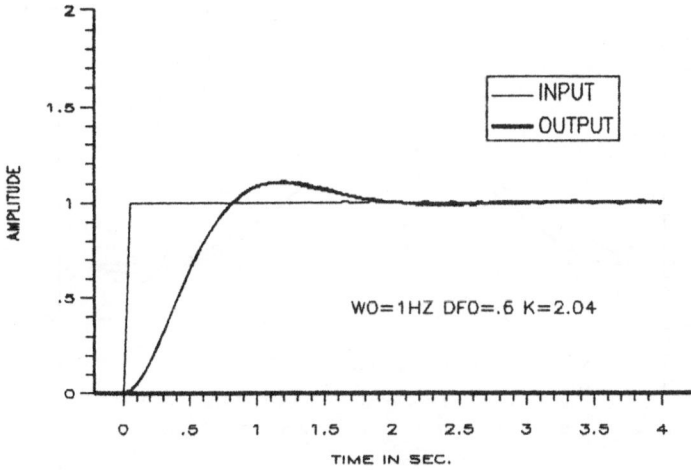

Fig. 15.8   Transient response to step input.

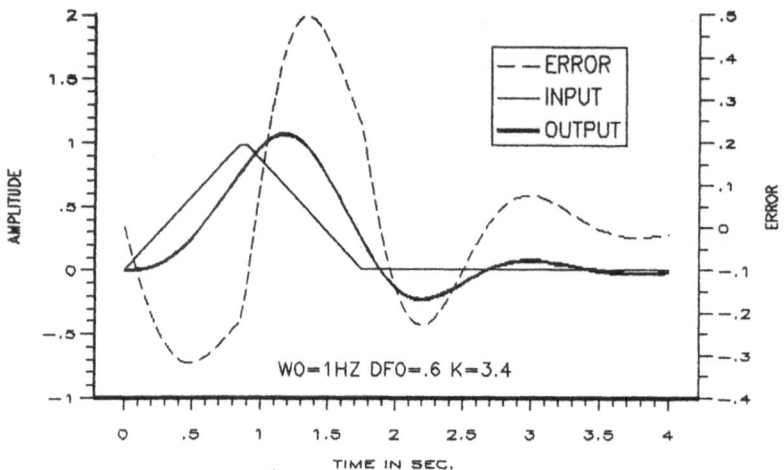

Fig. 15.9 Transient response to triangular input.

optimum operating condition are shown in Figs 15.9 to 15.13. Response to a single triangular input are plotted in Figs 15.9, 15.10 and 15.11 and to a continuous input in Figs. 15.12 and 15.13.

By rationalizing the transient response characteristics seen in the above graphs, we can establish some basic trends which are applicable to all types of control system. The effect of varying the system damping factor ζ_n on

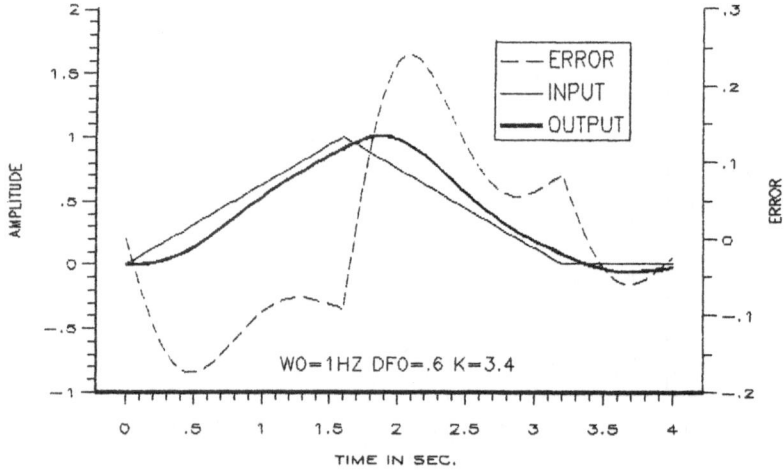

Fig. 15.10 Transient response to triangular input.

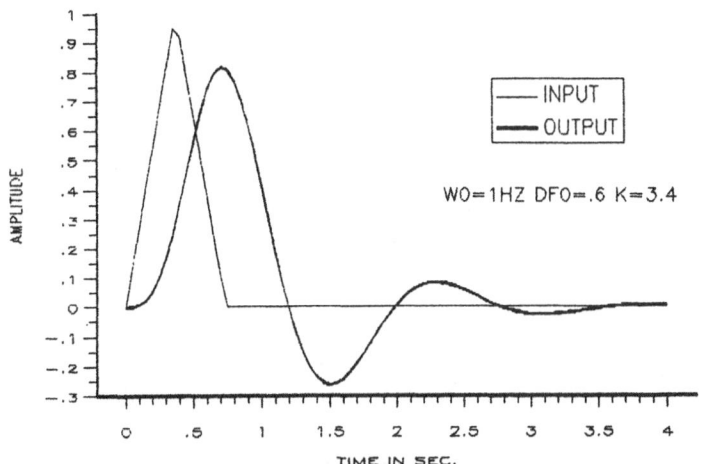

Fig. 15.11 Transient response to triangular input.

transient response to a step input is shown in Fig. 15.14. Between zero and 0·85 system damping, maximum overshoot changes from 100% down to zero, while the 90% response time increases at a lower rate up to approximately 0.85 damping and at a higher rate at larger system damping. The step response at zero system damping was given in Fig. 8.3 as an example of absolute stability.

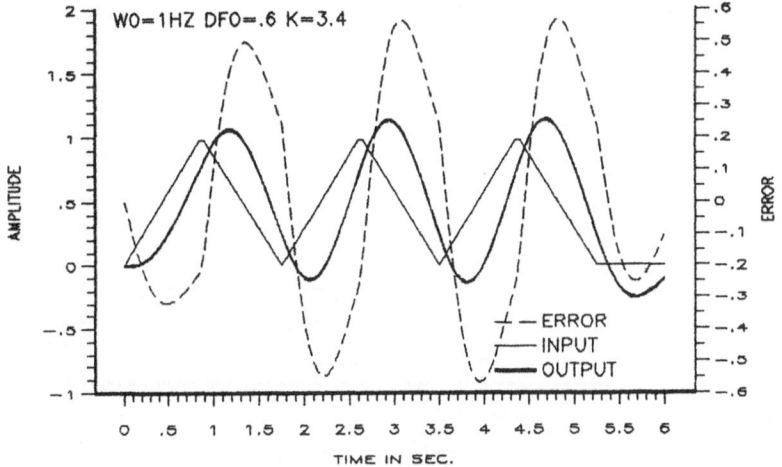

Fig. 15.12 Transient response to continuous input.

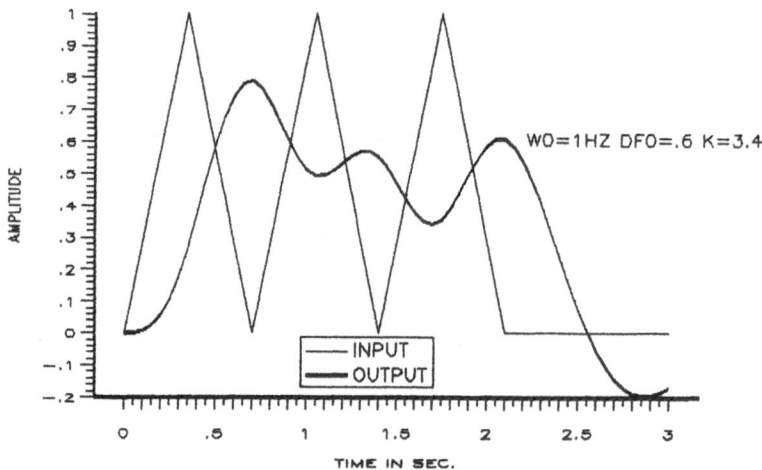

Fig. 15.13   Transient response to continuous input.

The effect of ramp time on overshoot and system reponse at the optimum operating condition is shown in Fig. 15.15. Above ramp times of, say, 1·5 s, overshoot is only marginally affected, response times increasing at a fairly constant rate over the entire range. Figure 15.15 is applicable to a system subjected to a ramp-step demand. Corresponding characteristics of systems subjected to a single triangular input, as shown in Figs 15.9 to 15.11, are

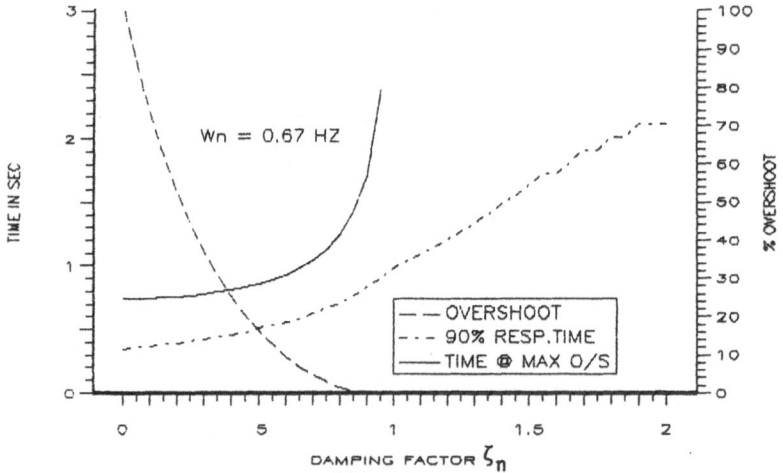

Fig. 15.14   Step response characteristics.

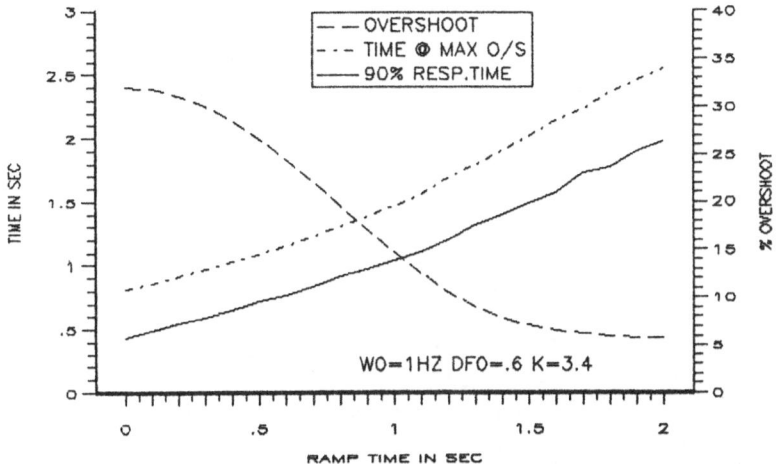

Fig. 15.15 Ramp-step response characteristics.

given in Fig. 15.16. Up to a ramp time of 0·5 s the output undershoots, between 0·5 and 1·5 s there is a slight overshoot and at ramp times higher than 1·5 s the system settles down to zero overshoot. The time taken for the output to reach maximum amplitude increases at a fairly linear rate with ramp time.

Transient response to a sinusoidal input stimulus is plotted in Fig. 15.17.

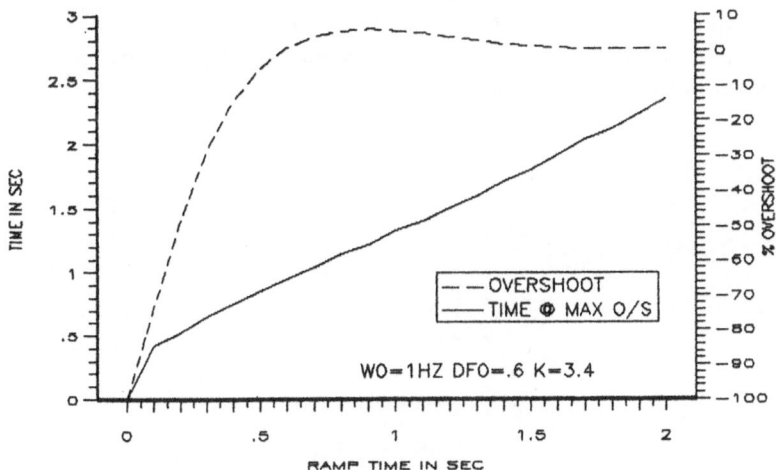

Fig. 15.16 Triangular response characteristics.

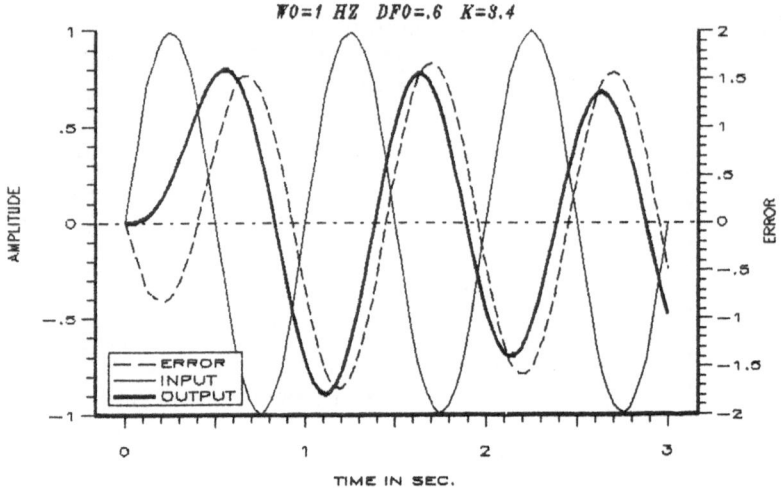

Fig. 15.17 Transient response to sinusoidal input.

It can be seen that initial transients have died down after approximately two cycles when steady-state conditions have been reached.

Since the characteristics illustrated in the above plots are applicable to a third order system having a hydraulic transmission natural frequency ω_0 of 1 Hz, all given parameters can be treated as non-dimensional, i.e. if all non-dimensional parameters are denoted by a prime sign ',

Loop gain $K = K'\omega_0$ Frequency $\omega = \omega'\omega_0$
Time $t = t'/\omega_0$ Bandwidth $\omega_C = \omega'_C\omega_0$
Natural frequency $\omega_n = \omega'_n\omega_0$

16

Further Case Studies

The investigation of the simple third order system in Chapters 11 and 15 can be extended to cover more complex systems, and we shall now proceed to analyse a number of typical system configurations.

(1) Third order system with flow feedback.
(2) Fourth order hydrostatic transmission.
(3) Fifth order system.
(4) Seventh order system with flow feedback.

16.1 THIRD ORDER SYSTEM WITH FLOW FEEDBACK

The system, introduced in Chapter 12, is represented by the block diagram Fig. 12.3, the system transfer function Fig. 12.4 and the mathematical models given in Figs 12.8 and 12.9. The methods described in the preceding chapters will be applied to the analysis of a simple control system incorporating flow feedback. The stability boundary compatible with the previously applied phase and gain margin criteria of 45° and 7 dB is given in Fig. 16.1, where the optimum loop gain is plotted against the hydraulic transmission damping factor, ζ. A comparison with the stability boundary of the simple third order system analysed in Chapter 11 and plotted in Fig. 11.5 shows a marked similarity between the two types of system, the only significant difference being the optimum operating condition, which has shifted from 0·6 to 0·32 transmission damping. There is consequently a similar analogy between the frequency response characteristics of the flow feedback system shown in Fig. 16.2 and the corresponding characteristics of the position control system Fig. 15.1. The same applies to the transient response characteristics which are plotted in Fig. 16.3 for the flow feedback

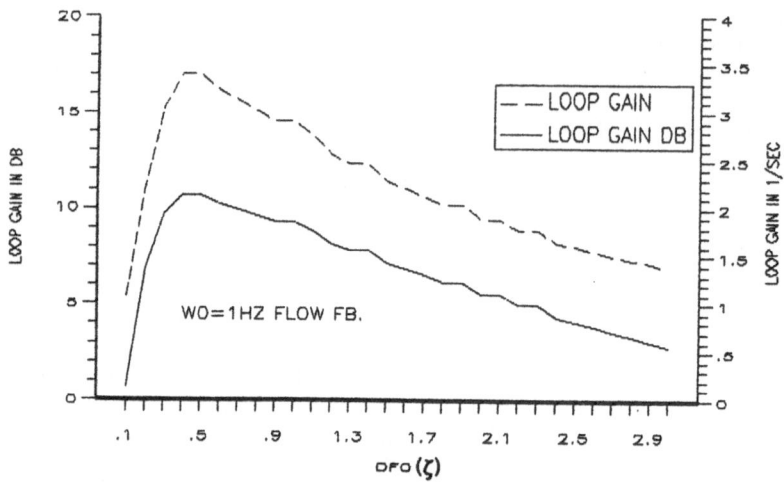

Fig. 16.1   Stability boundary.

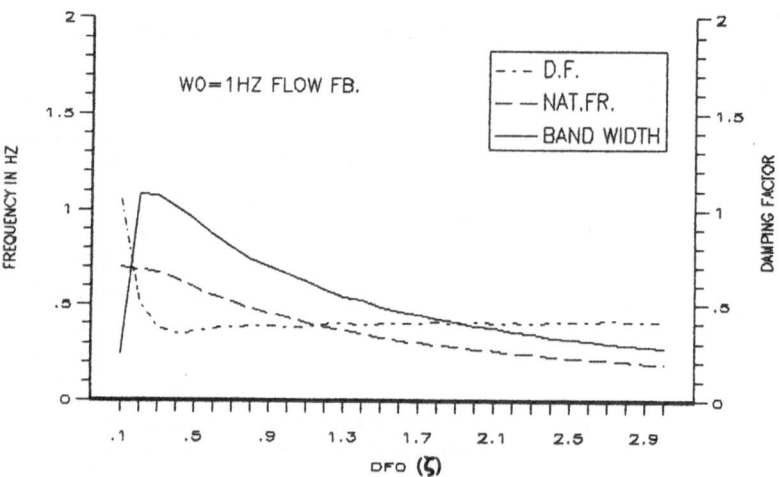

Fig. 16.2   Frequency response parameters.

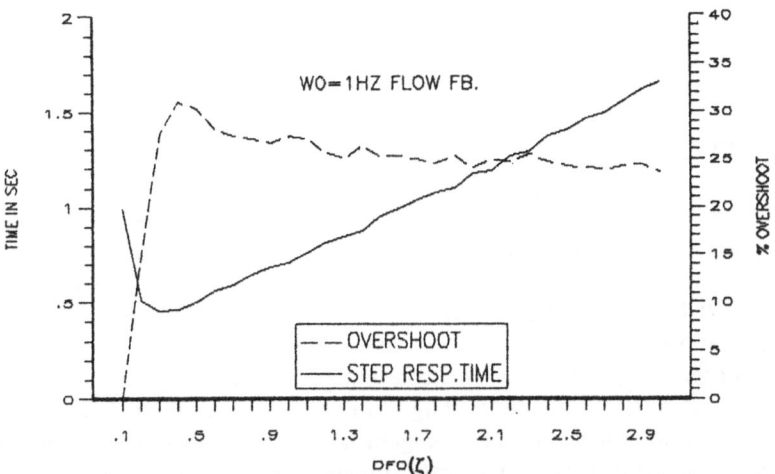

Fig. 16.3 Transient response parameters.

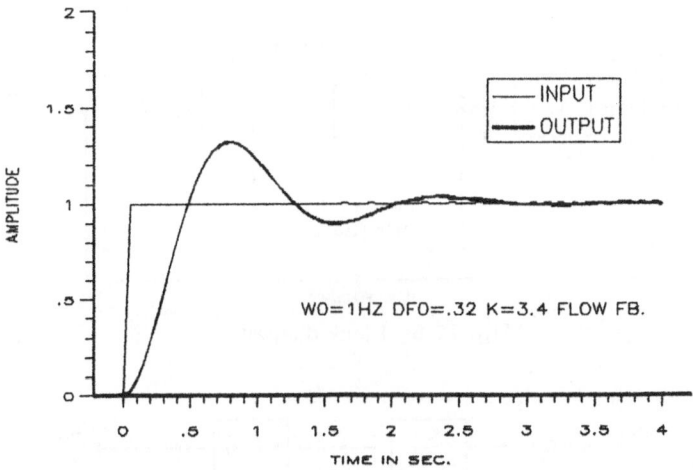

Fig. 16.4 Transient response to step input.

and in Fig. 15.2 for the position control system. Transient response to a step input at the optimum operating condition is virtually identical, as shown by graphs Figs 16.4 and 15.3 for the flow feedback and position control systems respectively.

16.2 FOURTH ORDER HYDROSTATIC TRANSMISSION

The hydrostatic transmission, shown as a block diagram in Fig. 5.5 and analysed in Chapter 12, can be represented by the simplified block diagram, Fig. 16.5. K_2 denotes the open loop gain of the swash control servo, K_1 the normalized swash angle, i.e. per unit error ε, where ε is referred to the output, and K_3 the output per unit swash angle. K_3 can also be expressed as the ratio of pump and actuator displacements, or for the system shown in Fig. 12.6, $K_3 = C_p/C$. The block diagram Fig. 16.5 can be further simplified to that of Fig. 16.6, where the loop gain $K = K_1 K_3$, and the swash control servo is represented by a first order transfer function with time constant $\tau = 1/K_2$. If the controlled output quantity is position, the system transfer function is similar to that of the third order system analysed in Chapters 11 and 15 but with the addition of a first order transfer function representing the swash angle control servo loop. The effect of the swash angle controller

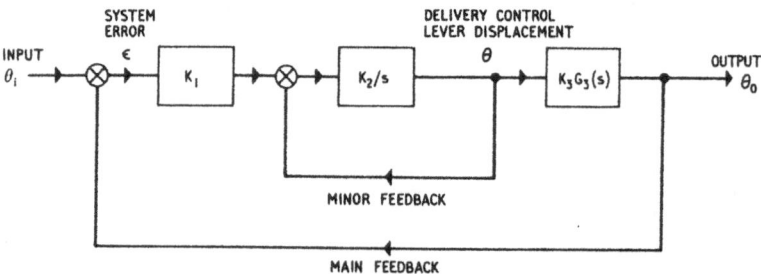

Fig. 16.5 Block diagram.

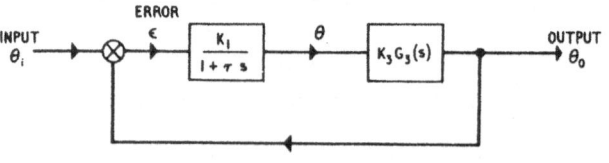

Fig. 16.6 Block diagram.

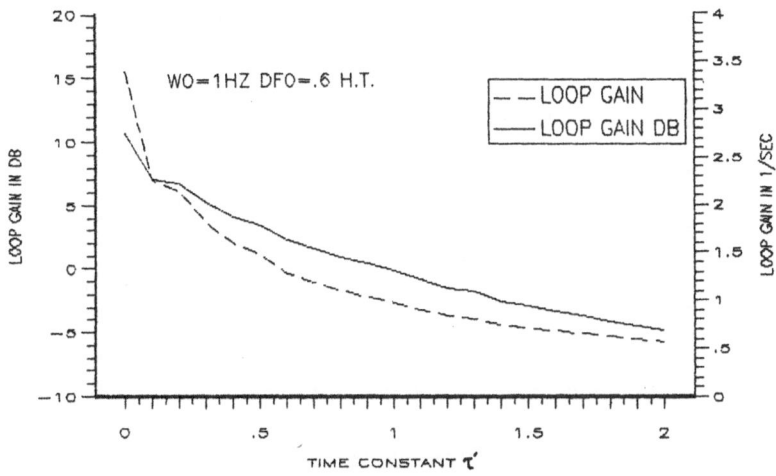

Fig. 16.7 Stability boundary.

on system performance is shown in Figs 16.7, 16.8 and 16.9, where the optimum loop gain, frequency and transient response parameters are plotted against the non-dimensional time constant τ'. The actual time constant is given by the ratio $\tau = \tau'/\omega_0$, where ω_0 is the open loop natural frequency of the hydrostatic transmission in hertz. At $\tau = 0$, performance parameters of the hydrostatic transmission and the third order position

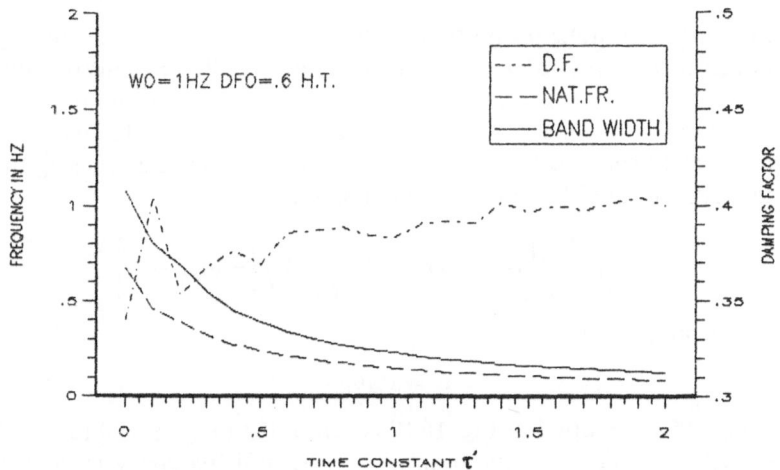

Fig. 16.8 Frequency response parameters.

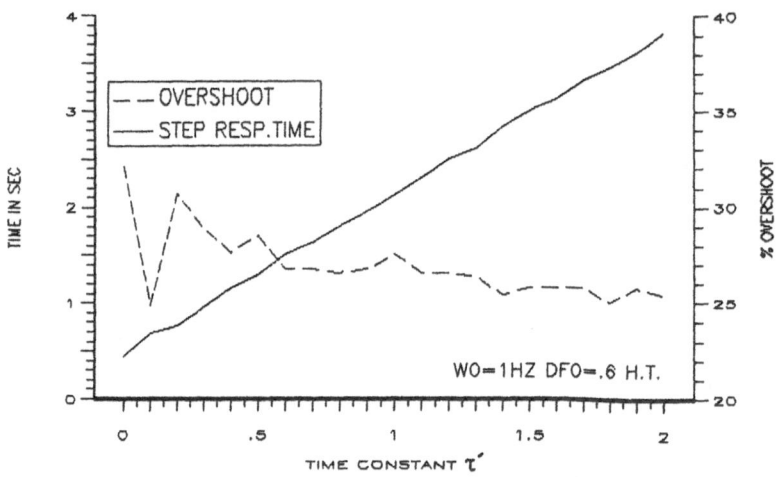

Fig. 16.9 Transient response parameters.

control system are, of course, identical. It can be seen that optimum loop gain and frequency bandwidth initially decrease fairly rapidly, but tend to level off as the time constant τ is increased, whereas the response time increases at a constant rate at an almost constant maximum overshoot.

16.3 FIFTH ORDER SYSTEM

Most electro-hydraulic proportional and servo valves can be adequately identified by a second order transfer function. Usually the control valve characteristics are experimentally established by frequency response testing, the 90° phase lag specifying the natural frequency, ω_n. The damping factor ζ_n can then be calculated from the following expression, using the bandwidth, ω_C, at either 3 or 4 dB attenuation.

$$\frac{\omega_C}{\omega_n} = \sqrt{\frac{1}{2}\left\{\sqrt{\left[(4\zeta_n^2 - 2)^2 + 4\left(\frac{1}{A_R^2} - 1\right)\right]} - (4\zeta_n^2 - 2)\right\}}$$

where the amplitude ratio

$$A_R = |\theta_o/\theta_i| \tag{16.1}$$

Equation 16.1 is plotted in Fig. 16.10 covering the range $\zeta_n = 0$ to $\zeta_n = 2$.

The effect of the hydraulic transmission natural frequency to control valve natural frequency ratio ω_0/ω_1 on system performance is plotted in

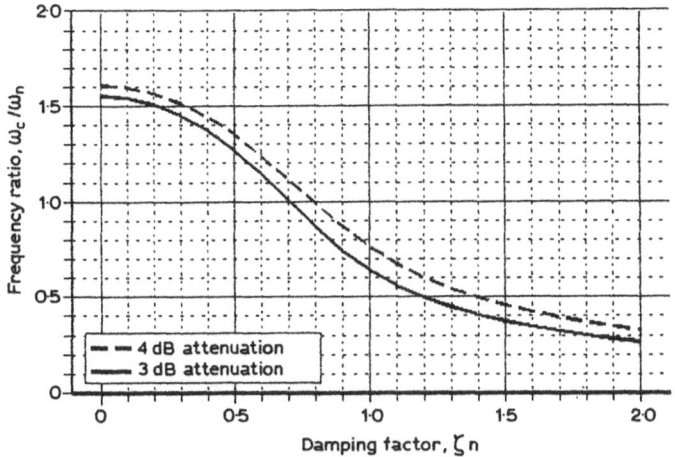

Fig. 16.10 Equation system characteristics.

Figs 16.11, 16.12 and 16.13. At $\omega_0/\omega_1 = 0$, the system becomes a third order system with performance characteristics identical to the previously analysed position control system. A control valve damping factor of $0\cdot8$ was arbitrarily chosen as a fairly typical value. A comparison with the hydrostatic transmission performance characteristics, Figs 16.7 to 16.9, shows a marked similarity between the two system configurations.

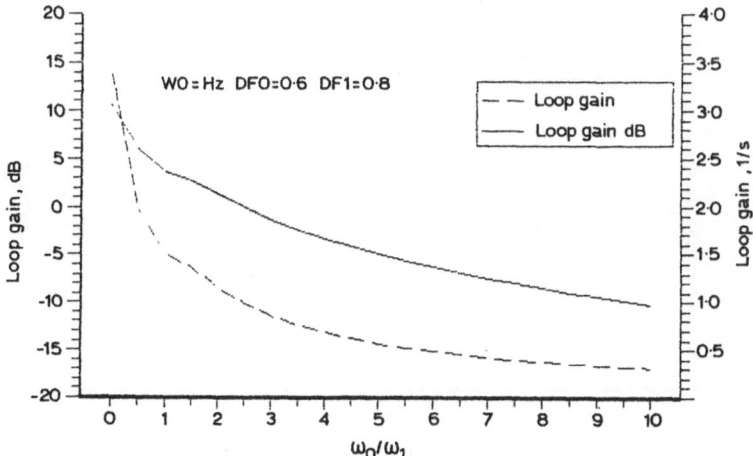

Fig. 16.11 Stability boundary.

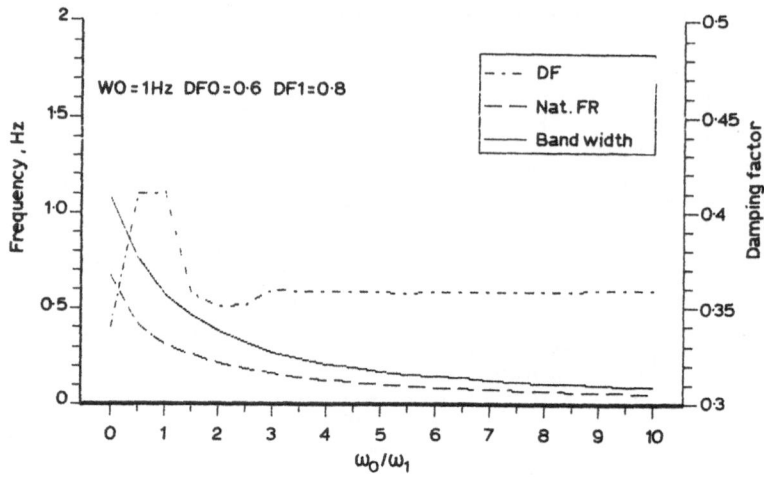

Fig. 16.12   Frequency response parameters.

Up to now we have investigated a number of control systems in the frequency and time domains, using non-dimensional notation, and have established some characteristic trends specific to the particular system configuration. We shall now analyse a typical fifth order position control system with the aid of the software package HYDRASOFT, described in Chapter 13. A print-out of the input and output parameters is given in Fig.

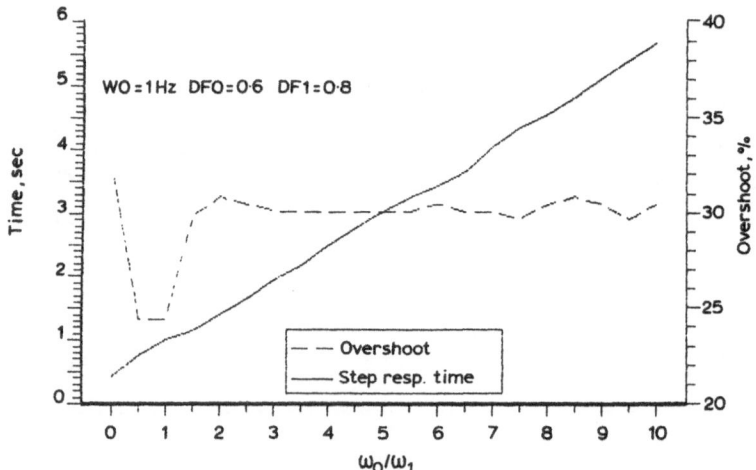

Fig. 16.13   Transient response parameters.

16.14. In accordance with the program structure, Fig. 13.2, the print-out is subdivided into three sections, corresponding to the hydraulic transmission identification module 'S', the system identification frequency response module 'F' and the output parameter statement module 'P'. The discerning reader will note that the actuator specified is an asymmetrical cylinder, and that one of the hydraulic transmission variables is the inlet to outlet orifice area ratio. So far our mathematical model derivation has been confined to symmetrical systems, but in the following chapter will be expanded to include non-symmetrical systems.

S Module

```
Closed loop system
valve operated system without flow feedback
4 way valve
Single ended extending Cylinder
  10 bore x  5 rod dia. x  300 stroke (cm)
inlet cylinder area= 78.53983 cm², outlet cylinder area= 58.90487 cm²
mass = 2000 kg
coefficient of viscous damping = 0 kp/cm per sec
load = 12 kN
inlet/outlet orifice area ratio= 1
equivalent valve flow rating= 210.7928 L/min
actuator velocity= 53.05174 cm/sec
inlet flow = 250 L/min
supply pressure = 70 bar,   valve pressure drop= 45.73751 bar
port diameter= 1.5 cm
trapped volume= 28.56195 L ( 23.56195 :Actuator +  5 :Piping)
actuator shunt coefficient = 0 L/min per bar
bulk modulus= 13793 bar
WHO= 11.46hz
DFO=  0.53
KO/K= 1
```

F Module

```
Closed loop system
Components selected:SH4-3090 SERVO VALVE
OPEN LOOP SYSTEM PARAMETERS
Frequencies in hz:
WHO= 11.46
WH1= 47.86
WH2=%100000.00
WH3=%100000.00
Damping factors:
DFO=  0.53
DF1=  0.79
DF2=  0.00
DF3=  0.00
Time constants in seconds:
t0= 0        t1= 0        t2= 0        t3= 0        t4= 0
free integrator:SINGLE
loop gain= 25.72  1/sec
```

P Module

```
Natural frequency WR=  5.94hz              CL Damping factor DFR=  0.44
Band width at 4 dB attenuation WC= 10.52hz
Max.Overshoot at  0.094sec = 21.07%
Step response time=  0.053sec
```

Fig. 16.14 System parameter print-out.

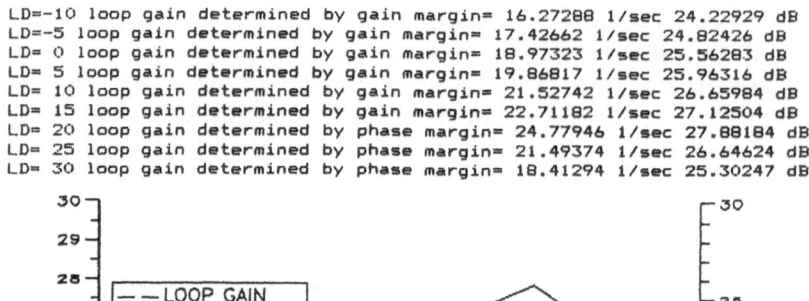

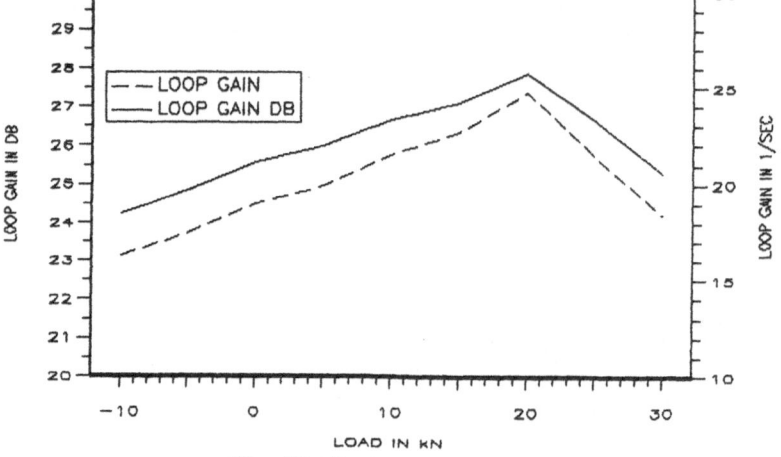

Fig. 16.15 Stability boundary.

Stability boundary plots are shown in Figs 16.15 and 16.16. In Fig. 16.15 optimum loop gain is plotted as a function of load at constant inlet flow of 150 litres/min, and in Fig. 16.16 as a function of flow at constant cylinder loading of 12 kN. In the first case, the optimized loop gain peaks at 20 kN loading, in the second case at an inlet flow of 200 litres/min. It is of interest to note that in both cases the peak loop gain occurs at the change-over from gain to phase margin as the critical stability criterion. This is equally true for systems previously analysed. Optimized frequency response parameters are plotted in Figs 16.17 and 16.18 and optimized transient response parameters in Figs 16.19 and 16.20. It is apparent from the plots that for the given specification, load variation has less effect on system performance than flow changes.

16.4 SEVENTH ORDER SYSTEM WITH FLOW FEEDBACK

To complete our initial set of case studies, we shall now consider a cylinder velocity control system comprising a two-stage, two-way valve and

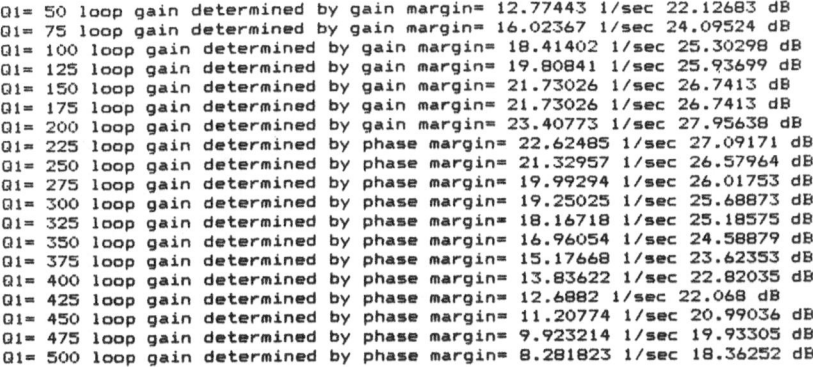

```
Q1= 50 loop gain determined by gain margin= 12.77443 1/sec 22.12683 dB
Q1= 75 loop gain determined by gain margin= 16.02367 1/sec 24.09524 dB
Q1= 100 loop gain determined by gain margin= 18.41402 1/sec 25.30298 dB
Q1= 125 loop gain determined by gain margin= 19.80841 1/sec 25.93699 dB
Q1= 150 loop gain determined by gain margin= 21.73026 1/sec 26.7413 dB
Q1= 175 loop gain determined by gain margin= 21.73026 1/sec 26.7413 dB
Q1= 200 loop gain determined by gain margin= 23.40773 1/sec 27.95638 dB
Q1= 225 loop gain determined by phase margin= 22.62485 1/sec 27.09171 dB
Q1= 250 loop gain determined by phase margin= 21.32957 1/sec 26.57964 dB
Q1= 275 loop gain determined by phase margin= 19.99294 1/sec 26.01753 dB
Q1= 300 loop gain determined by phase margin= 19.25025 1/sec 25.68873 dB
Q1= 325 loop gain determined by phase margin= 18.16718 1/sec 25.18575 dB
Q1= 350 loop gain determined by phase margin= 16.96054 1/sec 24.58879 dB
Q1= 375 loop gain determined by phase margin= 15.17668 1/sec 23.62353 dB
Q1= 400 loop gain determined by phase margin= 13.83622 1/sec 22.82035 dB
Q1= 425 loop gain determined by phase margin= 12.6882 1/sec 22.068 dB
Q1= 450 loop gain determined by phase margin= 11.20774 1/sec 20.99036 dB
Q1= 475 loop gain determined by phase margin= 9.923214 1/sec 19.93305 dB
Q1= 500 loop gain determined by phase margin= 8.281823 1/sec 18.36252 dB
```

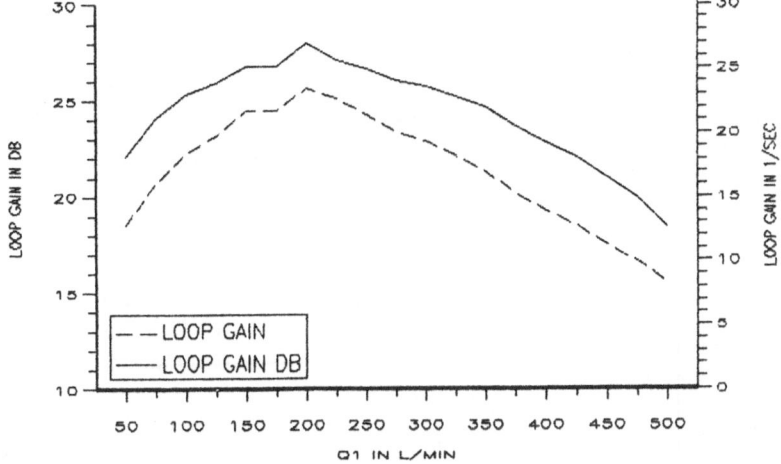

Fig. 16.16 Stability boundary.

incorporating a flow feedback transducer, as shown diagrammatically in Fig. 16.21 and as a circuit diagram in Fig. 16.22. The LVDT-type position transducer attached to the flow sensing element provides a flow feedback signal which is compared with the demand signal in the summing junction of the valve drive amplifier to produce an error signal which, after suitable amplification, drives the solenoid-operated pilot stage, which in turn controls the main stage two-way valve. An optional valve spool position feedback loop can be used to give enhanced fail-safe and large step response characteristics. In any velocity or flow feedback system, introducing a minor feedback loop, and thereby eliminating the free

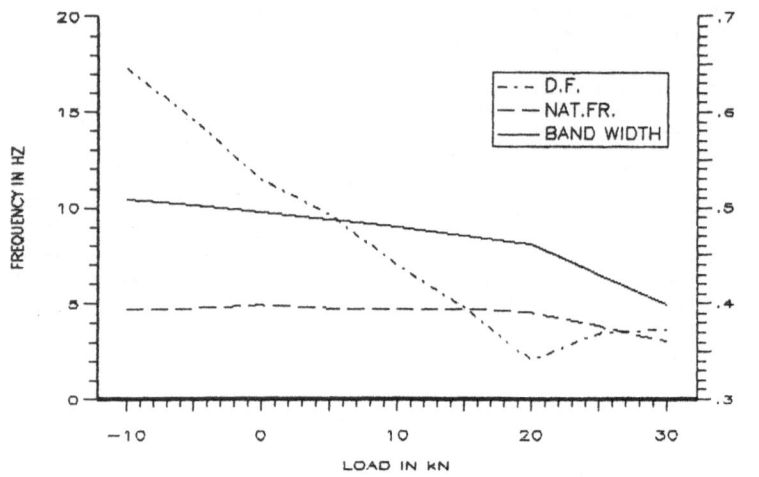

Fig. 16.17   Frequency response parameters.

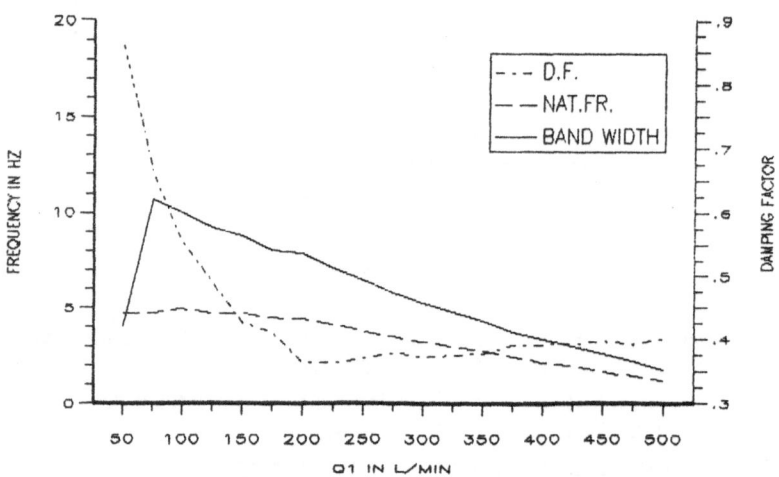

Fig. 16.18   Frequency response parameters.

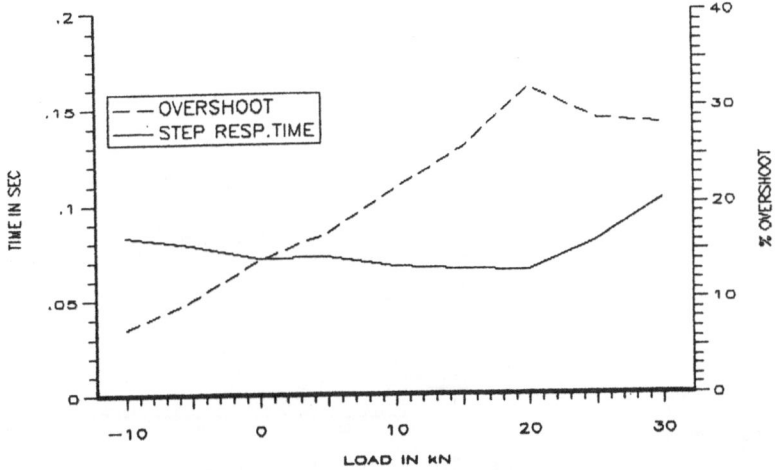

Fig. 16.19 Transient response parameters.

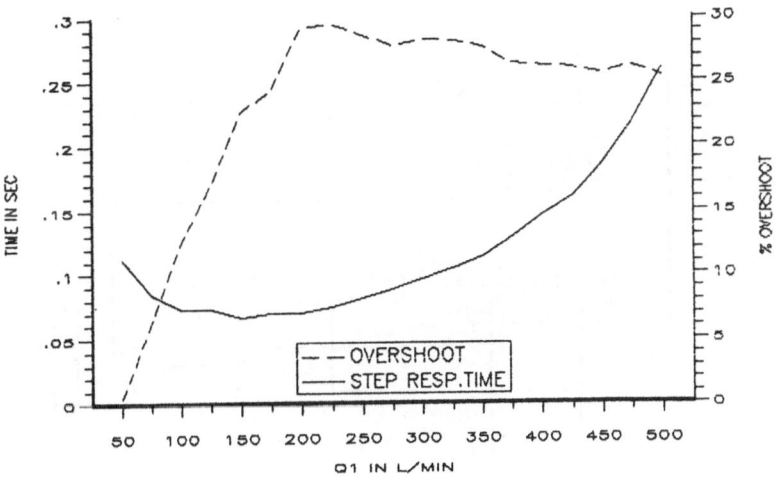

Fig. 16.20 Transient response parameters.

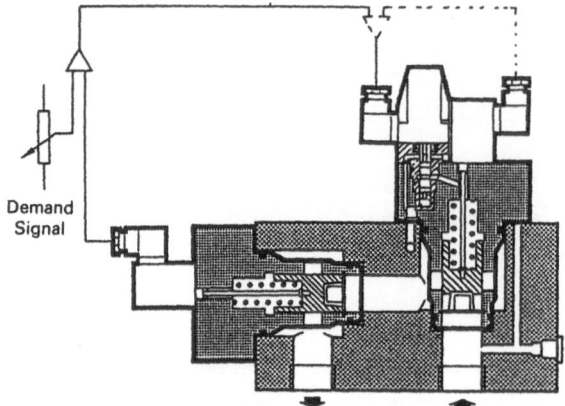

Demand
Signal

Fig. 16.21 System diagram.

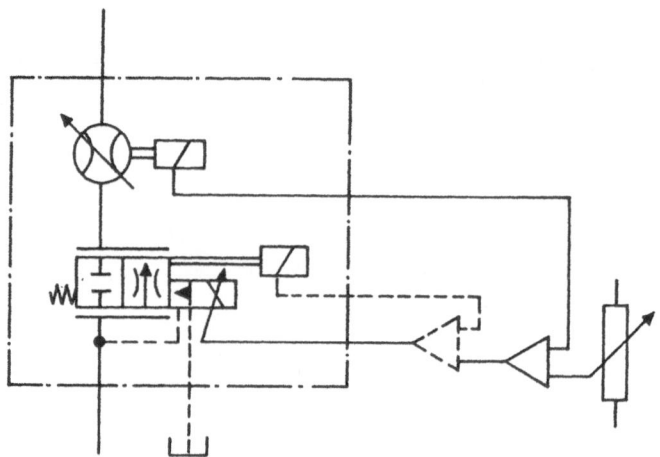

Fig. 16.22 Circuit diagram.

S Module

Closed loop system
valve operated system with flow feedback
2 way valve meter-in
Single ended extending Cylinder
  10 bore x 5 rod dia. x 300 stroke (cm)
inlet cylinder area= 78.53982 cm², outlet cylinder area= 58.90487 cm²
mass = 2000 kg
coefficient of viscous damping = 0 kp/cm per sec
load = 12 kN
actuator velocity= 53.05175 cm/sec
inlet flow = 250 L/min
supply pressure = 70 bar,   valve pressure drop= 45.73751 bar
port diameter= 1.5 cm
trapped volume= 28.56194 L (23.56194 :Actuator + 5 :Piping)
actuator shunt coefficient = 0 L/min per bar
bulk modulus= 13793 bar
WHO= 6.14hz
WH3= 6.14hz
DFO= 0.29
DF3= 0.00
KO/K= 1
K1= 1

F Module

Closed loop system
Components selected:KFDG 5V8   FLOW TRANSDUCER FT07
OPEN LOOP SYSTEM PARAMETERS
Frequencies in hz:
WHO= 6.14
WH1= 17.00
WH2= 45.00
WH3= 6.14
Damping factors:
DFO= 0.29
DF1= 0.88
DF2= 0.70
DF3= 0.00
Time constants in seconds:
t0= 0   t1= 0   t2= 0   t3= 0   t4= 0
free integrator:SINGLE
loop gain= 10.03 1/sec

P Module

Natural frequency WR= 2.58hz   CL Damping factor DFR= 0.54
Band width at 4 dB attenuation WC= 5.07hz
Max.Overshoot at 0.230sec = 13.55%
Step response time= 0.138sec

Fig. 16.23 System parameter print-out.

integrator, has, however, an adverse effect on system performance and should therefore be treated as an alternative rather than additional feedback signal. We are now dealing with a fairly complex control system, but having prepared the ground in previous chapters, the increased complexity does not complicate the analytical process. The mathematical model for the hydraulic transmission was derived in Chapter 12 and is summarized in Figs 12.3, 12.4, 12.8 and 12.9. Referring to the input parameter statement of Fig. 16.23 and the compound system block diagram Fig. 12.10, the hydraulic transmission is identified by the natural

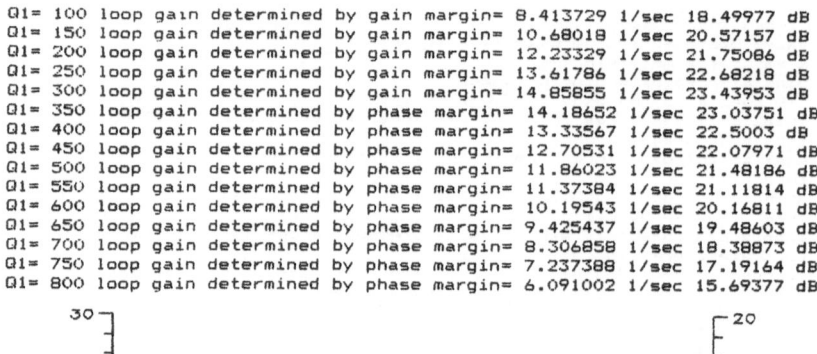

```
Q1= 100 loop gain determined by gain margin= 8.413729 1/sec 18.49977 dB
Q1= 150 loop gain determined by gain margin= 10.68018 1/sec 20.57157 dB
Q1= 200 loop gain determined by gain margin= 12.23329 1/sec 21.75086 dB
Q1= 250 loop gain determined by gain margin= 13.61786 1/sec 22.68218 dB
Q1= 300 loop gain determined by gain margin= 14.85855 1/sec 23.43953 dB
Q1= 350 loop gain determined by phase margin= 14.18652 1/sec 23.03751 dB
Q1= 400 loop gain determined by phase margin= 13.33567 1/sec 22.5003 dB
Q1= 450 loop gain determined by phase margin= 12.70531 1/sec 22.07971 dB
Q1= 500 loop gain determined by phase margin= 11.86023 1/sec 21.48186 dB
Q1= 550 loop gain determined by phase margin= 11.37384 1/sec 21.11814 dB
Q1= 600 loop gain determined by phase margin= 10.19543 1/sec 20.16811 dB
Q1= 650 loop gain determined by phase margin= 9.425437 1/sec 19.48603 dB
Q1= 700 loop gain determined by phase margin= 8.306858 1/sec 18.38873 dB
Q1= 750 loop gain determined by phase margin= 7.237388 1/sec 17.19164 dB
Q1= 800 loop gain determined by phase margin= 6.091002 1/sec 15.69377 dB
```

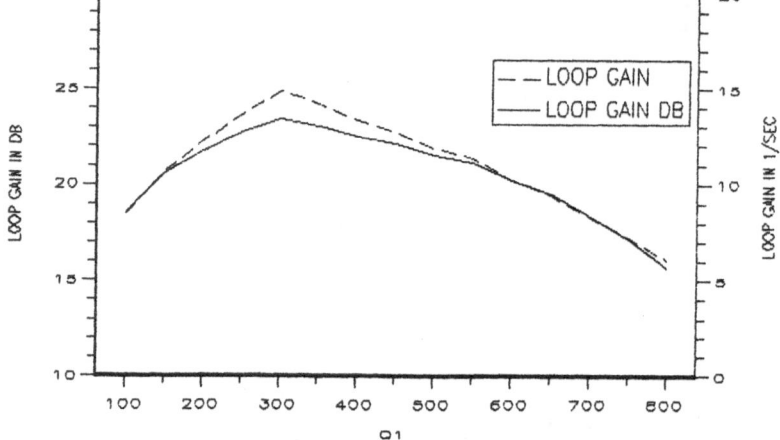

Fig. 16.24 Stability boundary in litres/min.

frequencies, $\omega_0(WH0) = 6\cdot14\,\mathrm{Hz}$ and $\omega_3(WH3) = 6\cdot14\,\mathrm{Hz}$, and the damping factors, $\zeta_0(DF0) = 0\cdot29$ and $\zeta_3(DF3) = 0$; the control valve by the natural frequency $\omega_1(WH1) = 17\,\mathrm{Hz}$ and damping factor $\zeta_1(DF1) = 0\cdot88$ and the flow transducer by the natural frequency $\omega_2(WH2) = 45\,\mathrm{Hz}$ and damping factor $\zeta_2(DF2) = 0\cdot7$.

The optimized loop gain for the given operating condition and corresponding output parameters are given in Fig. 16.23. Stability boundaries for two independent variables are plotted in Figs 16.24 and 16.25. In Fig. 16.24, where the optimum loop gain is shown as a function of inlet flow Q1, loop gain peaks at the gain/phase margin criterion change-over point at 300 litres/min. Figure 16.25 shows the effect of hydraulic transmission natural frequency variation on maximum permissible loop gain. As this frequency approaches that of the control elements, its effect progressively diminishes, the loop gain eventually levelling off at

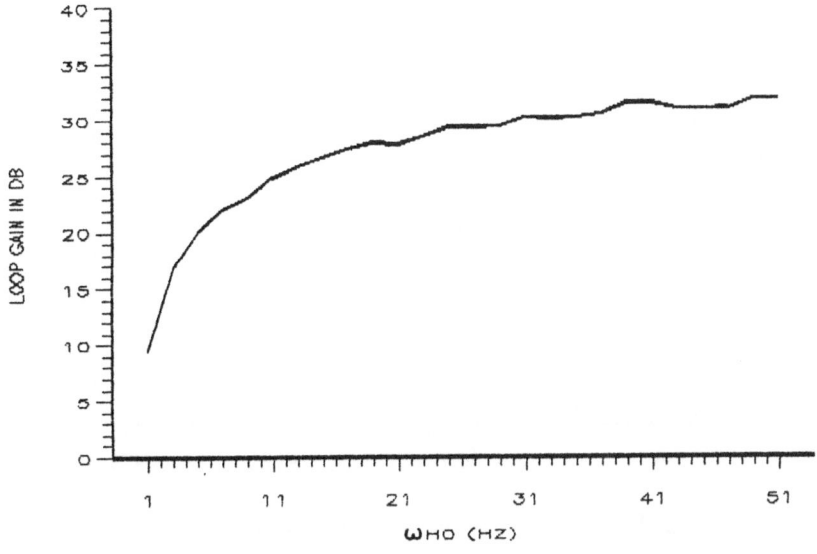

Fig. 16.25 Stability boundary.

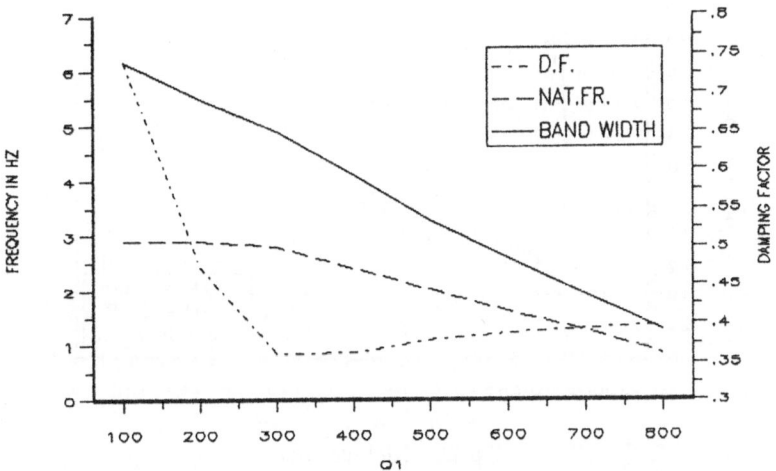

Fig. 16.26 Frequency response parameters in litres/min.

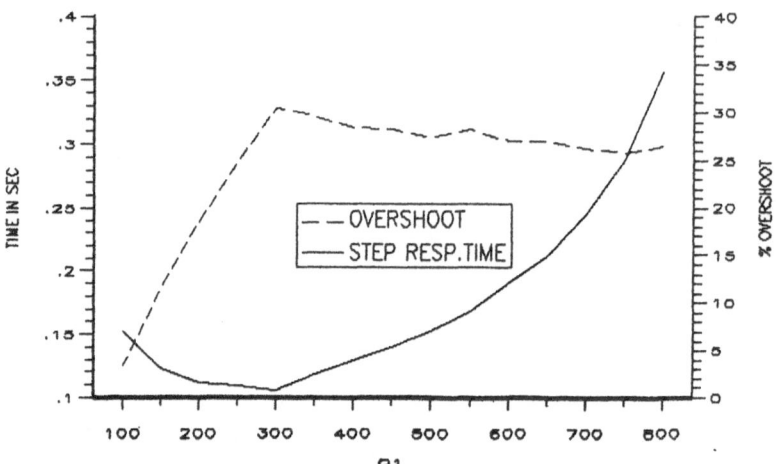

Fig. 16.27 Transient response parameters in litres/min.

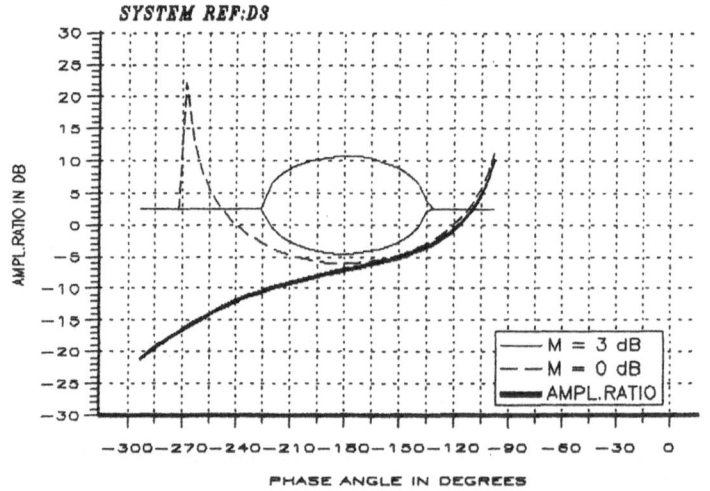

16.28 Nichols chart.

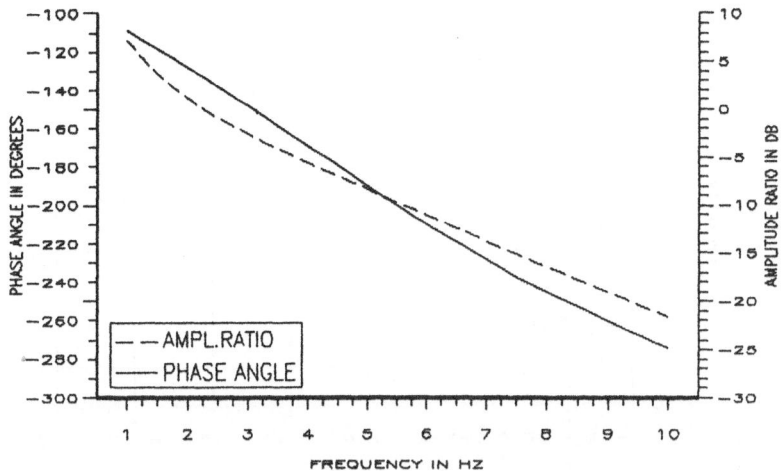

Fig. 16.29 Open loop Bode diagram.

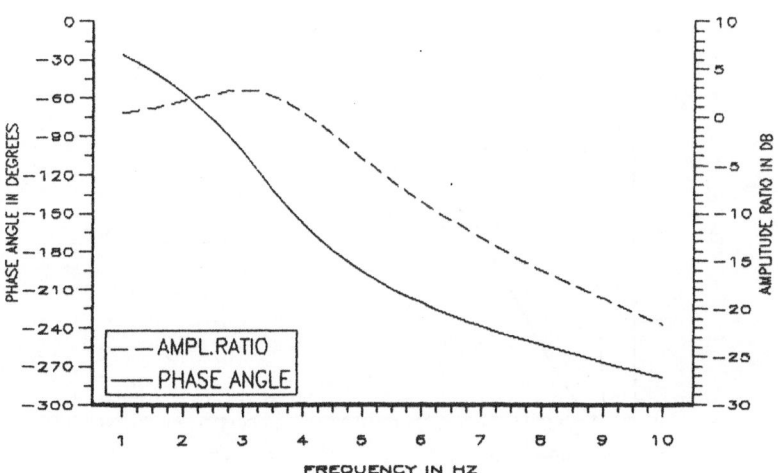

Fig. 16.30 Closed loop Bode diagram.

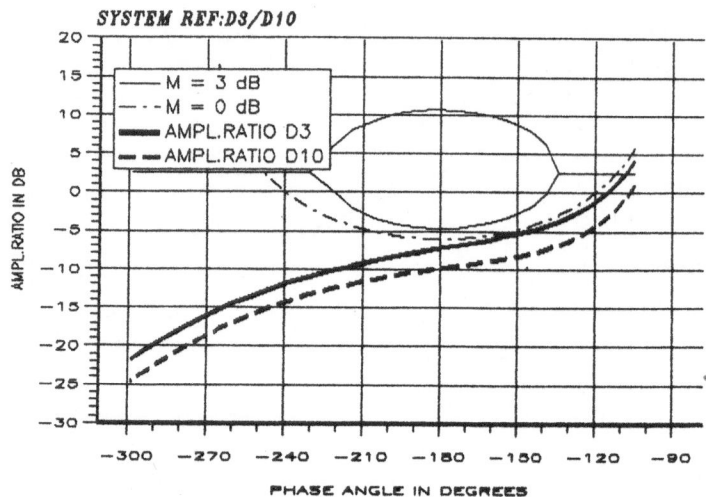

Fig. 16.31   Nichols Chart, dual plot.

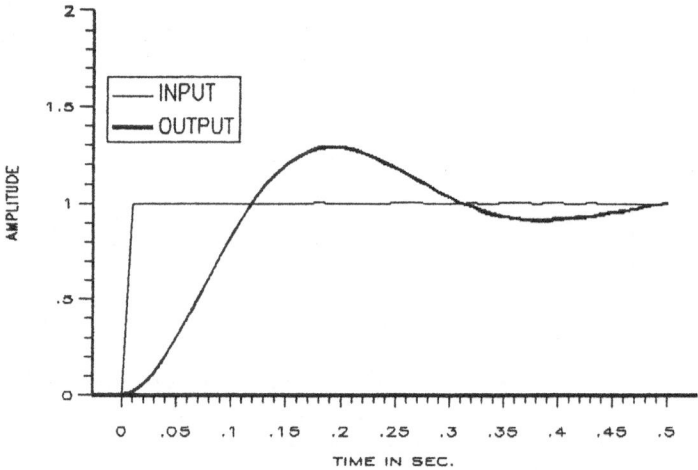

Fig. 16.32   Step response.

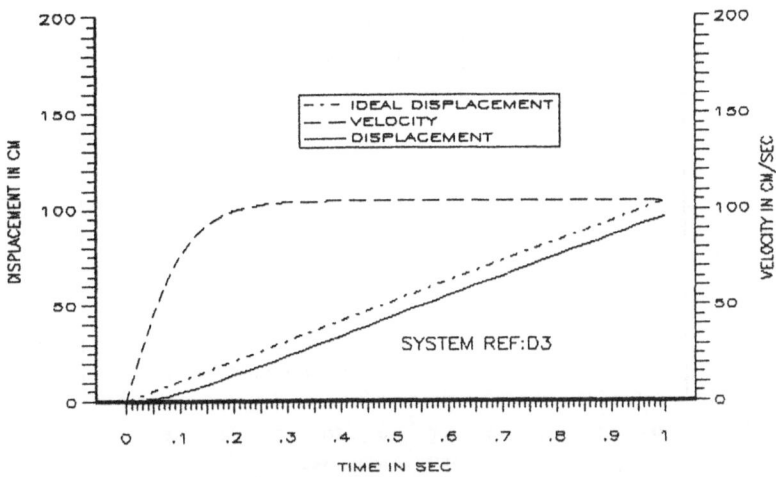

Fig. 16.33 Large step response.

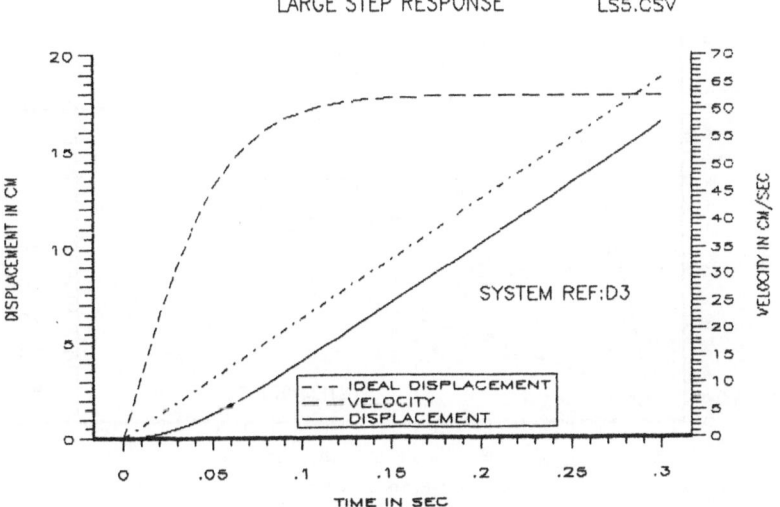

Fig. 16.34 Large step response.

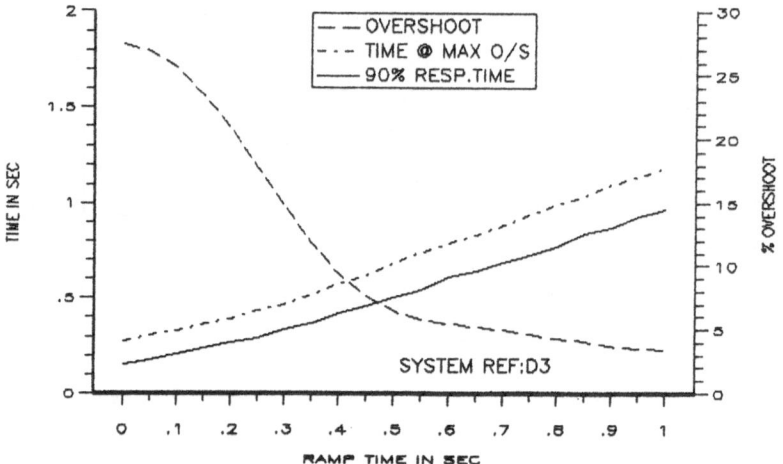

Fig. 16.35 Ramp-step response parameters.

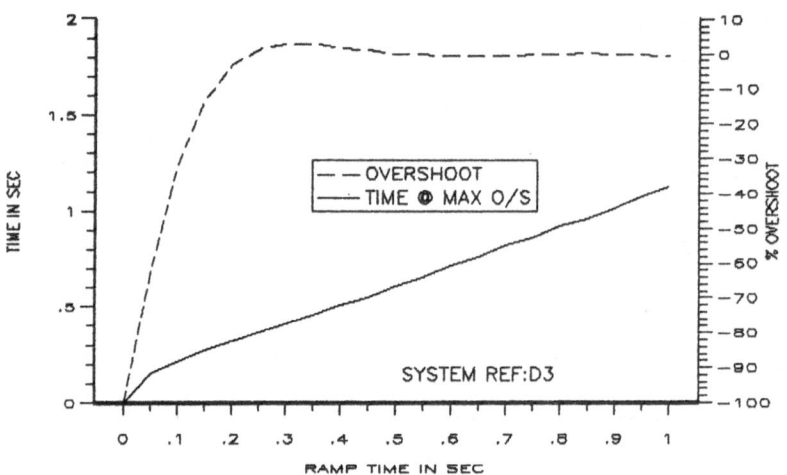

Fig. 16.36 Triangular response parameters.

approximately 32 dB as the hydraulic natural frequency reaches the natural frequency of the flow sensing element at 45 Hz. Optimized frequency and transient response parameters, plotted as a function of inlet flow, are shown in Figs 16.26 and 16.27. A Nichols chart, an open loop Bode diagram and a closed loop Bode diagram for the optimum operating condition corresponding to 300 litres/min are shown respectively in Figs 16.28, 16.29 and 16.30. A Nichols chart showing the effect of increasing stability margins is shown in Fig. 16.31 under reference D10. Transient response characteristics are plotted in Figs 16.32 to 16.35. Figure 16.32 shows the response to a step demand within the flow capacity of the pilot valve. The response to a large step demand, i.e. with the pilot valve saturated, is shown in Fig. 16.33 for a flow rate of 500 litres/min, and in Fig. 16.34 for a flow rate of 300 litres/min. The equations applicable to the response of systems operating under saturated conditions will be derived in Chapter 18. Ramp-step and triangular response characteristics at a flow rate of 500 litres/min are shown in Figs 16.35 and 16.36.

It should be noted that all performance curves are applicable to typical operating conditions whereas Figs 16.14 and 16.23 are the computer print-outs of a specific set of parameters.

17

Non-symmetrical Systems

Any control system employing a meter-in/meter-out flow control valve in which actuator inlet and outlet configurations are not identical falls into the category of non-symmetrical systems. Two alternative system configurations employing four-way directional and flow control valves are shown in Figs 17.1 and 17.2. An equivalent symmetrical system, as depicted in Fig. 9.1, is defined as a system having identical inlet operating conditions to the asymmetrical system. Although the majority of proportional and servo valves currently on the market incorporate

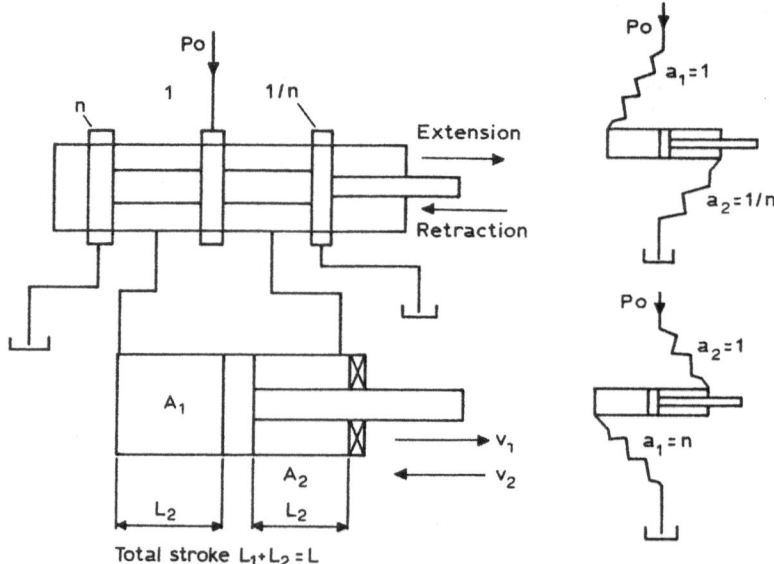

Fig. 17.1   Non-symmetrical system diagram.

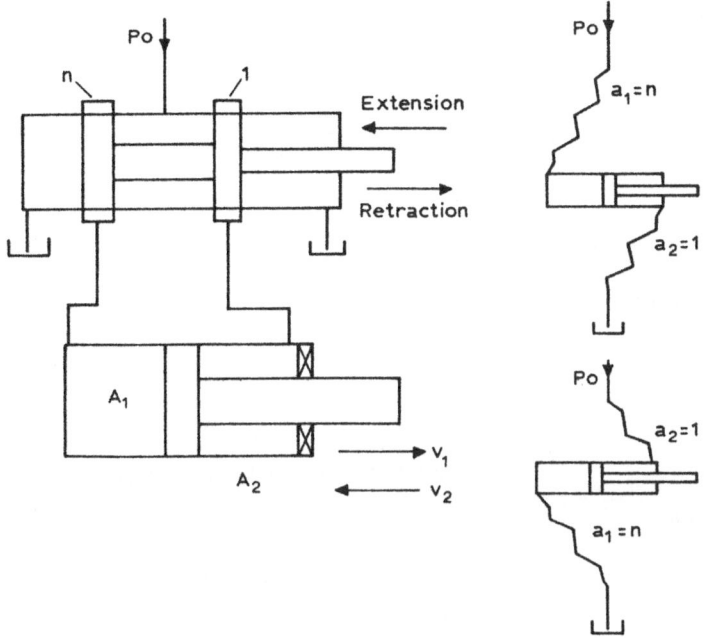

Fig. 17.2 Non-symmetrical system diagram.

symmetrical control orifices, and are perfectly capable of controlling non-symmetrical actuators, system performance can be enhanced by introducing non-symmetrical control orifice characteristics. The primary benefits obtained by using non-symmetrical control valves to control non-symmetrical actuators are: (1) increased effective flow rating of a valve controlling a retracting cylinder; (2) improved overrunning load capacity of an extending cylinder.

Let us first consider the system represented by the circuit of Fig. 17.1 when the cylinder is extending.

For the non-symmetrical system,

$$q_1 = C_d a_1 \sqrt{(2/\sigma)} \sqrt{(P_0 - P_1)} \qquad (17.1)$$

and

$$q_2 = C_d a_2 \sqrt{(2/\sigma)} \sqrt{(P_2)} \qquad (17.2)$$

For the equivalent symmetrical system,

$$q = C_d a_1 \sqrt{(2/\sigma)} \sqrt{(P_0 - P_1)} = C_d a_1 \sqrt{(2/\sigma)} \sqrt{(P_2)} \qquad (17.3)$$

hence $P_0 - P_1 = P_2$ and since $F/A_1 = P_L = P_1 - P_2$

$$q = \frac{C_d}{\sqrt{\sigma}} a_1 \sqrt{(P_0 - P_L)} \qquad (17.4)$$

For the non-symmetrical system, $F = P_1 A_1 - P_2 A_2$, let $F/A_1 = P_L$, then $P_L = P_1 - P_2(A_2/A_1)$

when $A_1/A_2 = \alpha$, $\qquad\qquad P_L = P_1 - \dfrac{P_2}{\alpha} \qquad (17.5)$

From eqns (17.1) and (17.2),

$$\frac{q_1}{q_2} = \alpha = \frac{a_1}{a_2} \sqrt{\left(\frac{P_0 - P_1}{P_2} \right)} \qquad (17.6)$$

$$q_2 = C_d a_2 \sqrt{\left(\frac{2}{\sigma} \right)} \sqrt{\left\{ \frac{(P_0 - P_L)}{(1/\alpha) + [\alpha^2 (a_2/a_1)^2]} \right\}} \qquad (17.7)$$

and since

$$v = \frac{q}{A_1} \qquad \text{and} \qquad v_1 = \frac{q_2}{A_2} = q_2 \frac{\alpha}{A_1}$$

combining eqns (17.7) and (17.4) yields

$$\frac{v_1}{v'} = \frac{q_1}{q'} = \sqrt{\left\{ \frac{2}{[(a_1/a_2)^2/\alpha^3] + 1} \right\}} \qquad (17.8)$$

$$\frac{v_2}{v'} = \frac{q_2}{q'} = \sqrt{\left\{ \frac{2}{[(a_2/a_1)^2 \alpha^3] + 1} \right\}} \qquad (17.9)$$

Equations (17.8) and (17.9) are plotted for orifice area ratios of 1, 1.5 and 3 in Fig. 17.3, and for orifice area ratios of 1, 2 and 4 in Fig. 17.4. The velocity and inlet flow of the asymmetrical cylinder are denoted by v and q, and velocity and flow of the equivalent symmetrical cylinder by v' and q' respectively. In accordance with our previously adopted definition of an equivalent cylinder having identical inlet conditions to the asymmetrical cylinder, the effective area of the extending equivalent cylinder is the piston area, A_1, and of the retracting equivalent cylinder the annular area, A_2. The stalling condition for the extending cylinder is reached when $F = P_0 A_1$ and for the retracting cylinder when $F = P_0 A_2$.

Equations (17.8) and (17.9) and Figs 17.3 and 17.4 can be used to size a four-way meter-in/meter-out flow control valve for non-symmetrical systems.

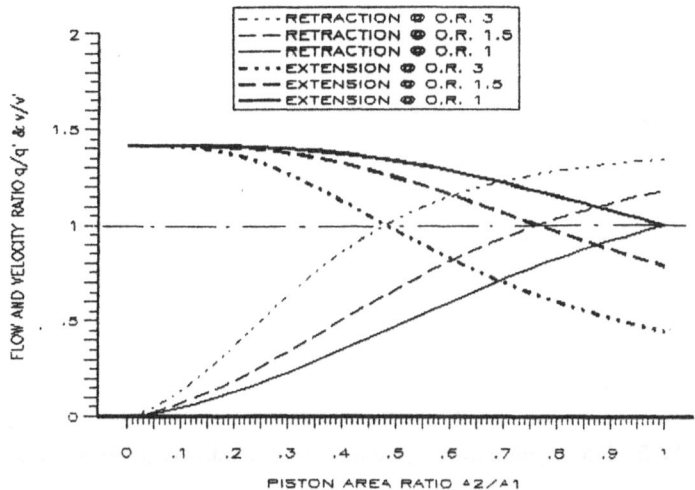

Fig. 17.3 Non-symmetrical system flow and velocity characteristics.

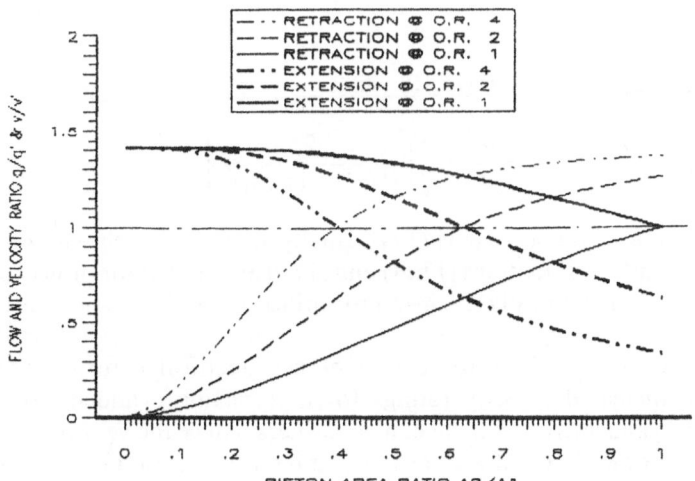

Fig. 17.4 Non-symmetrical system flow and velocity characteristics.

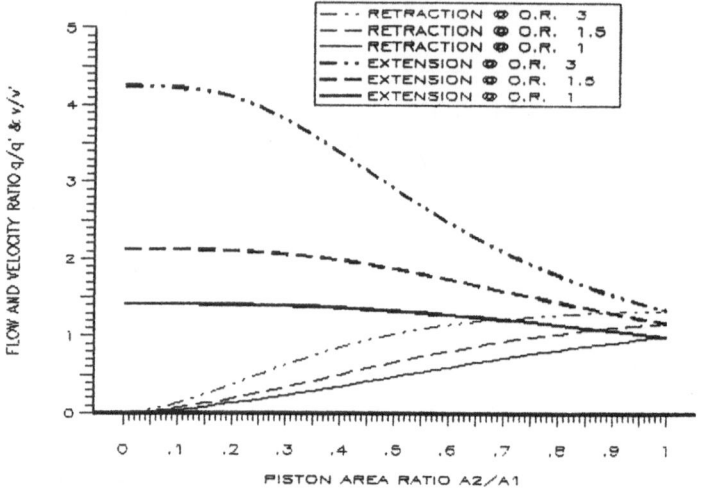

Fig. 17.5 Non-symmetrical system flow and velocity characteristics.

For the system represented by circuit Fig. 17.2, eqn (17.4) changes to

$$q = \frac{C_d}{\sqrt{\sigma}} a_2 \sqrt{(P_0 - P_L)} \qquad (17.10)$$

Combining eqns (17.7) and (17.10) yields

$$\frac{v_1}{v'} = \frac{q_1}{q'} = \sqrt{\left[\frac{2}{(1/\alpha^3) + (a_2/a_1)^2} \right]} \qquad (17.11)$$

For a retracting cylinder, eqn (17.9) is equally applicable to the circuits of Figs 17.1 and 17.2. Equations (17.11) and (17.9) are plotted for orifice area ratios of 1, 1·5 and 3 in Fig. 17.5, and for orifice areas of 1, 2 and 4 in Fig. 17.6.

For both circuits the introduction of asymmetrical control orifices results in increased velocity ratings for a retracting cylinder. For an extending cylinder asymmetrical control orifices reduce the velocity rating in the case of the circuit of Fig. 17.1, whilst increasing it in the case of the circuit of Fig. 17.2.

It should be noted that the orifice ratios referred to in Figs 17.3, 17.4, 17.5 and 17.6 are quoted as the ratio of the larger over the smaller for all flow paths.

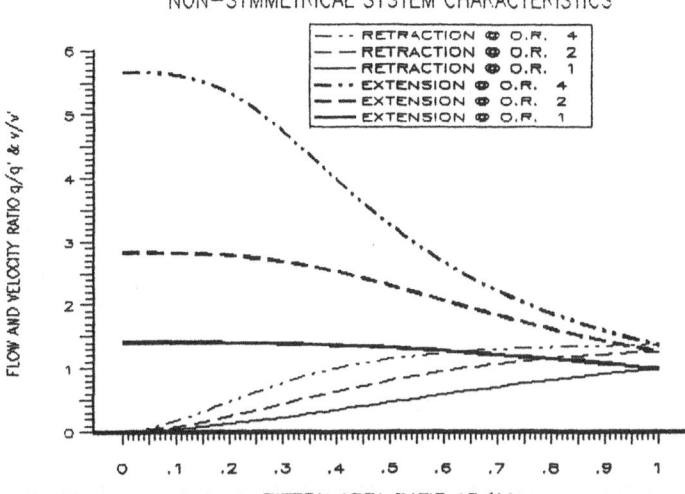

Fig. 17.6 Non-symmetrical system flow and velocity characteristics.

17.1 OIL COMPLIANCE

Referring to Fig. 17.1, the oil compliance of the cylinder on either side of the piston is given by the expressions:

$$L_1/(A_1 N) \quad \text{and} \quad L_2/(A_2 N)$$

giving a total compliance

$$\lambda_1 = \frac{1}{N[(A_1/L_1) + (A_2/L_2)]} \tag{17.12}$$

Let $L_1/L = \beta$, then eqn (17.12) can be re-written as

$$\lambda_1 = \frac{L}{N\{(A_1/\beta) + [A_2/(1 - \beta)]\}}$$

Differentiating λ_1 with respect to β gives the condition for maximum compliance, which occurs when

$$\beta = \frac{1}{1 + \sqrt{(A_2/A_1)}} \tag{17.13}$$

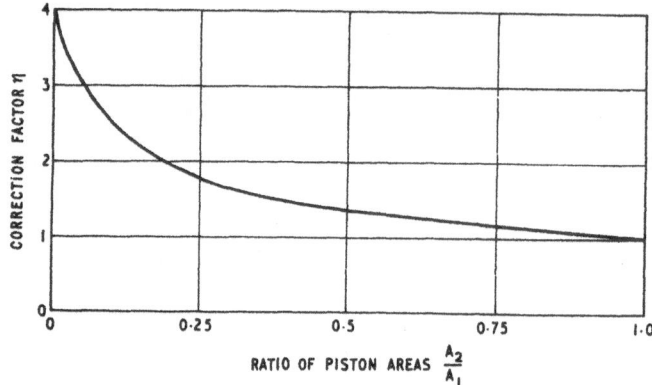

Fig. 17.7 Non-symmetrical system oil compliance correction factor.

By combining eqns (17.12) and (17.13), the maximum oil compliance of the cylinder is given by

$$\lambda_1 = \frac{L}{A_1 N} \frac{1}{[\sqrt{(A_2/A_1)} + 1]^2} \qquad (17.14)$$

The maximum oil compliance of a non-symmetrical cylinder can then be espressed as a ratio of the maximum oil compliance of an equivalent symmetrical cylinder of equal stroke and effective area, A_1. By combining eqns (17.14) and (11.4), the correction factor to be applied to convert the compliance of a symmetrical cylinder to that of an equivalent non-symmetrical cylinder is given by the expression

$$\mu = \frac{4}{[\sqrt{(A_2/A_1)} + 1]^2} \qquad (17.15)$$

Equation (17.15) is plotted in Fig. 17.7.

17.2 CAVITATION EFFECTS OF OVERRUNNING LOADS

Overrunning loads can cause the inlet chamber pressure to fall below zero thereby inducing cavitation conditions which are detrimental to system performance and integrity.

Let us consider an extending cylinder subject to an overrunning load, F.

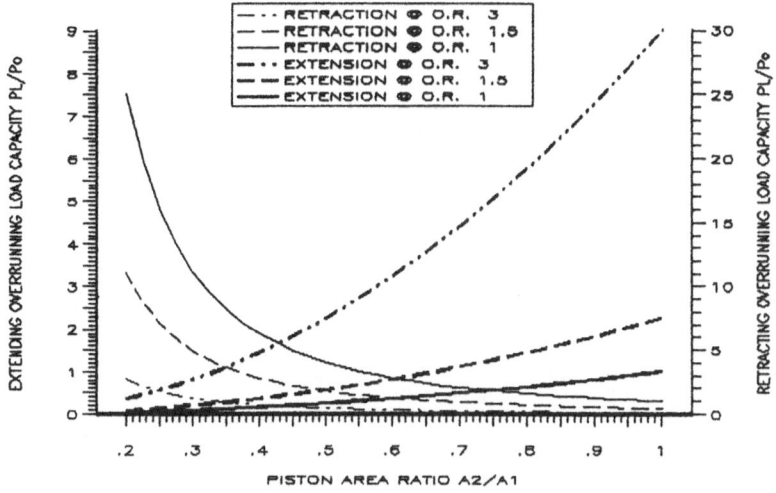

Fig. 17.8 Overrunning load capacity.

Cavitation will occur when the inlet pressure $P_1 = 0$, then $q_1 = C_d a_1 \sqrt{(2/\sigma)} \sqrt{P_0}$ and $q_2 = C_d a_2 \sqrt{(2/\sigma)} \sqrt{P_2}$; also $F = P_2 A_2$, and combining the above equations, the limiting overrunning load

$$F = P_0 A_2 \left(\frac{a_1}{a_2}\right)^2 \left(\frac{A_2}{A_1}\right)^2$$

or when $F/A_2 = P_L$,

$$\frac{P_L}{P_0} = \left(\frac{a_1}{a_2}\right)^2 \left(\frac{A_2}{A_1}\right)^2 \qquad (17.16)$$

For a retracting cylinder,

$$\frac{P_L}{P_0} = \left(\frac{a_2}{a_1}\right)^2 \left(\frac{A_1}{A_2}\right)^2 \qquad (17.17)$$

where the overrunning load $F = P_L A_1$.

Limiting overrunning load capacities for both extending and retracting cylinders are plotted in Figs 17.8 and 17.9. The graphs show that the overrunning load capacity of extending asymmetrical cylinders can be considerably improved by using valves with unequal inlet and outlet control orifice areas.

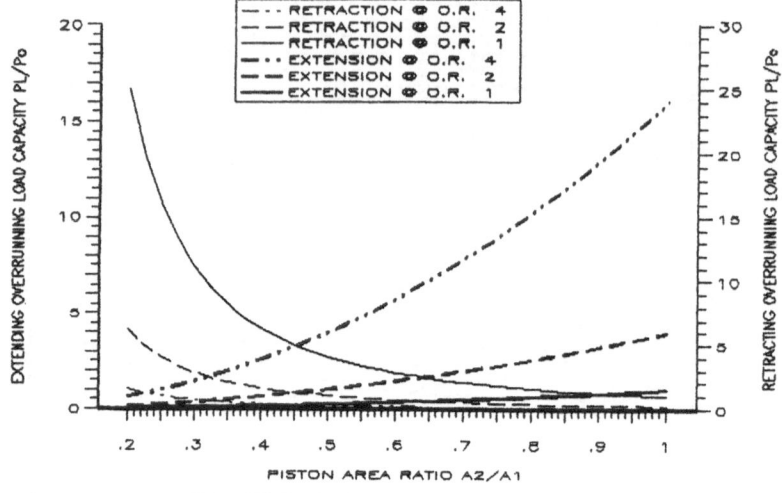

Fig. 17.9 Overrunning load capacity.

17.3 WORKED EXAMPLE

The worked example will illustrate how the equations and graphs can be applied to a typical system.

Specification

Cylinder piston area:	$100\,cm^2$
Cylinder annular area:	$50\,cm^2$
Cylinder stroke:	$100\,cm$
Supply pressure:	$150\,bar$
Extending cylinder loading:	$10\,000\,kp$
Retracting cylinder loading:	$5\,000\,kp$
Mass referred to cylinder:	$2\,000\,kg$
Symmetrical valve characteristics:	$100\,litres/min$ @ 50 bar valve pressure drop

From eqn (17.8) and Fig. 17.3, $q_1/q' = 1\cdot333$ for a symmetrical control valve when $A_2/A_1 = 0\cdot5$ and $q_1/q' = 1\cdot25$ for an asymmetrical control valve of the type shown in Fig. 17.1 when the orifice area ratio is $1\cdot5\!:\!1$. Similarly from eqn (17.8) and Fig. 17.4, $q_1/q' = 1\cdot155$ for an orifice area ratio of $2\!:\!1$. The corresponding flow ratios for a retracting cylinder can be established from eqn (17.9) and Figs 17.3 and 17.4 as:

$q_2/q' = 0.47$ for a symmetrical valve
$q_2/q' = 0.66$ for a 1·5:1 orifice area ratio
$q_2/q' = 0.816$ for a 2:1 orifice area ratio

Extending cylinder flow ratios for an asymmetrical valve of the type represented by Fig. 17.2 obtained from eqn (17.11) and Figs 17.5 and 17.6 are:

$q_1/q' = 1.874$ for a 1·5:1 orifice area ratio
$q_1/q' = 2.31$ for a 2:1 orifice area ratio

Retracting flow ratios for the circuit of Fig. 17.2 are identical to those of Fig. 17.1.

Maximum cylinder velocities can now be calculated from $v_1 = q_1/A_1$ and $v_2 = q_2/A_2$. For cylinder extension $P_L = 10\,000/100 = 100$ bar and for retraction $P_L = 5000/50 = 100$ bar, hence valve pressure drop $P_V = 150 - 100 = 50$ bar and $q' = 100$ litres/min for both extension and retraction. For example, for a symmatrical valve $q_1 = 133$ litres/min and $q_2 = 47$ litres/min, hence $v_1 = 133 \times 1000/(60 \times 100) = 22.17$ cm/s and $v_2 = 47 \times 1000/(60 \times 50) = 15.67$ cm/s. Velocities pertaining to other operating conditions are tabulated in the summary of results.

From eqn (17.16) and Figs 17.8 and 17.9 overrunning load capacity of an extending cylinder $P_L/P_0 = 0.25$ for a symmetrical valve, 0·562 for an asymmetrical valve with 1·5:1 orifice area ratio and 1·0 for a 2:1 orifice area ratio, giving limiting overrunning loads of 1875 kp, 4219 kp and 7500 kp respectively. For a retracting cylinder corresponding values can be obtained from eqn (17.17) and Figs 17.8 and 17.9, i.e. $P_L/P_0 = 4.0$ for a symmetrical valve, 1·778 for a 1·5:1 orifice area ratio and 1·0 for a 2:1 orifice area ratio, giving limiting overrunning loads of 60 000 kp, 26 670 kp and 15 000 kp respectively.

From eqn (17.15) and Fig. 17.7 the oil compliance correction factor to be applied to an equivalent symmetrical cylinder of area A_1 is 1·3726, giving an equivalent trapped cylinder volume of $10 \times 1.3726 = 13.726$ litres. The hydraulic actuator natural frequency can be calculated using either eqn (11.17) or eqn (12.9). In the absence of external damping, the natural frequency is given by the expression

$$\omega_0 = 2\sqrt{\left[\left(\frac{N}{V}\right)\left(\frac{A^2}{m}\right)\right]}$$

Before the natural frequency can be calculated, all values have to be expressed in SI units. Taking the bulk modulus for mineral oil as 13 793 bar, $N = 13\,793 \times 10^5$ Pa, $A = 0\cdot01\,\mathrm{m}^2$, $V = 13\cdot726 \times 10^{-3}\,\mathrm{m}^3$ and $m = 2000\,\mathrm{kg}$; then

$$\omega_0 = 2\sqrt{\left[\frac{(13\,793 \times 100^2)}{(13\cdot726 \times 2000)}\right]} = 141\cdot77 \text{ radians/second}$$

or $22\cdot56$ Hz. The results are summarized in Table 17.1.

Table 17.1 Summary of results

Circuit	Orifice ratio	Max. velocities (cm/s)		Limiting o/r load (kp)	
		Extend	Retract	Extend	Retract
	1:1	22·17	15·67	1 875	60 000
17·1	1·5:1	20·83	22·0	4 219	26 670
17·2	1·5:1	31·23	22·0	4 219	26 670
17·1	2:1	19·25	27·2	7 500	15 000
17·2	2:1	38·5	27·2	7 500	15 000

18

Response to Large Step Demand

In considering the response of a system to a step demand, a clear distinction has to be drawn between small and large step response. The dynamic equations derived so far assume small perturbations about a fixed steady-state operating point, and are therefore applicable to transient response to a small step demand. Since small and large step response are relative terms, a definition providing a clear distinction is required.

When a system performs within the confines of its limiting operating conditions, i.e. solenoid or force motor effort of single-stage valves, pilot valve displacement of multi-stage valves, main stage travel, its transient response characteristics can be described by a small step response to a unit input.

A system required to perform beyond the confines of at least one of its limiting operating conditions has to be described by a large step response to a given demand. Large step response can also be applied to bang–bang (on–off) valves.

Let us consider an actuator, operating under valve saturated conditions, subjected to inertia and resistance loading. If the pressure drop across the actuator due to the resistance loading is denoted by P_F and that due to inertia forces by P_I,

$$Av = ka\sqrt{(P_0 - P_F - P_I)} \tag{18.1}$$

and

$$m\frac{dv}{dt} = P_I A \tag{18.2}$$

where A is the effective cylinder area, v the actuator velocity, P_0 the supply pressure, a the control orifice area at maximum valve opening and m the mass referred to the actuator; hence

$$t = m/A \int_0^{v_m} \frac{dv}{P_0 - P_F - [Av/(ka)]^2}$$

$$= \frac{v_0 m}{2(P_0 - P_F)A} \log_e \left[\frac{1 + (v_m/v_0)}{1 - (v_m/v_0)} \right] \qquad (18.3)$$

where v_m is the maximum actuator velocity attained, and the steady-state velocity,

$$v_0 = ka \frac{\sqrt{(P_0 - P_F)}}{A}$$

Equation (18.2) can be re-written in terms of the actuator travel, θ_0, as

$$mv \frac{dv}{d\theta_0} = P_1 A \qquad (18.4)$$

whence

$$\theta_0 = \frac{m}{A} \int_0^{v_m} \frac{v}{P_0 - P_F - [Av/(ka)]^2} dv$$

$$= \frac{v_0^2 m}{2(P_0 - P_F)A} \log_e \left[\frac{1}{1 - (v_m/v_0)^2} \right] \qquad (18.5)$$

Equations (18.3) and (18.5) can be expressed non-dimensionally as

$$T = \log_e \left[\frac{1 + v_m/v_0}{1 - v_m/v_0} \right] \qquad (18.6)$$

and

$$\theta = \log_e \left[\frac{1}{1 - (v_m/v_0)^2} \right] \qquad (18.7)$$

where the non-dimensional time

$$T = \frac{2(P_0 - P_F)A}{v_0 m} t$$

and the non-dimensional actuator travel

$$\theta = \frac{2(P_0 - P_F)A}{v_0^2 m} \theta_0$$

The space–time equation for saturated throttled flow can now be obtained by combining eqns (18.6) and (18.7), and hence

$$\theta = 2\log_e\left[\frac{e^T + 1}{2}\right] - T \qquad (18.8)$$

Equation (18.6) can be transposed to express the velocity ratio v_m/v_0 as a function of the non-dimensional time T, or

$$v_m/v_0 = \frac{e^T - 1}{e^T + 1} \qquad (18.9)$$

Equations (18.8) and (18.9) are plotted in Fig. 18.1.

Corresponding velocity and displacement time characteristics in respect of non-throttled flow at constant and infinite acceleration are also shown for comparison with throttled flow characteristics. Figure 18.1 uses non-dimensional units; large step response characteristcs for the system analysed previously in Section 16.4 are plotted in Figs 16.33 and 16.34.

Since for any given valve configuration and actual time, t, the non-dimensional time, T, is directly proportional to the ratio A^2/m, small masses

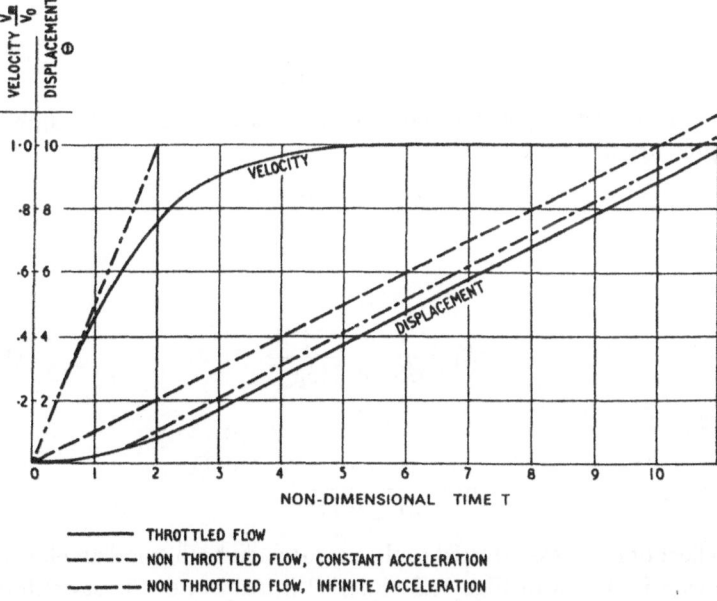

Fig. 18.1   Non-dimensional large step response characteristics.

referred to large cylinder areas can lead to fairly high values of T, which in turn will result in very high exponentials, e^T. To prevent overflow errors, it is advisable to modify eqns (18.8) and (18.9) for higher values of T when used in computer programs. When t is greater than say, 10, giving an exponential $e^T > 22\,026$ no discernible errors are introduced by changing eqn (18.8) to $\theta = T - 2\log_e 2 = T - 1\cdot3863$ and eqn (18.9) to $v_m/v_0 = 1$.

It is obviously of interest to be able to quantify the relative significance of large to small step response. A comparison of the response plots to a large step demand shown in Figs 16.33 and 16.34 with the response plot to a small ramp input shown in Fig. 14.3 demonstrates the similarity of output characteristics. In both cases the output settles down to follow the demand at a constant following or velocity error. In a closed loop system containing a free integrator, the steady-state velocity error is given by the velocity to loop gain ratio, or $\varepsilon_v = v_0/K$. The corresponding error to a large step demand is $2\log_e 2$ when expressed non-dimensionally, which can be converted to an actual velocity error of

$$\varepsilon_v = \frac{\log_e 2mv_0^2}{(P_0 - P_F)A} \tag{18.10}$$

or to an equivalent loop gain

$$K_L = \frac{(P_0 - P_F)A}{\log_e 2mv_0} \tag{18.11}$$

alternatively, the loop gain can be expressed as a function of the flow, q, or the pressure factor, R_V, as

$$K_L = \frac{(P_0 - P_F)\,A^2}{q}\frac{1}{m}\frac{1}{\log_e 2} \tag{18.12}$$

or

$$K_L = \frac{1}{R_V}\frac{A^2}{m}\frac{1}{2\log_e 2} \tag{18.13}$$

where

$$R_V = \frac{1}{2}\frac{q}{(P_0 - P_F)}$$

The effect of the mass referred to the actuator on both small and large step loop gain is shown in Figs 18.2 and 18.3 in respect of the seventh order system analysed in Chapter 16 for flow rates of 500 and 300 litres/min

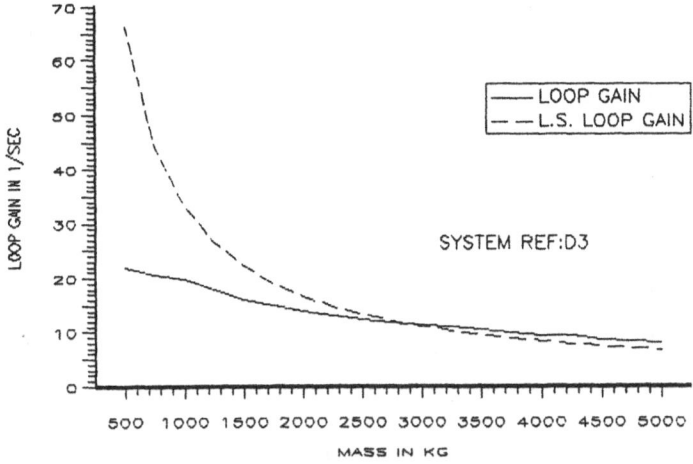

Fig. 18.2 Stability boundary.

respectively. Both graphs show that the large step loop gain varies exponentially with mass, with a considerably reduced effect on small step loop gain. Figure 18.2 shows a distinct cross-over point at around 3000 kg, where a small ramp input and a large step demand would exhibit identical steady-state velocity errors. Response to a ramp input under identical operating conditions to the large step response shown in Fig. 16.33 is

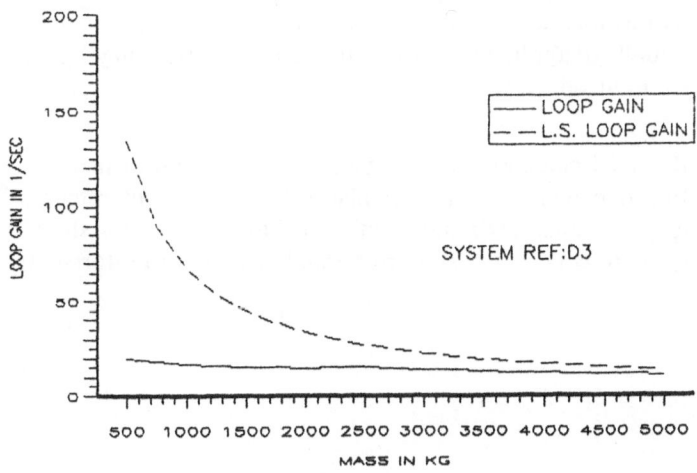

Fig. 18.3 Stability boundary.

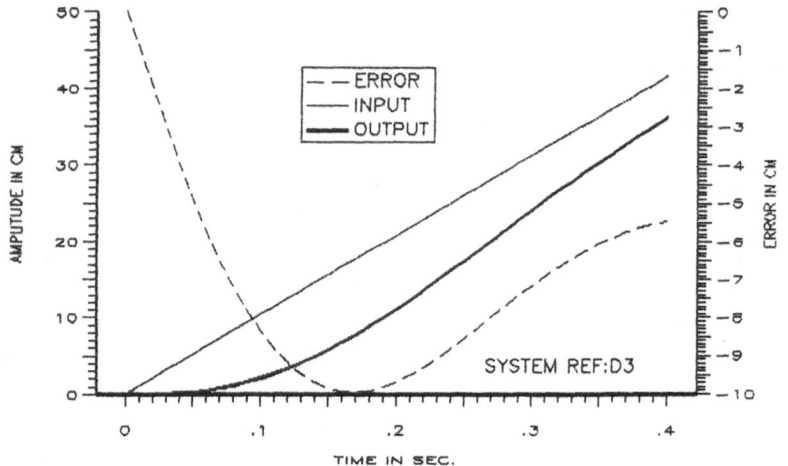

Fig. 18.4   Response to ramp input.

plotted in Fig. 18.4. The response to the ramp input is under closed loop control with the control valve in its operative condition, resulting in an initial dynamic error before settling down to a constant steady-state following error. Figure 16.33 depicts the response to a velocity step demand under saturated operating conditions, i.e. with the control valve at maximum opening, the gradual velocity build-up being due to the interaction of valve throttling and inertia loading.

The expressions derived above for cylinder-operated systems can, of course, equally be applied to motor-driven systems, providing the following changes are introduced:

(1)   Referred mass, m, replaced by referred moment of inertia, I.
(2)   Effective cylinder area, A, replaced by motor displacement, C.
(3) v_m represents maximum rotational velocity of motor shaft.
(4) v_0 represents steady-state rotational velocity of motor shaft.

19

Valve Operating Forces

In Chapter 5 we reviewed several types of control valve. In considering valve operating forces, we have to clearly differentiate between spool, poppet and flapper-nozzle valve constructions.

19.1 SPOOL VALVES

In balanced spool valves, valve operating forces can be considered to be predominantly frictional and hydrodynamic; the frictional forces can be reduced by keeping the area of contact between spool and bore to a minimum and introducing balancing grooves on the spool lands. The hydrodynamic, or Bernoulli forces, arise due to the change of momentum of the fluid being throttled. The direction of the velocity of the fluid is not radial but inclined at approximately 69° relative to the axis of the spool, and the axial component of the velocity produces a force which always tends to centralize the spool. This force is directly proportional to the product of the control orifice area and the corresponding valve pressure drop, or

$$F_B = 2C_d a \cos 69° \, \delta P \qquad (19.1)$$

where F_B is the Bernoulli force and δP the pressure drop across an orifice of area a.

For a four-way valve,

$$F_B = 2C_d a \cos 69°(P_0 - P_L) \qquad (19.2)$$

and when the coefficient of discharge is taken as 0·8,

$$F_B = 0·57a(P_0 - P_L)$$

169

It is often more convenient to express the Bernoulli force in terms of the valve flow and pressure drop. By combining eqns (19.2) and (9.8), Bernoulli force for a four-way valve,

$$F_B = kq\sqrt{(P_0 - P_L)} \qquad (19.3)$$

The corresponding equation for a three-way valve can be derived from eqns (19.1) and (9.14) as

$$F_B = = kq\sqrt{\left(\frac{P_0}{2} - P_L\right)} \qquad (19.4)$$

For mineral oil and rectangular orifices, the constant k can be taken as 0·0076 and 0·0054 for four-way and three-way valves respectively. Corresponding constants for valves with round control orifices are 0·0097 and 0·0069. The above constants are applicable when q is in litres per minute, P_0 and P_L in bars and F_B in kp.

For valve constructions where the output available from the force/torque motor or solenoid is insufficient to overcome valve operating loads, hydrodynamic forces can be reduced by special shaping of the valve spool, but effective compensation is difficult to achieve over a wide operating range. In most cases a better solution is the introduction of a second stage. Various types of multi-stage valves are described in Chapter 5.

19.2 FLAPPER-NOZZLE VALVES

Forces associated with flapper-nozzle arrangements consist of three components as shown in Fig. 19.1.

(1)   Pressure forces due to fluid pressure in the nozzle acting on the projected flapper-nozzle area.
(2)   Momentum forces due to fluid velocity in the nozzle.
(3)   Bernoulli forces, i.e. negative forces due to the gradually increasing area between nozzle and flapper as the fluid moves radially outward.

An exhaustive analysis of flapper-nozzle forces is quite complex and outside the scope of this book, but a few general statements can be useful in appreciating the effect of the three components within the normal range of flapper operation.

(1)   Pressure forces are usually dominant.

(2) Momentum forces become more significant at larger flapper openings.
(3) Bernoulli forces become more significant at smaller flapper openings.
(4) All forces are directly proportional to control pressure P_C.

The relative magnitude of the three force components can be gauged by the following expressions:

$$F_P = \frac{\pi}{4} d_n^2 P_C K_S \tag{19.5}$$

$$F_M = 8\pi (C_d y)^2 P_C K_S \tag{19.6}$$

$$F_B = \pi (C_d d_n)^2 \left\{ \log_e \left(\frac{d_e}{d_n} \right) - \frac{1}{2} \left[1 - \left(\frac{d_n}{d_e} \right)^2 \right] \right\} P_C K_S \tag{19.7}$$

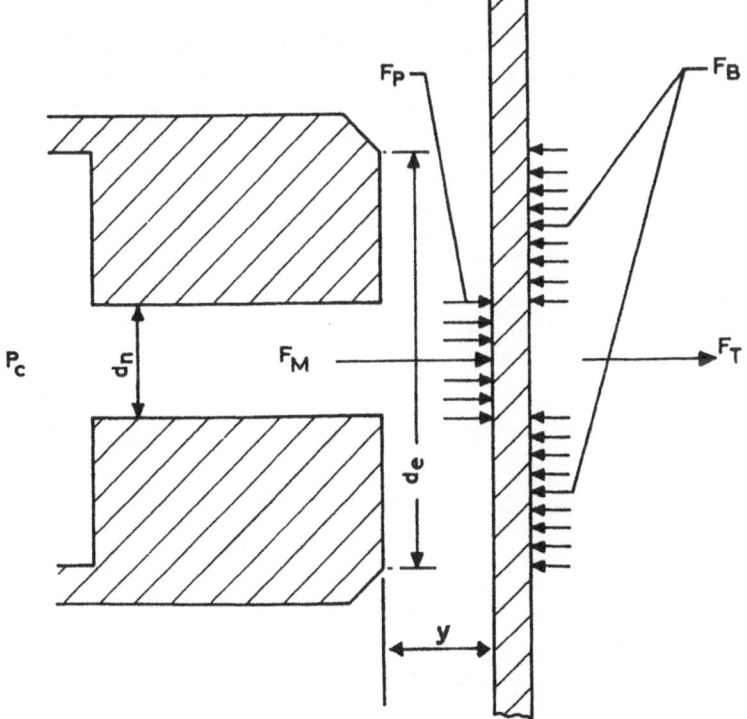

Fig. 19.1 Flapper-nozzle arrangement.

Where the saturation factor

$$K_S \simeq \frac{1}{1 + 16(Y/d_n)^2}$$

The total force acting on the flapper is then given by

$$F_T = F_P + F_M - F_B \qquad (19.8)$$

19.3 POPPET VALVES

In poppet valves dominant force components are usually pressure and Bernoulli forces. If the seat diameter is represented by d_n, eqn (19.5) can be used to calculate the pressure force acting on the poppet. By combining eqns (9.1) and (19.1), Bernoulli force is given by the equation

$$F_B = kq\sqrt{\delta P} \qquad (19.9)$$

Assuming a coefficient of discharge of 0·8, the constant, k, applicable to mineral oil can be taken as 0·0054 when q is in litres per minute, δP in bars and F_B in kp. It must be stressed, however, that this value should be treated with caution and may have to be modified in accordance with poppet and seat configuration. The total force acting on the poppet is then given by

$$F_T = F_P - F_B \qquad (19.10)$$

20

The Electronic Interface

The electronic interface, briefly referred to in Chapter 7, will now be discussed in more detail. The role of the electronic or electro-hydraulic interface in the overall control system is indicated in the block diagrams, Fig. 7.2. Its function is to process the low power signal generated by the controller in order to provide a signal compatible with the control element input characteristics. To be able to carry out the above stated function, some ancillary electronic circuit elements may have to be included:

(1) A stabilized DC power supply.
(2) A digitial–analogue converter.

The majority of electronic amplifiers require a stabilized 24V DC power supply, which, in applications other than mobile, is not usually readily available. Modern control systems will almost invariably utilize digital controllers, thus requiring a digital–analogue converter. D–A converters can be either binary or BCD (binary–decimal coded). To express 10 numbers requires a 4-bit D–A converter and a four-digit decimal number can be produced either by a 14-bit binary or by a 16-bit BCD converter. Circuits for 4-bit and 16-bit BCD converters are shown in Fig. 20.1.

A typical power amplifier circuit, specifically configured for a proportional flow control valve, is shown in Fig. 20.2. The features included in this circuit are common to most commercially available proportional valve amplifiers and will now be summarized to define their basic functions.

(1) Gain adjustment: the gain, i.e. the slope of the output/input signal characteristics can be individually adjusted. It can be used to compensate for asymmetrical cylinders or loading.
(2) Ramp setting: acceleration and deceleration adjustment can be used to optimize system performance. It is covered in Section 21.1, 'input shaping'.

173

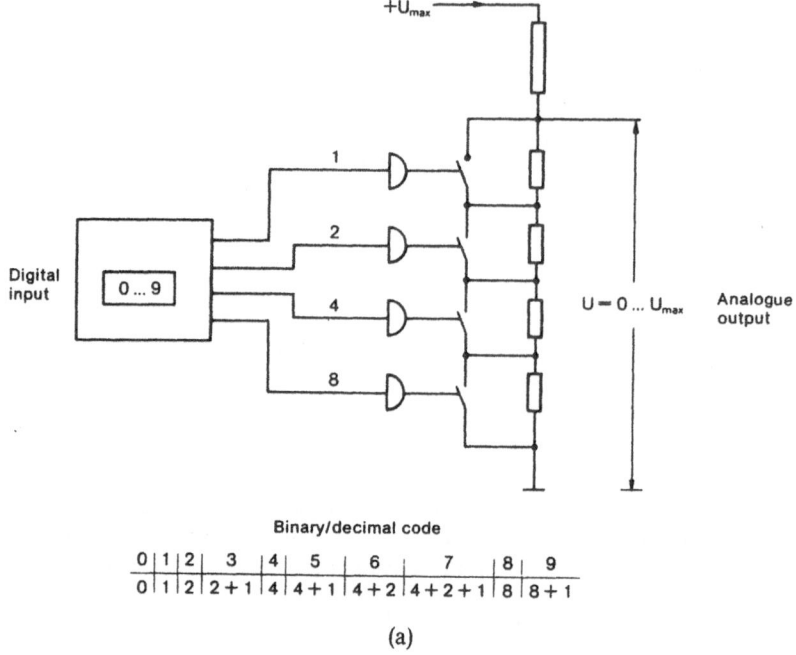

Binary/decimal code

0	1	2	3	4	5	6	7	8	9
0	1	2	2 + 1	4	4 + 1	4 + 2	4 + 2 + 1	8	8 + 1

(a)

Fig. 20.1   BCD digital–analogue converter.

(3) Deadband compensation: most proportional control valves incorporate overlapped spools, thus creating a deadband over which signal variation produces no output flow. In order to improve transient response when a directional flow change is demanded by a signal polarity reversal, a controlled jump can be introduced around the null region. This adjustment should be used with caution, since deadband compensation often extends into the operational range and can therefore turn a proportional into a 'bang-bang' valve.

(4) Current feedback: current feedback ensures that the output to the solenoid is a current rather than a voltage signal, thereby maintaining a magnetic force substantially unaffected by coil temperature and hence resistance variations.

(5) Dither: a sinusoidal or square wave superimposed dither signal has the effect of reducing valve hysteresis caused by frictional forces acting on the valve spool. Dither frequency and amplitude has to be carefully chosen to be effective in reducing valve hysteresis without disturbing the controlled output. Normally dither frequency is set at

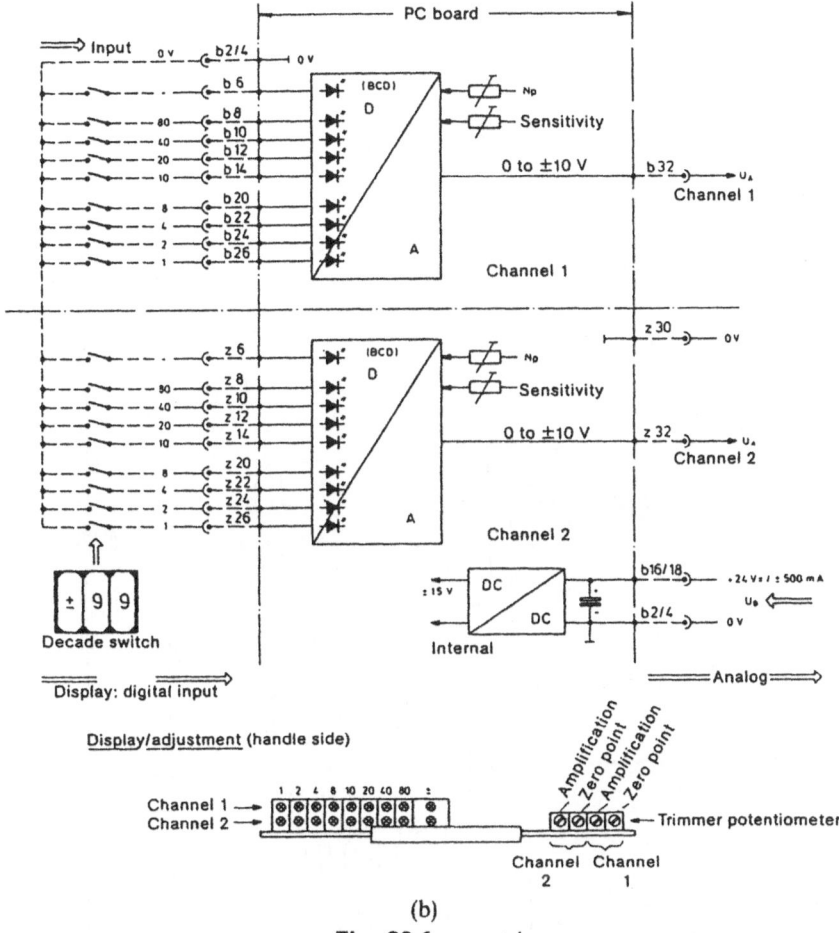

(b)

Fig. 20.1—*contd.*

a value below the natural frequency of the valve, or pilot valve in the case of a two-stage valve, and above the bandwidth of the control system. In some amplifiers dither frequency is pre-set while dither amplitude is adjustable.

(6) Modulator: the majority of power amplifiers driving proportional solenoids employ pulse width modulation. This is a more efficient way of controlling solenoids than the conventional proportional drive amplifier, resulting in less heat dissipation requiring smaller heat sinks. The modulator generates a square wave at constant

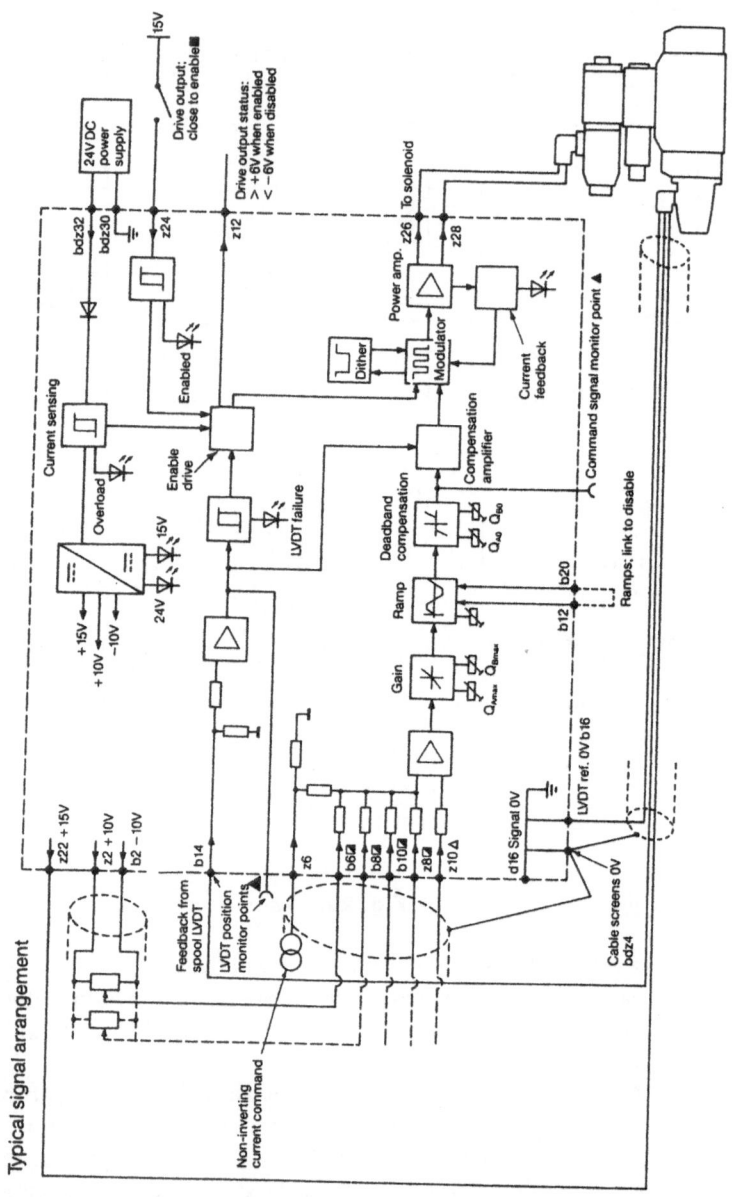

Command signals and outputs

Command signals	Input pins		Secondary pins ref.	Valve flow
Type	Ref.	Signal polarity		
Non-inverting voltages	b6/8/10 or z8	+		P-B
		–	d16	P-A
Non-inverting current	z6	+	d16	P-B
		–		P-A
Inverting voltage	z10	–	Link one of b6/8/10 or z8 to d16	P-B
		+		P-A
Differential voltage	z10	–	One of b6/8/10 or z8	P-B
		+		P-A
	One of b6/8/10 or z8	+	z10	P-B
		–		P-A

LVDT connections

LVDT plug pin	Amplifier pin
1	b14
2	z22
3	b16
4	–

◄ On front panel
■ 10 to 40V to enable. ≤ 0,8V or open circuit to disable
■◪ Non-inverting voltage commands
◁ Inverting or differential voltage command

Fig. 20.2 Power amplifier circuit.

frequency, the output signal being controlled by varying the pulse width as a function of the demand signal level.

(7) Summing junction: summing junctions provide a means for comparing two signals and generating an error signal between demand and output, thus catering for negative feedback loops. In the circuit shown in Fig. 20.2 two feedback loops can be accommodated: an internal spool position and an external output signal feedback loop.

21

System Enhancement

System enhancement can be attained by several means, the more important being shaping of the demand signal, passive network compensation, adaptive control, multiple feedback loops and the use of three term controllers.

21.1 INPUT SHAPING

The effect of changing the input signal from a step to a ramp-step demand has been demonstrated in preceding chapters, specifically for a non-dimensional system in Chapter 15 and for a typical flow feedback system in Chapter 16. In both cases it could be seen that the amount of overshoot was considerably reduced at the expense of increased response times.

We shall now investigate the effect of various forms of input shaping on the transient response of a fifth order closed loop control system described by the system parameter print-out, system reference D8. (see Fig. 16.14)

21.1.1 Ramp-Step Demand

In analogue circuits, ramps are generated by means of an RC network. The simplest form of network, comprising a resistor and a capacitor, produces an exponential demand as shown in Fig. 21.1(a); a true ramp can be generated by the circuit shown in Fig. 21.1(b), the latter being a more effective method of input shaping. The effect of varying the ramp time on system overshoot and response is illustrated in Fig. 21.2. A 0·15 s ramp reduces the overshoot from 22·6% to approximately 6% whilst increasing the 90% response time from 0·051 s to 0·15 s. At larger ramp times, the law of diminishing returns applies, as is clearly indicated by Fig. 21.2. Transient response plots for step and ramp-step inputs are shown in Fig. 21.4.

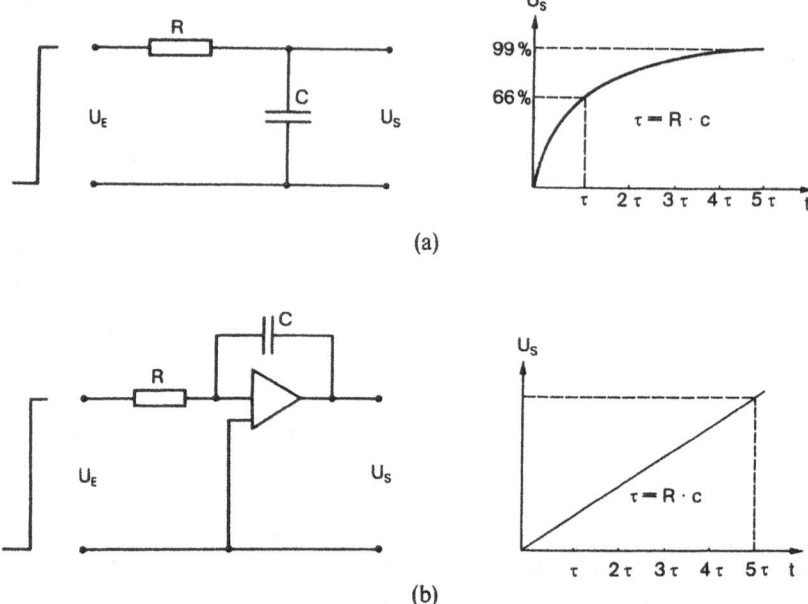

(a)

(b)

Fig. 21.1   Ramp generator circuits.

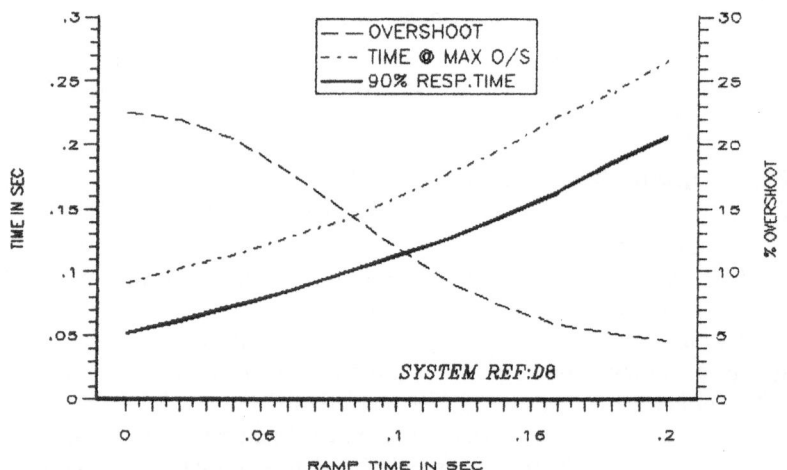

Fig. 21.2   Ramp-step response parameters.

21.1.2 Negative Ramping

A negative ramp can be introduced over an initial portion of the duty cycle in order to reduce the amount of system overshoot. Transient response of the non-dimensional system analysed in Chapter 15 when subjected to such a modified step demand is shown in Fig. 21.5. A comparison with the step response plotted in Fig. 15.3 shows that the overshoot has been reduced from 32% to 5% at some increase in response time.

21.1.3 Superimposed Negative Impulse

An alternative approach to negative ramping is the introduction of an impulse function of negative polarity to the step demand signal. The response of a non-dimensional system to a unit impulse function is shown in Fig. 21.6. Figure 21.7 shows the transient response of the system to a step demand with a 50% negative impulse applied at 0·4 s. The effect of applying a negative impulse to the closed loop control system D8 is shown in Fig. 21.4.

21.2 PASSIVE NETWORKS

In some cases improvement of system performance can be achieved by incorporating compensating networks in the amplifier. To increase high

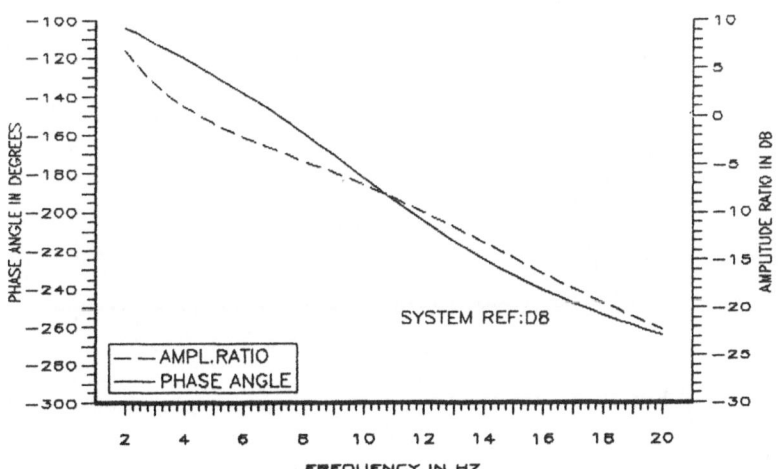

Fig. 21.3 Open loop Bode diagram, fifth order system.

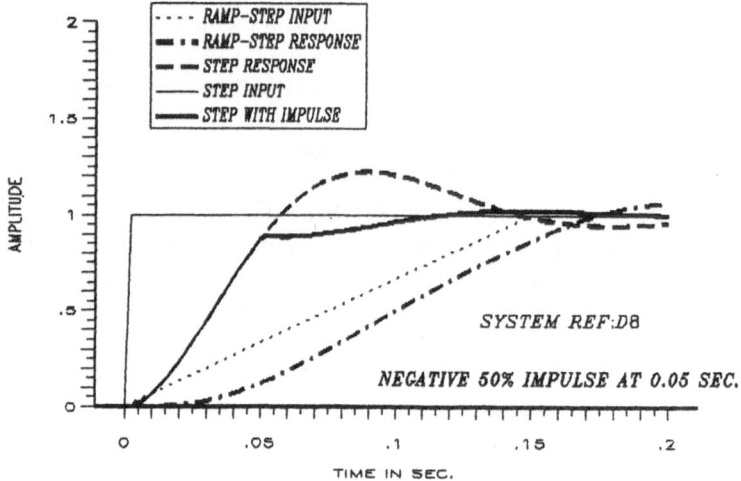

Fig. 21.4 Multiple transient response plots.

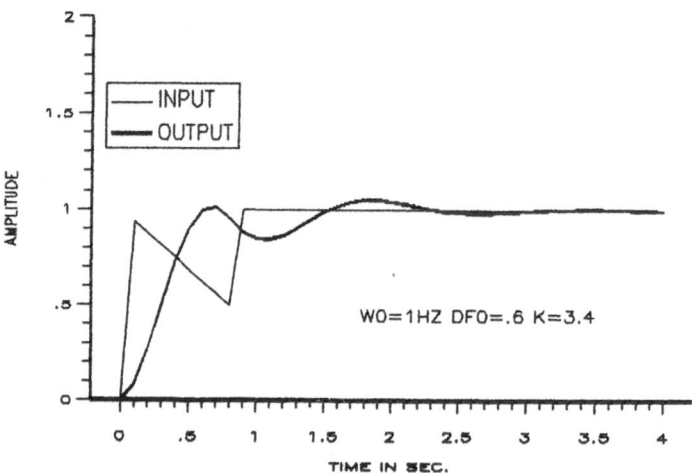

Fig. 21.5 Transient response to shaped step demand.

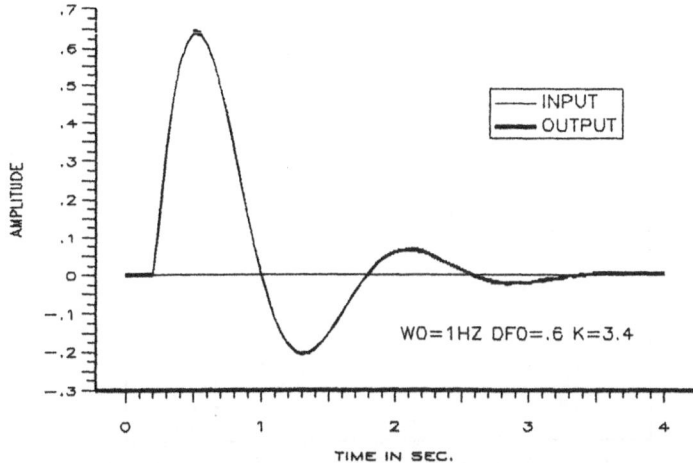

Fig. 21.6 Impulse function.

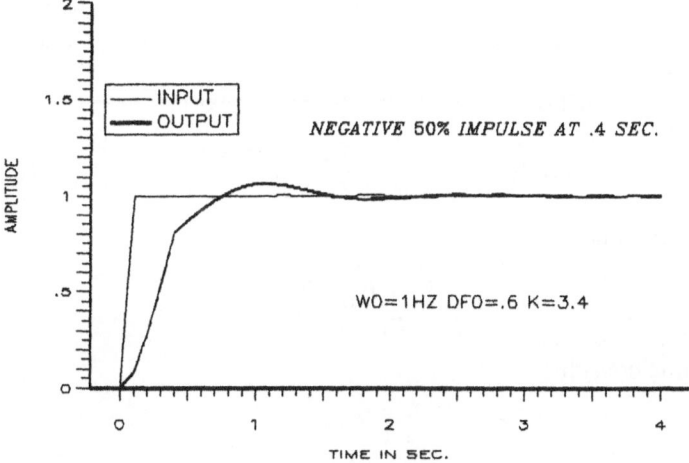

Fig. 21.7 Transient response with negative impulse.

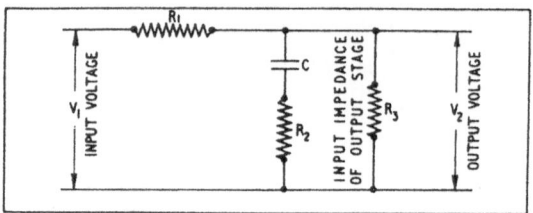

Fig. 21.8   Circuit diagram of integral network.

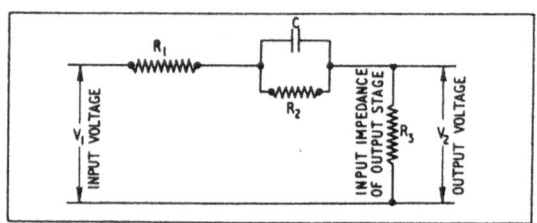

Fig. 21.9   Circuit diagram of phase advance network.

frequency or transient response a phase advance type network would normally be used, while low frequency or steady-state performance can be enhanced by means of an integral-type network. Circuit diagrams of integral and phase advance networks are shown in Figs 21.8 and 21.9 respectively.

The transfer function of the phase advance network,

$$K_N G_N(s) = \frac{V_2}{V_1} = \frac{R_3}{R_1 + R_2 + R_3} \frac{\alpha \tau s + 1}{\tau s + 1} \qquad (21.1)$$

where the amplification factor

$$\alpha = \frac{R_1 + R_2 + R_3}{R_1 + R_3}$$

and the time constant

$$\tau = C \frac{R_2}{\alpha}$$

The transfer function of the integral network is given by,

$$K_N G_N(s) = \frac{V_2}{V_1} = \frac{R_3}{R_1 + R_3} \frac{\tau s + 1}{\alpha \tau s + 1} \qquad (21.2)$$

where the attenuation factor

$$\alpha = 1 + \frac{R_1}{R_2} \frac{1}{1 + (R_1/R_3)}$$

and the time constant $\tau = CR_2$.

The time-dependent portion of the transfer function can be expressed in terms of the non-dimensional frequency $u = \tau\omega$.

Then for the phase advance network,

$$G_N(ju) = \frac{j\alpha u + 1}{ju + 1} \tag{21.3}$$

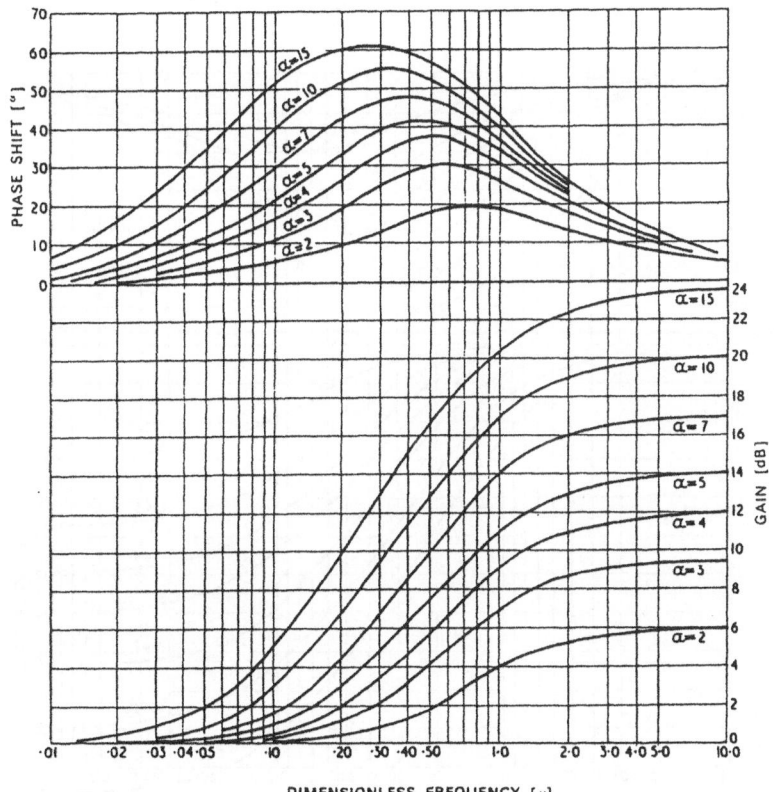

DIMENSIONLESS FREQUENCY [μ]

Fig. 21.10 Frequency response of phase advance network.

and for the integral network,

$$G_N(ju) = \frac{ju + 1}{j\alpha u + 1} \tag{21.4}$$

The above transfer functions are plotted in Figs 21.10 and 21.11.

By applying an integral network to our closed loop system D8, we can substantially improve its steady-state performance. Choosing an attenuation factor, α, of 5 should enable us to increase the loop gain by an equal factor without introducing any instability. Care has to be taken, however, to match the time constant, τ, to the system dynamics, since selecting an incorrect time constant can be counter-productive. This is illustrated in the stability boundary plot Fig. 21.12, which shows an initial drop in the permissible loop gain. The optimum loop gain is reached when the network

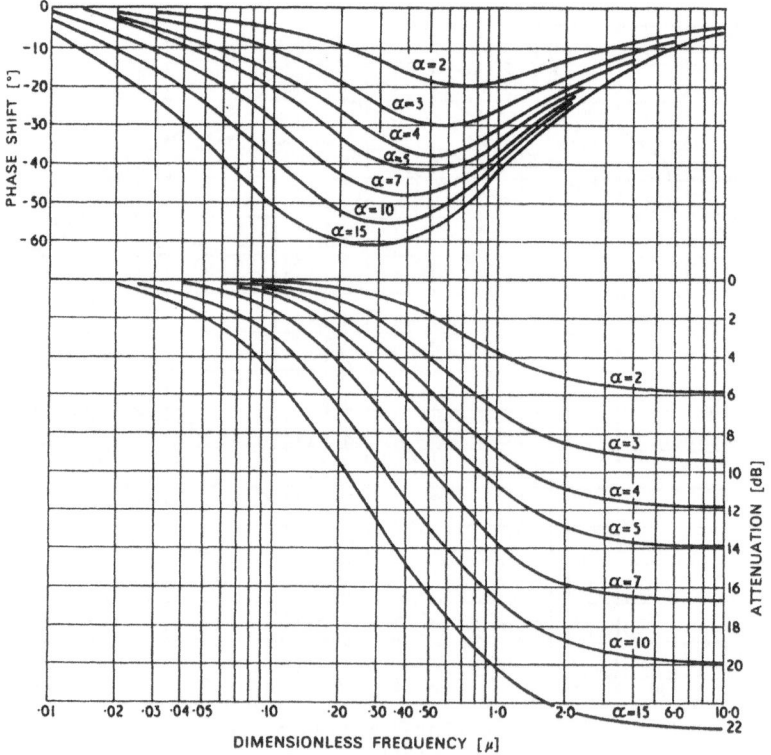

Fig. 21.11 Frequency response of integral network.

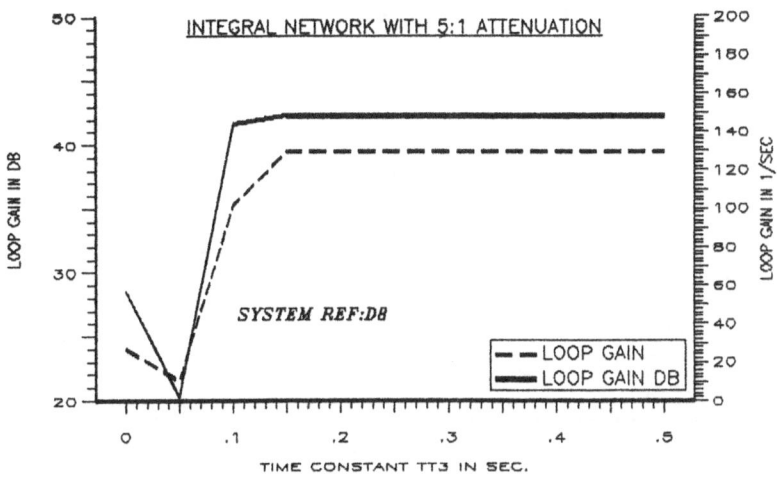

Fig. 21.12 Stability boundary with passive network.

time constant is greater than 0·15 s. In practice it is advisable to choose a slightly larger value to allow for parameter variations. The parameters for a compensated system with a network time constant $\tau = 0.23$ and attenuation factor $\alpha = 5$ are summarised in Table 21.1. It can be seen that, although the loop gain and hence steady-state performance characteristics have increased by a factor of 4·73:1, dynamic performance

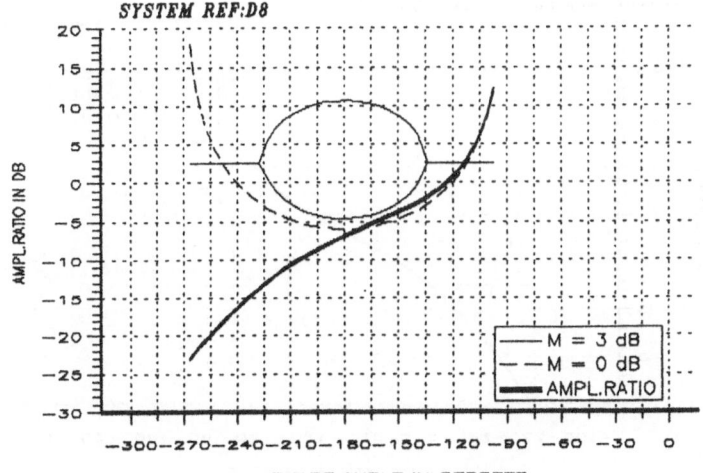

Fig. 21.13 Nichols chart, $K = 27\,\text{s}^{-1}$.

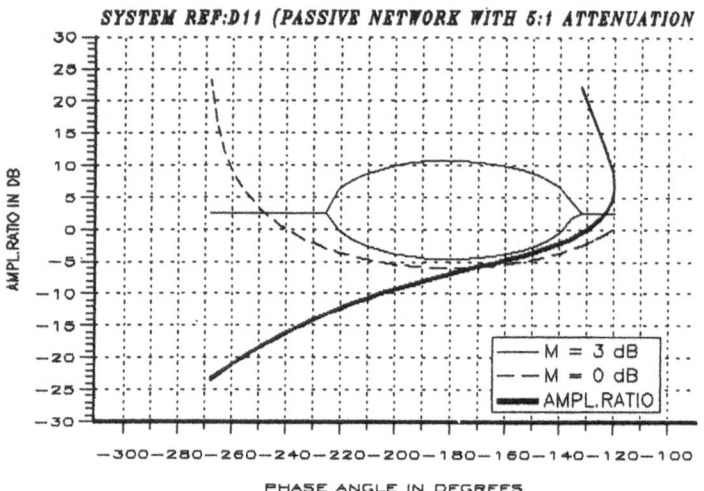

Fig. 21.14 Nichols chart, $K = 127 \cdot 7\,s^{-1}$, with network.

of the compensated system is virtually identical to the original system. Nichols charts for the original and compensated systems, plotted in Figs 21.13 and 21.14 respectively, show that both systems have almost identical stability margins, whereas an uncompensated system operating at the higher loop gain would be violently unstable, as clearly shown by the

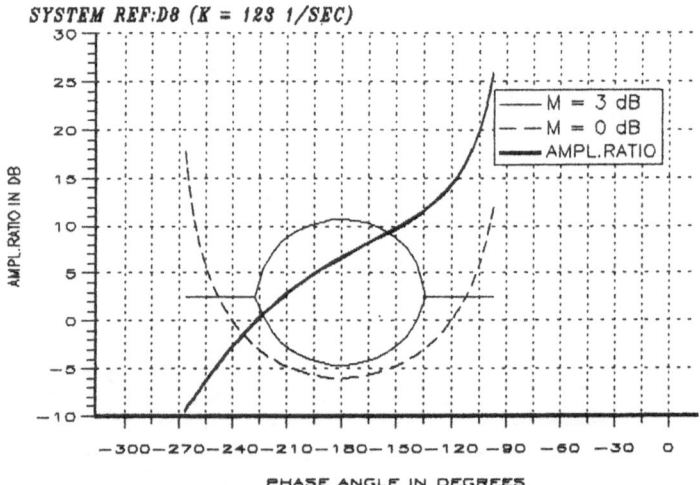

Fig. 21.15 Nichols chart, $K = 127 \cdot 7\,s^{-1}$, without network.

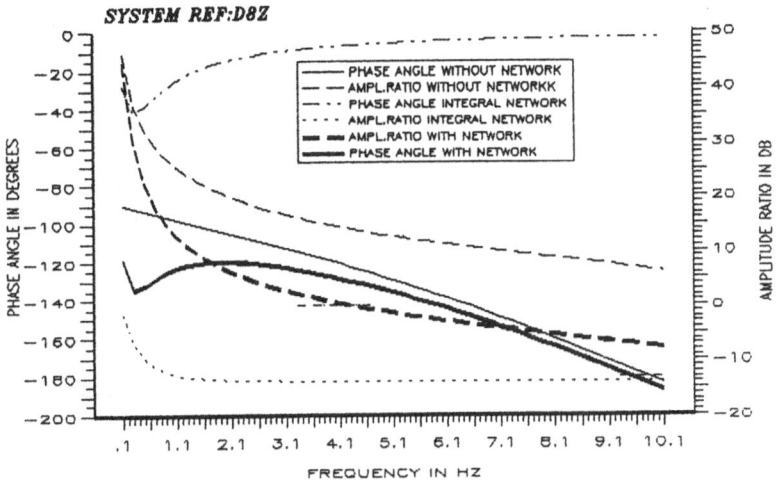

Fig. 21.16 Open loop Bode diagrams.

Nichols plot of Fig. 21.15. An open loop Bode diagram for the original system reference D8 is given in Fig. 21.3, and Bode diagrams for the network, and open loop plots for the compensated and uncompensated systems are shown in Fig. 21.16. Corresponding closed loop system Bode diagrams are plotted in Fig. 21.17. (For steady-state characteristics see Sec. 27.4).

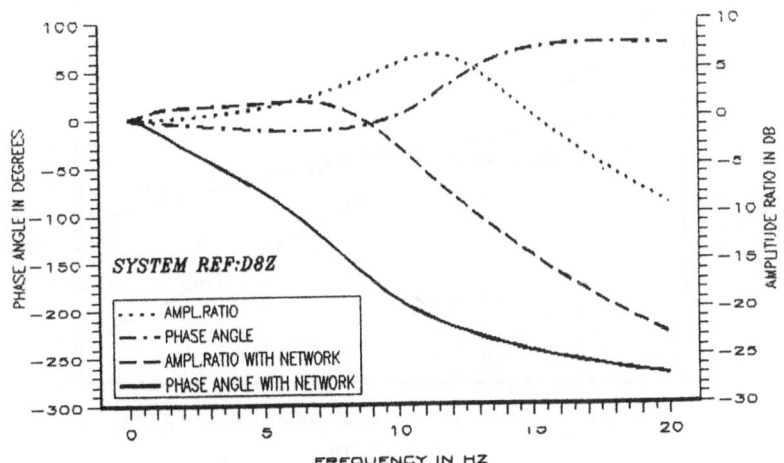

Fig. 21.17 Closed loop Bode diagrams.

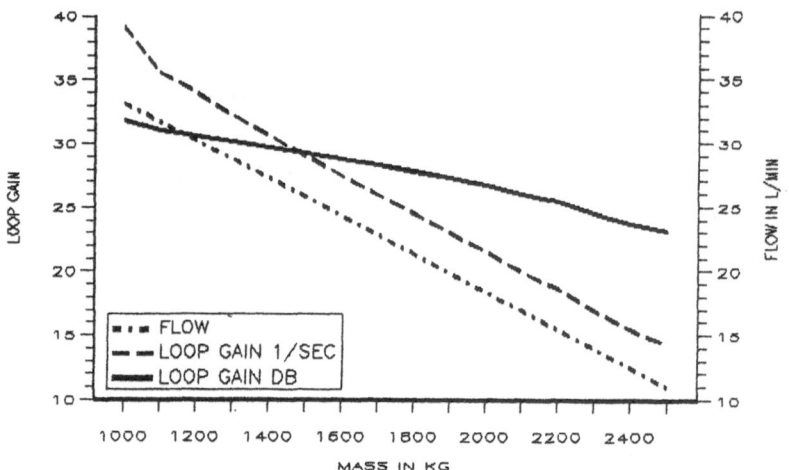

Fig. 21.18   Stability boundary with adaptive control.

21.3 ADAPTIVE CONTROL

In adaptive control systems, one or more system parameters are automatically adjusted in relation to other system variables to obtain enhanced system performance. As an example we shall consider a position control system in which there is a known relationship between actuator

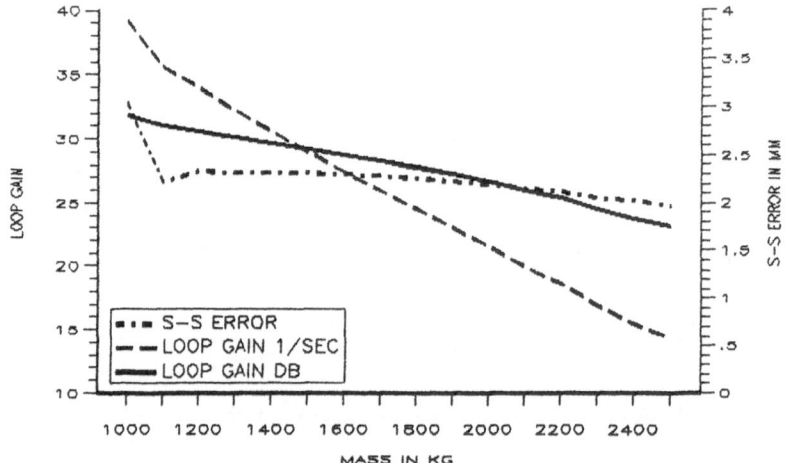

Fig. 21.19   Stability boundary with adaptive control.

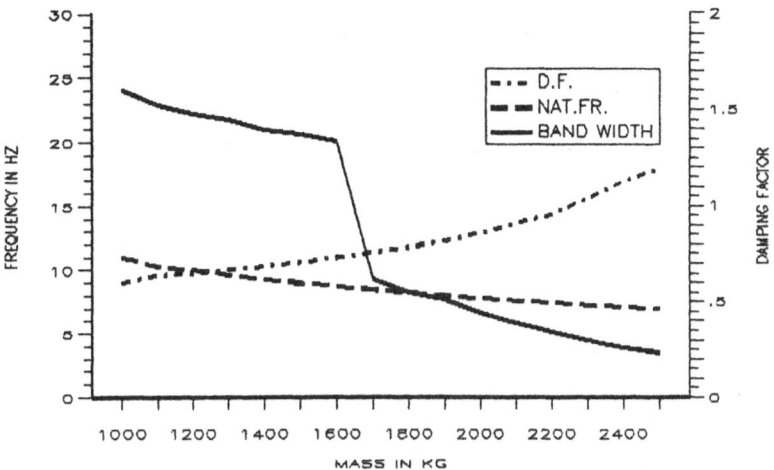

Fig. 21.20 Frequency response with adaptive control.

velocity, and hence flow rate, and the mass to be moved by the actuator. The stability boundary for this system is plotted in Fig. 21.18, which shows that an increasing mass is associated with a reducing actuator velocity and flow rate, which in turn leads to an optimum loop gain, decreasing from 39 s^{-1} or $31 \cdot 8 \text{ dB}$ at 1000 kg mass to 14 s^{-1} or $22 \cdot 9 \text{ dB}$ at 2500 kg mass. By varying the amplifier gain, and hence the loop gain, as a function of mass and

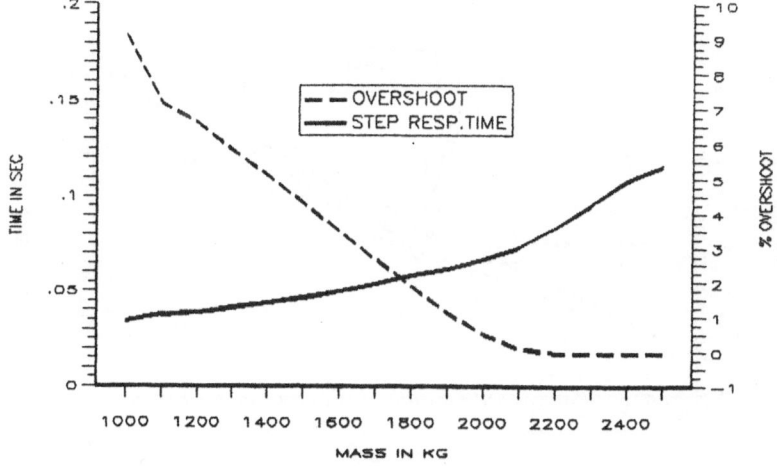

Fig. 21.21 Transient response with adaptive control.

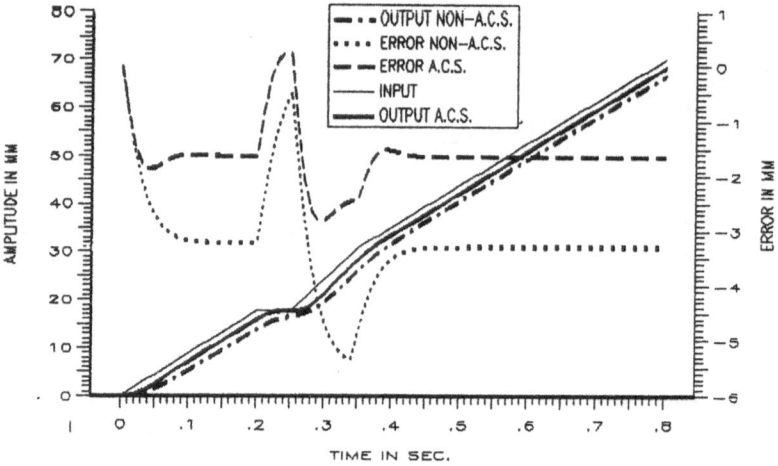

Fig. 21.22 Transient response with adaptive control.

associated flow rate, improved system performance, both steady-state and dynamic, can be achieved. Figure 21.19 shows that the steady-state following error can be maintained at a virtually constant level of around two millimetres over the entire operating range. Similarly the variation of frequency response and transient response parameters is reduced by adopting an adaptive control system, as shown by the curves plotted in Figs 21.20 and 21.21.

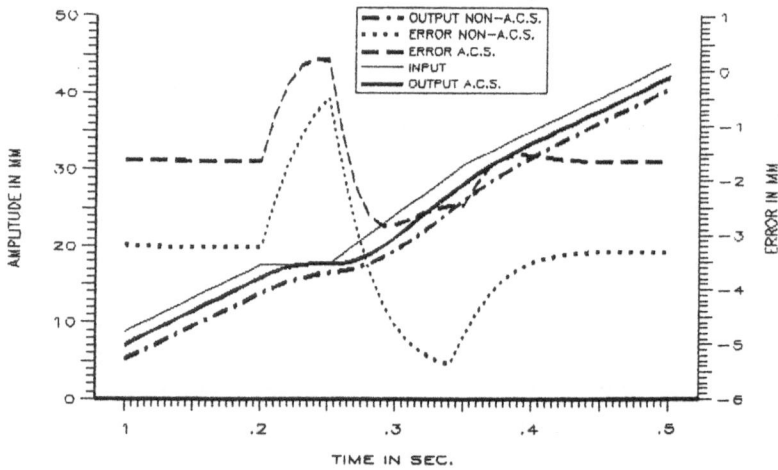

Fig. 21.23 Transient response with adaptive control.

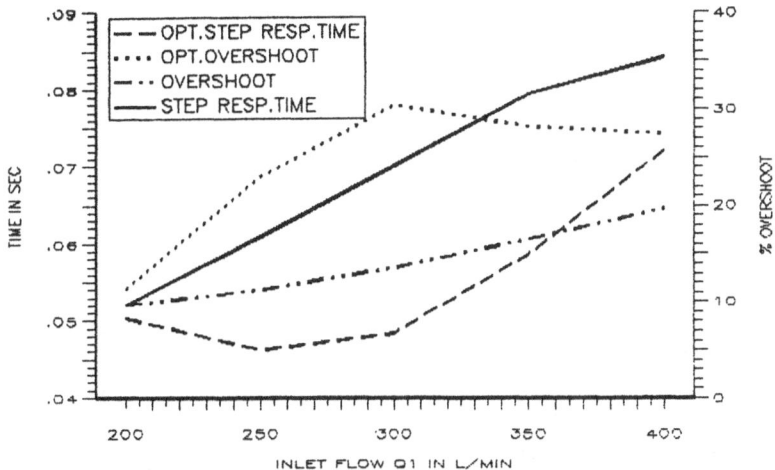

Fig. 21.24 Transient response parameters.

The transient response to a typical duty cycle is compared with the response of an equivalent system without adaptive control in Figs 21.22 and 21.23. It can be clearly seen that both steady-state and dynamic performance is enhanced by adopting adaptive control techniques.

Transient response parameters of a typical closed loop position control system both with and without adaptive control are plotted in the graph of Fig. 21.24 for flow ratings over the range 200 to 400 litres/min. Response times for the system optimized to conform to stability criteria of 45° phase margin and 7 dB gain margin are considerably faster than corresponding response times for the non-optimized system. For the chosen stability criteria, the optimized system exhibits higher overshoots, although these can of course be reduced at the expense of increased response times by selecting larger stability margins.

21.4 MULTIPLE FEEDBACK

An alternative method for system enhancement is the use of additional feedback loops, specifically the introduction of a minor velocity or flow feedback loop within a major position feedback loop. This requires additional feedback transducers of the type described in Chapter 6.

The computer analysis program described in Chapter 13 can easily be extended to cover multiple feedback loops, e.g. the frequency response

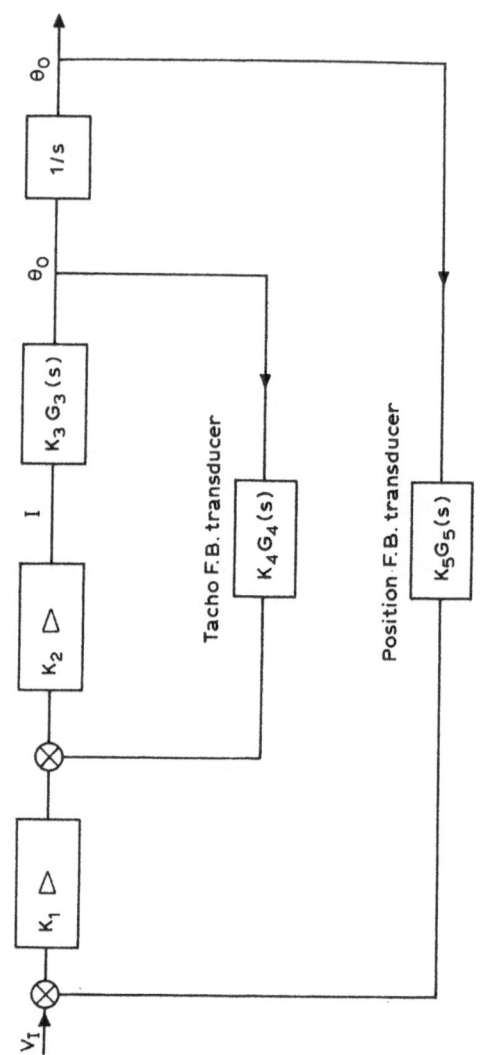

Fig. 21.25   Block diagram, position control with velocity feedback.

algorithm, Fig. 13.5, can be modified to contain any number of loops. To nest a velocity or flow feedback loop within a position feedback loop, the free integrator factor n would initially be set at zero and the loop closed. The closed loop system would then be cascaded with a function containing the major loop gain, K, and a free integrator, $n = 1$. Loop gain optimization can again be carried out by applying the algorithms, Figs 13.6 to 13.9, to the open loop system transfer function, and the closed loop system frequency response can then be obtained by closing the major position feedback loop. A block diagram of such a system is shown in Fig. 21.25.

The specific steps taken to analyse this system are:

(1) Optimize inner loop without free integrator and establish inner loop gain $K_1 = K_2 K_3 K_4$.
(2) Close inner loop and cascade with position transducer and free integrator to formulate major loop transfer function.
(3) Optimize outer loop to determine major loop gain, K. $K = K_1 K_2 K_3 K_5$.

In the performance summary in Section 21.6 the performance parameters of two alternative systems are listed. In the first arrangement the inner loop gain was optimized in accordance with a 45° phase margin/7 dB gain margin stability criterion, and for the second arrangement a somewhat higher inner loop gain was chosen. The closed loop Bode diagram of the inner loop of the first arrangement, plotted in Fig. 21.26,

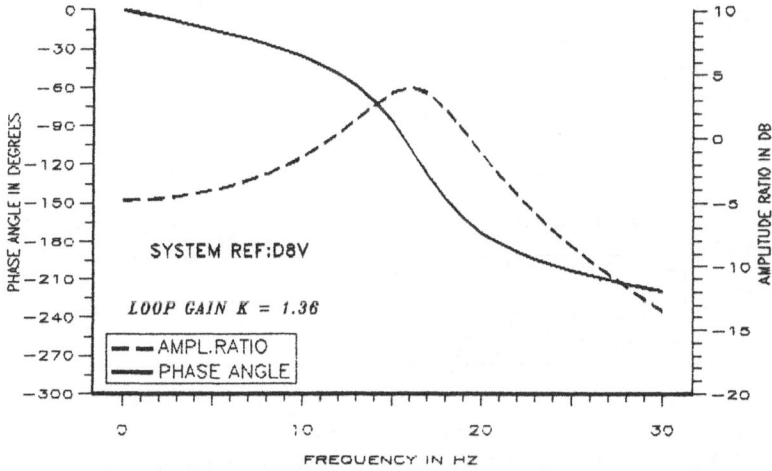

Fig. 21.26 Closed loop Bode diagram, minor loop.

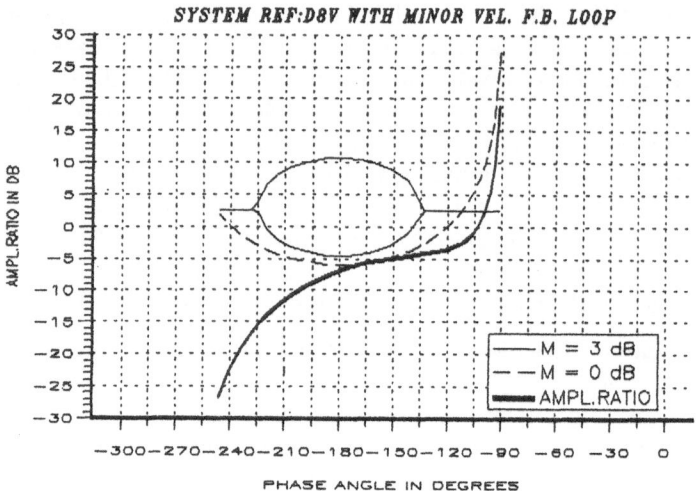

Fig. 21.27 Nichols chart, position control with velocity feedback.

shows that the lack of a free integrator severely limits the permissible loop gain, in this instance to 1·36. It can also be seen that the velocity loop has a residual amplitude error $= K_1/(1 + K_1) = 1·36/2·36 \equiv -4·78$ dB. Velocity feedback loops are one of the system configurations covered in Chapter 8 and included in Fig. 8.9 and Table 8.1. Frequency response characteristics

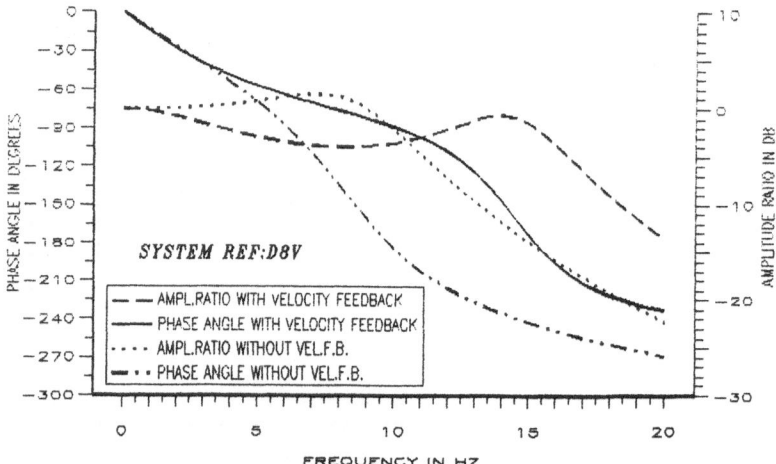

Fig. 21.28 Closed loop Bode diagram, position control with velocity feedback.

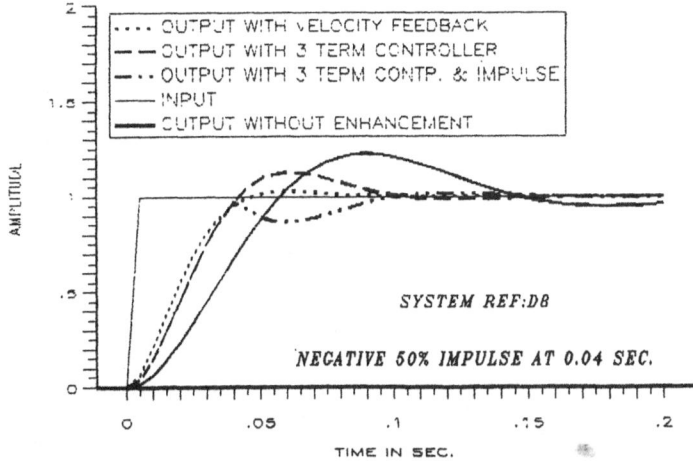

Fig. 21.29 Transient response with PID or velocity feedback

of the total system are shown in a Nichols chart, Fig. 21.27, and a closed loop Bode diagram, Fig. 21.28. Figure 21.27 shows that the gain margin of 7 dB was the critical criterion setting the value of the major loop gain, K. Figure 21.28 shows that the introduction of the minor feedback loop has considerably improved the dynamic performance of the system. This can also be seen in the transient response plot of Fig. 21.29. Both response times and overshoot have been reduced.

21.5 THREE-TERM CONTROLLER

In Chapter 11 the open loop transfer function of a hydraulic control system was described by a constant, K, and a time dependent or dynamic term, $G(s)$. In a three-term or PID (proportional, integral, derivative) controller, the constant, K, is replaced by a dynamic term $G_1(s) = K + K_D s + K_I/s$, which can be expressed as $G_1(s) = (1/s)(Ks + K_D s^2 + K_I)$, or alternatively as a function of frequency as

$$G_1(j\omega) = 1/j\omega(j\omega K + K_I - K_D\omega^2) \qquad (21.5)$$

The function $G_1(j\omega)$, added to the frequency response algorithm Fig. 13.5, is incorporated in the software package HYDRASOFT, described in Chapter 13. The effect of a PID controller on the fifth order control system, reference D8, is shown in the plots of Figs 21.30, 21.31 and 21.29. The

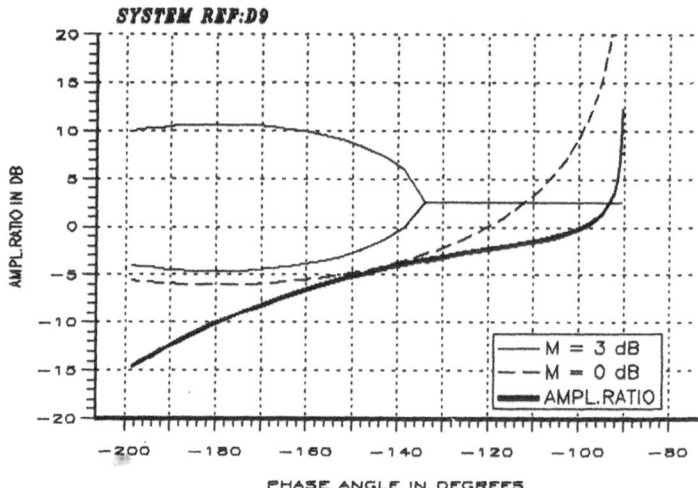

Fig. 21.30 Nichols chart with three-term controller.

differential and integral terms, K_D and K_I, were set at 50% of the proportional loop gain term, K. A comparison of Nichols plots, Fig. 21.13, for the original system and Fig. 21.30 for the system with PID control show that both phase and gain margins have been increased. Improvement of frequency response can be seen in the plot of Fig. 21.31, and that of transient response in Fig. 21.29. PID control has reduced the response time

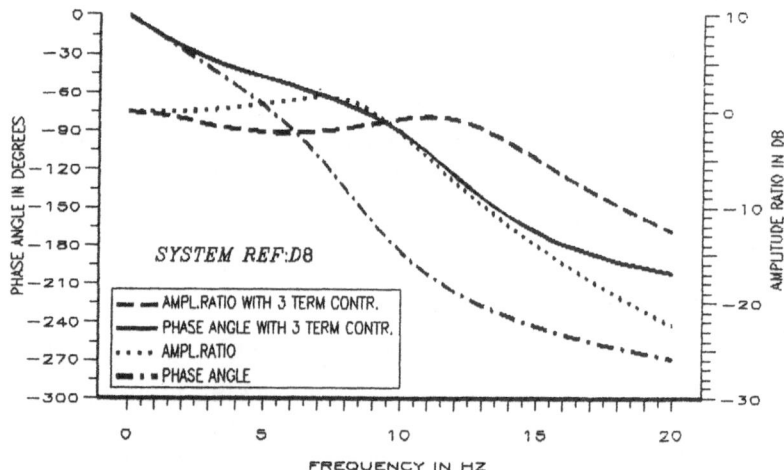

Fig. 21.31 Closed loop Bode diagram with three term controller.

and the amount of overshoot to a step demand. Figure 21.19 also demonstrates the effect of a 50% negative impulse superimposed on the PID control system. It should be noted that the improved stability margins and dynamic performance parameters were achieved by maintaining the optimized loop gain of the original system. Optimizing the PID control system would have only marginally reduced the response time while considerably increasing the overshoot; this can be seen in Section 21.6, 'Performance summary'.

21.6 PERFORMANCE SUMMARY

To give an overview of the effects of the various methods of system enhancement on steady-state and dynamic performance, the results obtained in the chosen example will now be summarized. It is worth noting that the parameter which has a direct bearing on steady-state performance is the loop gain, since it is one of the parameters affecting system output stiffness, hysteresis and following, or velocity, errors. Furthermore, for the adaptive control system, it can be seen from Figs. 21.22 and 21.23 that for the given duty cycle maximum steady-state following errors have been reduced from 3·3 mm to 1·6 mm, whilst maximum dynamic errors are reduced from 5·4 mm to 2·9 mm.

Table 21.1 Summary of the results obtained in the chosen example

System configuration	Loop gain (s^{-1})	Bandwidth (Hz)	Response time (s)	Overshoot (%)
Ref. D8, step demand	27	10·7	0·051	22·6
Ref. D8, ramp-step	27	10·7	0·15	6·0
Negative impulse	27	10·7	0·051	5·0
Passive network	127·7	10·4	0·055	24·4
PID control	27	14·5	0·036	12·7
PID control, optimized	38	15·4	0·031	27·6
PID with negative impulse	27	14·5	0·036	1·7
Adaptive control	21–39	7–24	0·06–0·035	0·6–9
(mass: 2 000–1 000 kg; flow: 18–33 litres/min)				
Velocity feedback (1·36)	27·4	15·8	0·034	7·3
Velocity feedback (1·50)	23·8	16·1	0·036	2·7

22

Analysis of Pressure Control System

We shall now proceed to analyse a system controlled by a pressure control valve of the type described in Section 5.1. Circuit and block diagrams are shown in Fig. 22.1.

We shall again consider small perturbations about a fixed operating condition. Since the controlled pressure $P = f(P_\varepsilon, Q_R)$,

$$P = \frac{\delta P}{\delta P_\varepsilon} P_\varepsilon + \frac{\delta P}{\delta Q_R} Q_R$$

$$= K P_\varepsilon + R_0 Q_R \tag{22.1}$$

where

$$K = \frac{\delta P}{\delta P_\varepsilon} \quad \text{and} \quad R_0 = \frac{\delta P}{\delta Q_R}$$

$$Q_P = Q_R + Q_C + Q + Q_S \tag{22.2}$$

$$P = \frac{Q}{A^2}(sm + D) \tag{22.3}$$

$$Q_C = sP \frac{V}{N} \tag{22.4}$$

$$Q_S = cP \tag{22.5}$$

and combining the above equations, yields the open loop transfer function:

$$\frac{P}{P_\varepsilon} = \frac{(K/R_0)[(sm/D) + 1]}{(mV/DN)s^2 + [(m/DR_0) + (V/N) + (cm/D)]s + (1/R_0) + (A^2/D) + c} \tag{22.6}$$

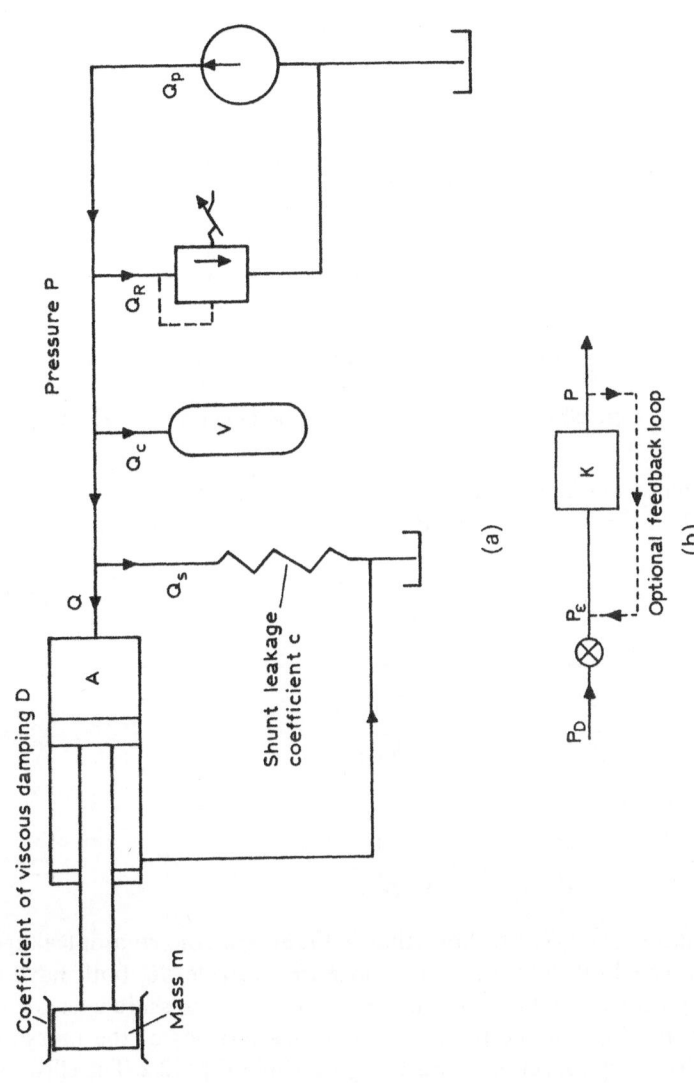

Fig. 22.1 Circuit diagram (a); and block diagram (b).

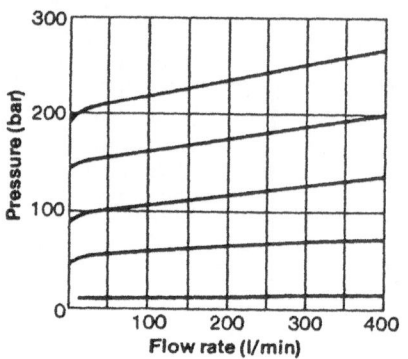

Typical overrides of a valve at various control settings. Back pressure at tank port = 0

Residual pressure in zero-current mode
All valves.................... 9 bar approx

Fig. 22.2   Pressure override characteristics.

The system transfer function can be simplified by considering a dead-headed cylinder, whence $D = \infty$. Equation (22.6) then becomes:

$$\frac{P}{P_\varepsilon} = \frac{K_0}{\sigma s + 1} \qquad (22.7)$$

where

$$K_0 = \frac{K}{(cR_0 + 1)} \qquad (22.8)$$

and the time constant

$$\sigma = \frac{V}{N(1/R_0 + c)} \qquad (22.9)$$

An examination of eqn (22.9) shows that, in the absence of actuator leakage flow, compressed oil volume, V, and pressure override, R_0 both have a direct proportional effect on the time constant, σ. The pressure override is defined as the slope of the pressure/flow characteristics of the pressure control valve, a typical set of curves being shown in Fig. 22.2. The effect of compressed oil volume on pressure build-up at a pressure override of 0·2 bar per litre/min is shown in Fig. 22.3. Transient response to a step demand of a proportional pressure control valve, having a natural frequency of 6 Hz

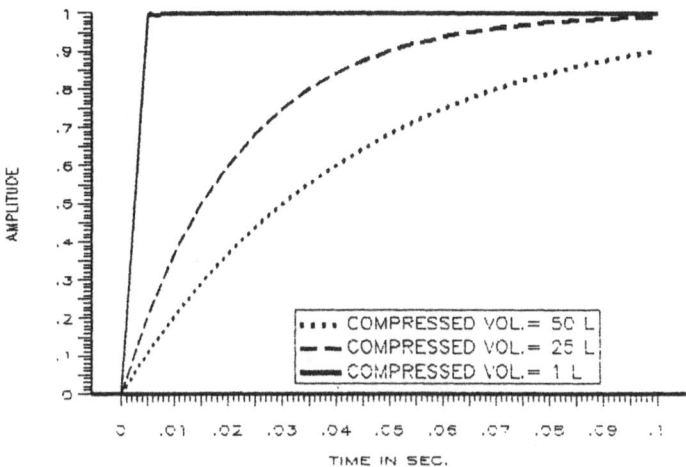

Fig. 22.3 Compressed oil volume effect.

at a damping factor of 0·4, is plotted for three oil volumes in Fig. 22.4, and corresponding transient response to a sudden flow disturbance is plotted in Fig. 22.5. The effect of the same pressure override of 0·2 bar per litre/min on a valve with doubled dynamic performance is shown in Figs 22.6 and 22.7.

A number of useful conclusions can be drawn from a study of the response curves.

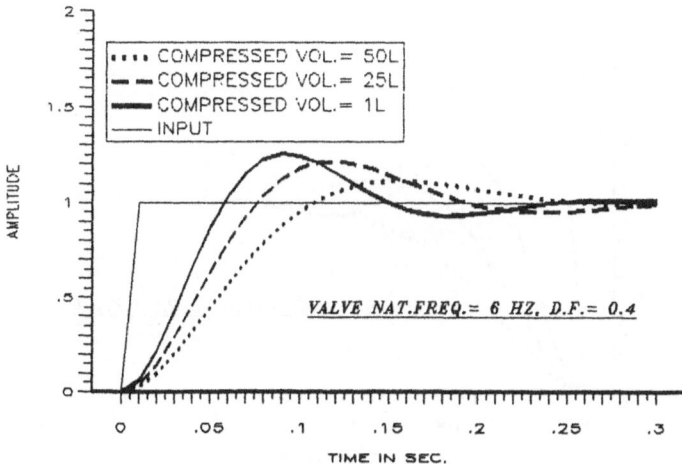

Fig. 22.4 Transient response to step demand.

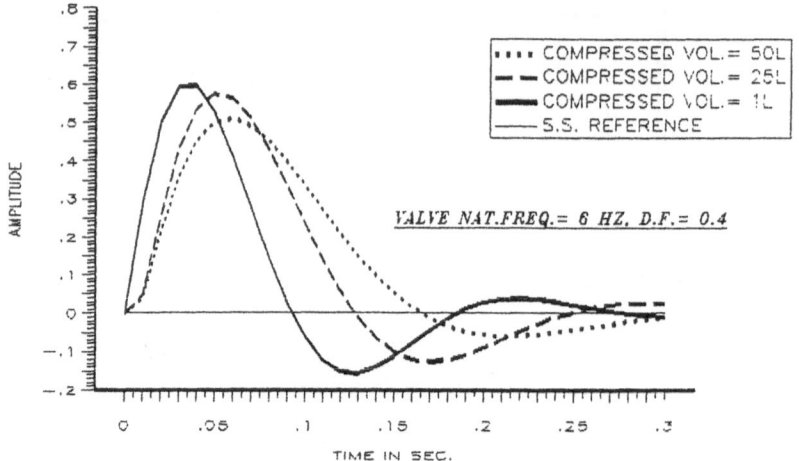

Fig. 22.5   Transient response to flow variation.

(1) Compressed oil volume has a damping effect, reducing pressure overshoots and increasing response and settling times.
(2) Positive pressure override has a similar damping effect on system response.
(3) Negative pressure override has a destabilizing effect.
(4) The effective damping due to oil compressibility and pressure override is more pronounced with fast-acting control valves.

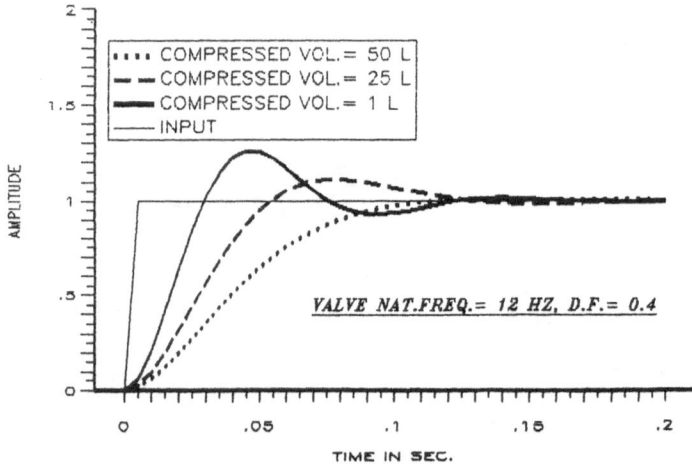

Fig. 22.6   Transient response to step demand.

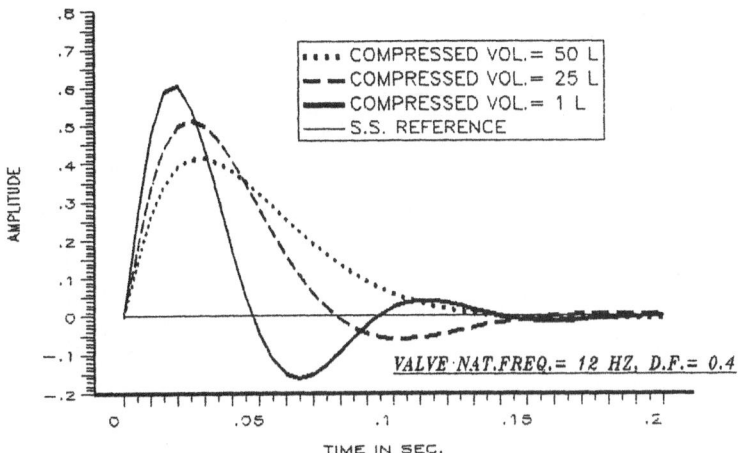

Fig. 22.7 Transient response to flow variation.

The closed loop option was previously mentioned in Section 7.2, but is less frequently applied to pressure controls than to motion control systems. The modular optimized system simulation concept discussed in Chapter 13 as well as the system analysis in the time domain, covered in Chapters 14 and 15, is of course equally applicable to open and closed loop pressure control systems. As opposed to position control systems, pressure control systems do not contain a free integrator in the form of the actuator, and it is

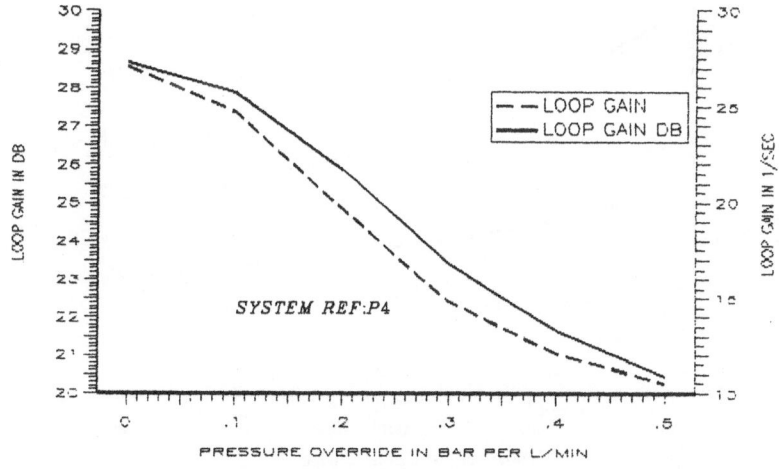

Fig. 22.8 Stability boundary.

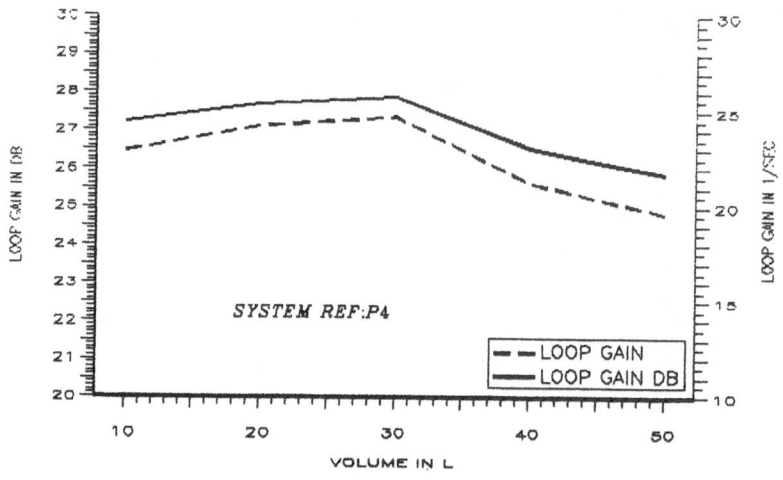

Fig. 22.9 Stability boundary.

therefore advisable to consider the introduction of a free integrator or an integral type network to the electronic control circuit to enhance system performance if a closed loop approach is chosen.

The dynamic performance characteristics of the closed loop equivalent of the open loop system whose transient response was shown in the graphs of Figs 22.6 and 22.7 are plotted in Figs 22.8 to 22.12. The maximum permissible loop gain compatible with a 45° phase margin per 7 dB gain

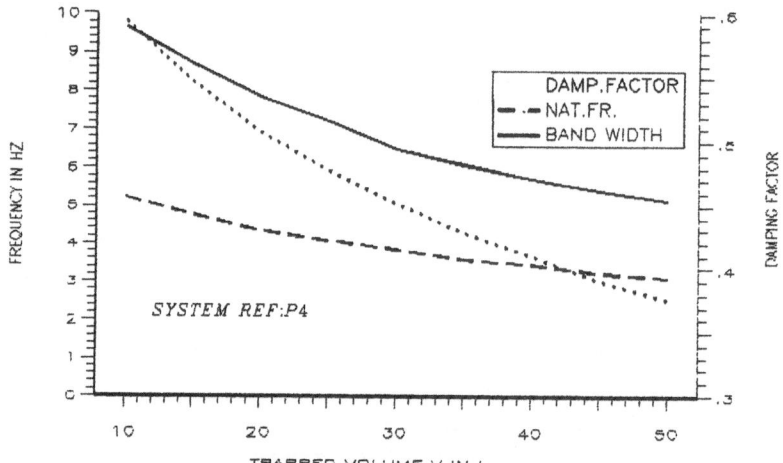

Fig. 22.10 Frequency response parameters.

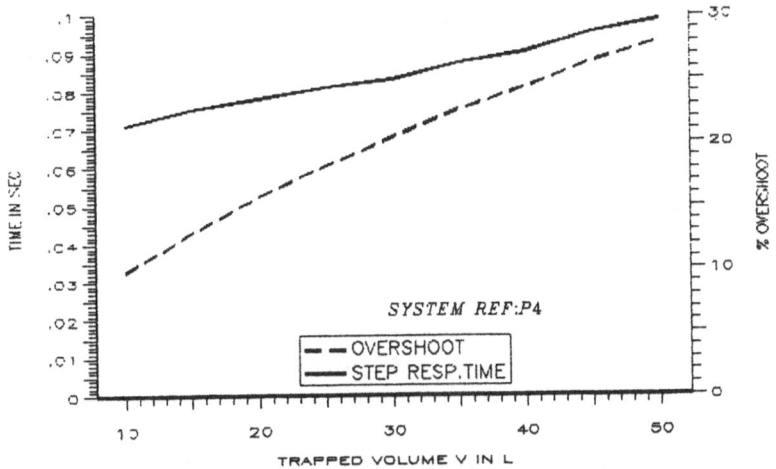

Fig. 22.11 Transient response parameters.

margin stability criterion is plotted as a function of pressure override in Fig. 22.8 and as a function of compressed oil volume in Fig. 22.9. In Fig. 22.8 the stability boundary is plotted for a compressed oil volume of 50 litres, whilst the stability boundary, Fig. 22.9, is applicable to a pressure override of 0·2 bar per litre/min. The effect of compressed oil volume on dynamic performance is plotted in Figs 22.10 and 22.11 for a system with constant loop gain of $19.6\,s^{-1}$ and a pressure override of 0·2 bar per litre/min. It can

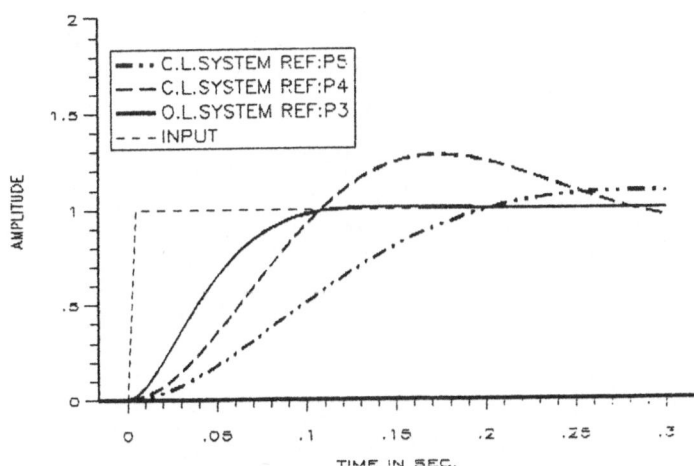

Fig. 22.12 Transient response comparison.

be seen that an increasing compressed oil volume degrades dynamic performance. A comparison of transient response of three system arrangements, one open loop and two closed loop, is shown in the graph of Fig. 22.12. The plots show that the open loop system for the given circuit conditions has a higher dynamic performance rating than the corresponding closed loop systems. The only justification for closing the loop in this instance would be to achieve enhanced steady-state performance. A summary of system parameters is given in Table 22.1.

Table 22.1 Summary of system parameters

Designation	OL/CL	Loop gain (s^{-1})	Bandwidth (Hz)	Response time (s)	Overshoot (%)
P3	OL	—	5·85	0·082	0·96
P4	CL	19·65	5·16	0·099	27·9
P5	CL	10·0	2·83	0·170	8·8

23

Efficiencies and Power Dissipation

As a rule, hydrostatic transmissions are more efficient than equivalent valve-operated flow controls. The choice of pump configuration can, however, significantly affect power efficiency. In Section 5.2, four basic valve control circuits were described; meter-in, meter-out, meter-in/meter-out and bleed-off. In the bleed-off circuit, the pump operates as a flow source, in the other three cases the pump acts as a pressure source. We shall now examine several types of pump arrangement with special emphasis on power dissipation and efficiency. The five basic circuit arrangements are:

(1) Fixed displacement pump at constant supply pressure.
(2) Pressure-compensated variable displacement pump.
(3) Fixed displacement pump with pressure match.
(4) Variable displacement pump with power match.
(5) Fixed displacement pump with bleed-off.

Circuits (3) and (4) can further be broken down into either fixed or variable pressure or power match.

Circuit diagrams and corresponding equations for efficiency and power dissipation covering the five basic circuit arrangements are tabulated in Fig. 23.1.

(1) This is the simplest and least efficient arrangement. The pump supplies fluid at a flow rate directly proportional to input speed and at substantially constant pressure, the supply pressure setting of the relief valve having to cater for maximum load and valve pressure drop requirements. It can be seen from the graph and equations that the high corner horsepower can result in large power dissipation and low efficiency. Since power dissipation creates heat, systems using constant-pressure fixed displacement pumps often require oil coolers to prevent excessive fluid temperatures.

System	Circuit application	Comments	Graphical representation	Efficiency	Power dissipation
Fixed displacement pump	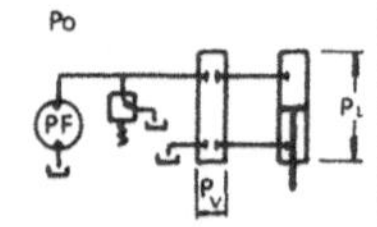 Po	Power wastage is a function of both excess pressure and flow above that required by the system to do work.	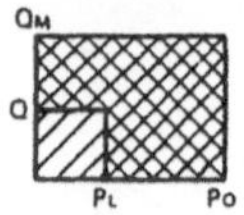	$\eta = \frac{P_L Q}{P_o Q_M}$	$(P_o Q_M - P_L Q) \times .00163$
Pressure match	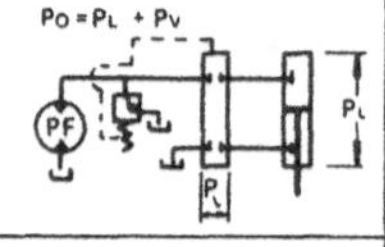 $P_o = P_L + P_v$	Power wastage is a function of the excess flow at the required system pressure, this results in a loss at the relief valve.	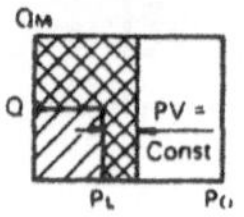	$\eta = \frac{Q}{Q_M} \times \frac{1}{1 + \frac{P_v}{P_L}}$	$[(P_L + P_v)Q_M - P_L Q] \times .00163$
Pressure compensated	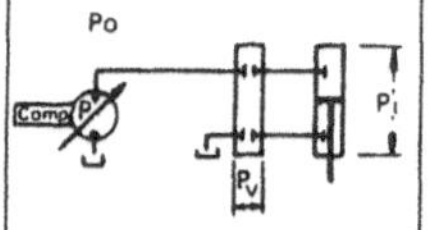 Po	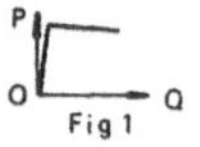 Fig 1. Power wastage is a function of $(P_o - P_L)$ at a particular flow output. Compensated pump characteristic giving variable volume at compensator pressure setting.	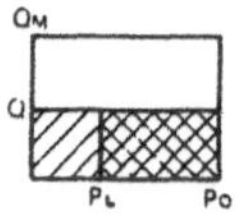	$\eta = \frac{P_L}{P_o}$	$Q(P_o - P_L) \times .00163$
Power match	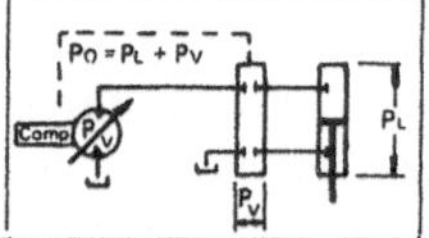 $P_o = P_L + P_v$	Power wastage is purely due to (P_v) in the valve package. (Requires power match compensator)	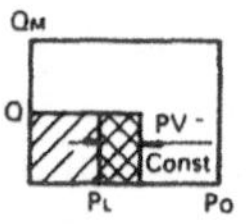	$\eta = \frac{1}{1 + \frac{P_v}{P_L}}$	$P_v Q \times .00163$
Bleed-off	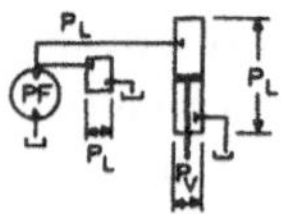	Since load determines system operating pressure, power wastage is a function of excess flow $(Q_M - Q)$ at the given load pressure, P_L.	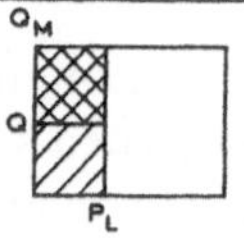	$\eta = \frac{Q}{Q_M}$	$P_L(Q_M - Q) \times .00163$

Units: Pressure in bar   Flow in l/min   Power in K watts

Fig. 23.1 Summary of efficiencies and power dissipation.

(2) A pressure-compensated variable displacement pump supplies fluid at a flow rate matched to system requirements at substantially constant supply pressure determined by the pressure compensator setting, thus providing a more efficient power source. Pressure-compensated pumps are particularly effective for systems operating over a wide flow range and narrower load pressure range, although by no means restricted to such a system.

(3) Fixed displacement pumps can be run more efficiently by employing load sensing and adjusting the pressure setting of the reflief valve in line with load pressure variations. Valve pressure drop and other pressure losses have to be allowed for. The pressure allowance over and above the sensed load pressure can be either a fixed or a variable

amount, and pressure matching can be achieved by hydraulic or electro-hydraulic means. In the latter case pressure transducers would sense the load pressure, and a processed signal would then be applied to a proportional (electrically modulated) pressure relief valve. This type of pump arrangement is particularly effective when wide load variations are encountered, but the effectiveness is reduced by large flow changes.

(4) The most efficient power source is a variable displacement pump with power match. Power-matching is accomplished by applying load-sensing to vary the setting of the pressure compensator. As in the case of pressure match, power-matching can be achieved by either hydraulic or electro-hydraulic means. Contrary to the previous arrangements, power dissipation is unaffected by load variations and less sensitive to flow changes than circuit arrangements (1) and (3).

(5) A fixed displacement pump supplies fluid at a pressure directly related to the external load. This arrangement provides a very efficient power source for systems operating at relatively constant velocity and varying actuator loading.

The relative efficiencies and power dissipation of the five basic types of power source, applied to a typical system, are plotted in the graphs of Figs 23.2 to 23.9.

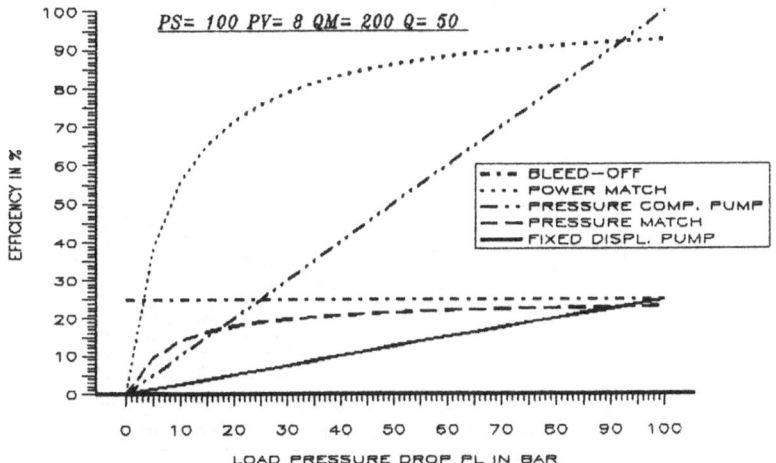

Fig. 23.2   Power efficiencies, $Q = 50$ litres/min.

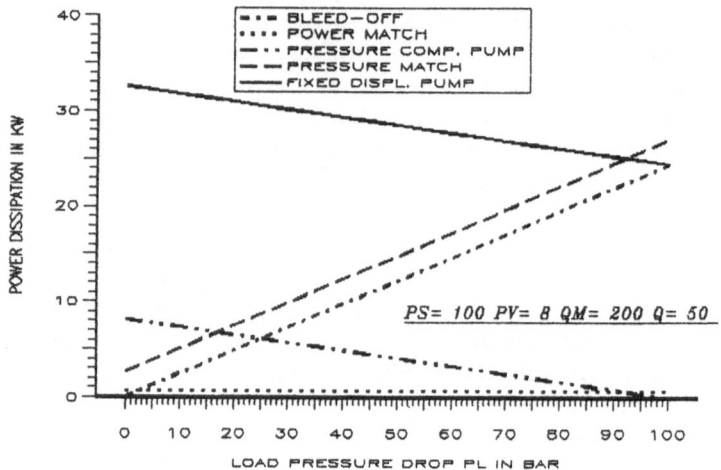

Fig. 23.3 Power dissipation, $Q = 50$ litres/min.

The following parameters were chosen for the example:

Supply pressure, P_S:	100 bar
Pressure and power match setting, P_V:	8 bar
Maximum flow rating, Q_M:	200 litres/min
Rated flow, Q:	0 to 200 litres/min
Load pressure, P_L:	0 to 100 bar

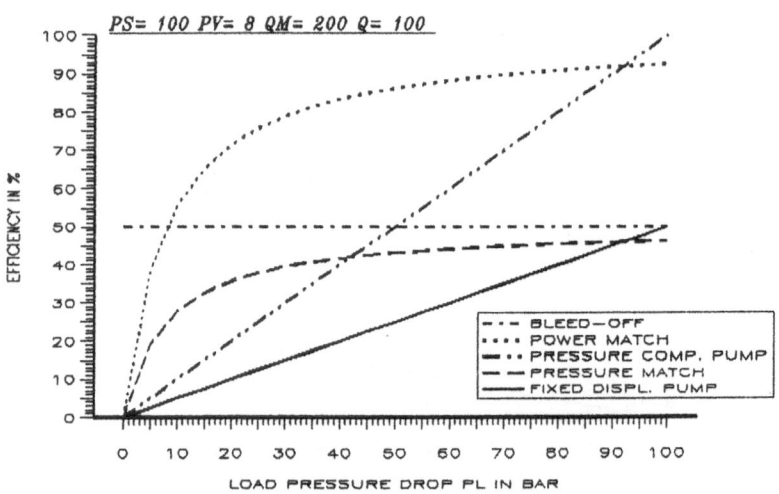

Fig. 23.4 Power efficiencies, $Q = 100$ litres/min.

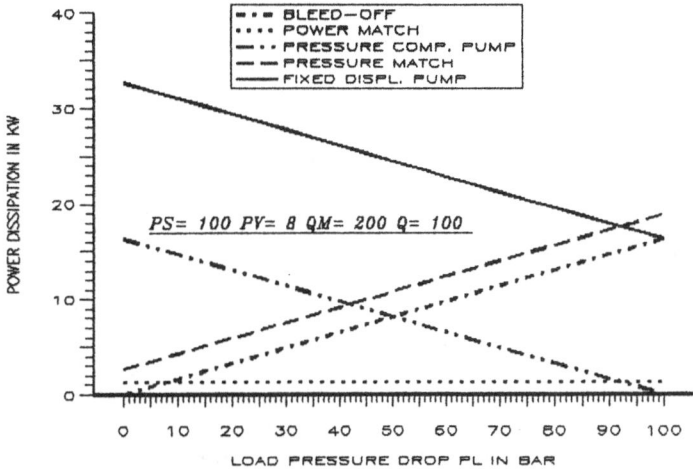

Fig. 23.5 Power dissipation, $Q = 100$ litres/min.

Efficiencies and power dissipation at a rated flow of 50 litres/min are shown in Figs 23.2 and 23.3, at a rated flow of 100 litres/min in Figs 23.4 and 23.5, and at 150 litres/min rated flow in Figs 23.6 and 23.7. Corresponding graphs at 50 bar load pressure, plotted as a function of flow rate, are shown in Figs 23.8 and 23.9.

In all cases, a fixed displacement pump operating at constant supply

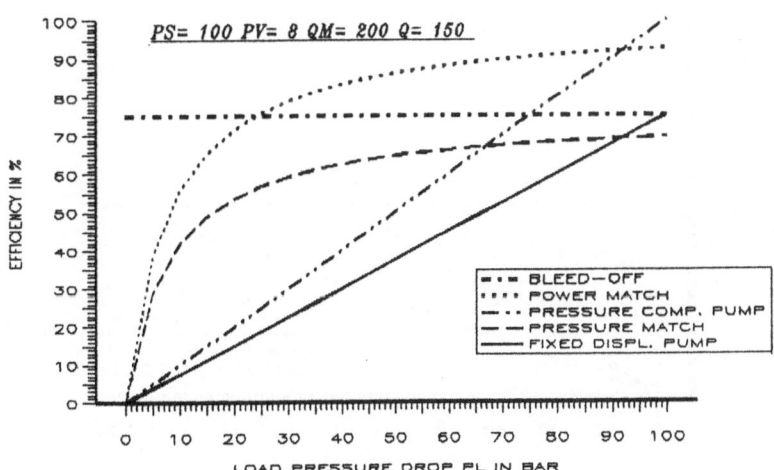

Fig. 23.6 Power efficiencies, $Q = 150$ litres/min.

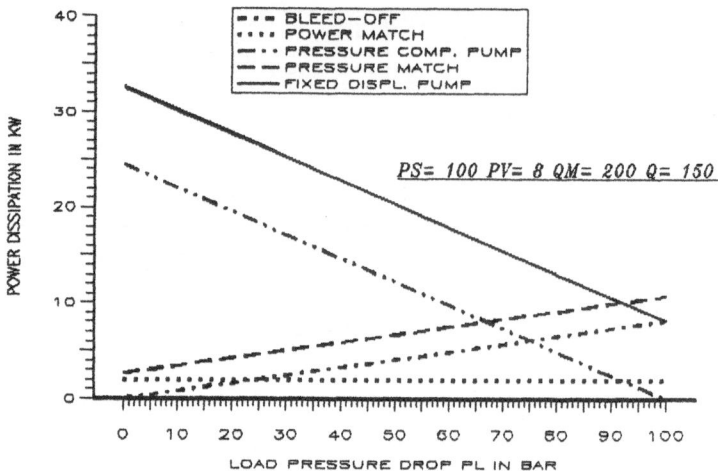

Fig. 23.7   Power dissipation, $Q = 150$ litres/min.

pressure is the least efficient power source. The plots clearly indicate the improvements which can be achieved by selecting one of the alternative circuit arrangements. The operating conditions strongly influence the choice of the most beneficial circuit configuration, although a few general guidelines can be stipulated, i.e.

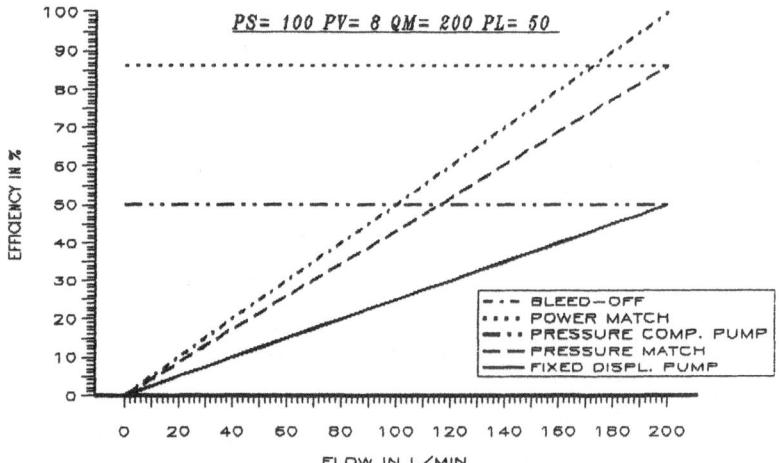

Fig. 23.8   Power efficiencies, $P_L = 50$ bar.

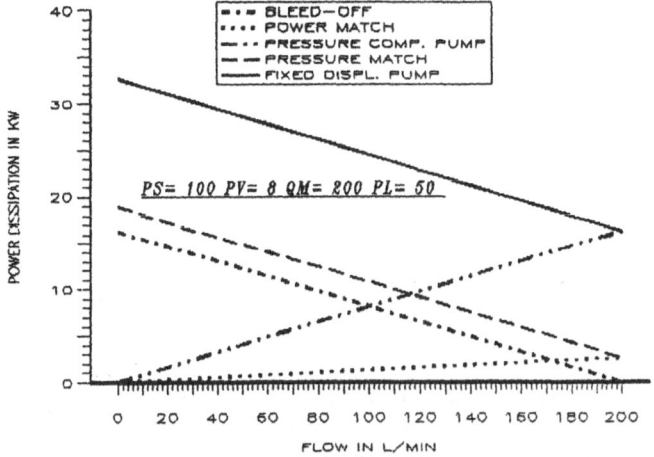

Fig. 23.9 Power dissipation, $P_L = 50$ bar.

(1) Power match is more efficient than either pressure match or a pressure-compensated pump.
(2) The efficiency of a bleed-off circuit is not affected by load pressure variations.
(3) At flow rates approaching the maximum flow rating, a bleed-off circuit can be more efficient than a power-matched variable displacement pump, particularly at lower load pressures.
(4) The efficiency of a pressure-compensated or power-matched pump is unaffected by flow variations.

24

Elastically Mounted Mass Systems

In the systems analysed so far, the mass was assumed to be rigidly attached to the actuator. Although this is a valid assumption for the majority of industrial applications, there are some cases in which an allowance for the elasticity of either the mass itself or the member attaching it to the actuator has to be made. Aircraft powered flying controls are a typical example of an elastic mass system.

The derivation of a system transfer function can be simplified by representing the system by a rigidly mounted mass m_R, and an elastically mounted mass m_E, attached to the actuator by means of a spring of compliance, σ, as shown in Fig. 24.1. The equation of motion, assuming pure inertia loading is then given by:

$$F = m_E s^2 \theta_E + m_R s^2 \theta_0 \qquad (24.1)$$

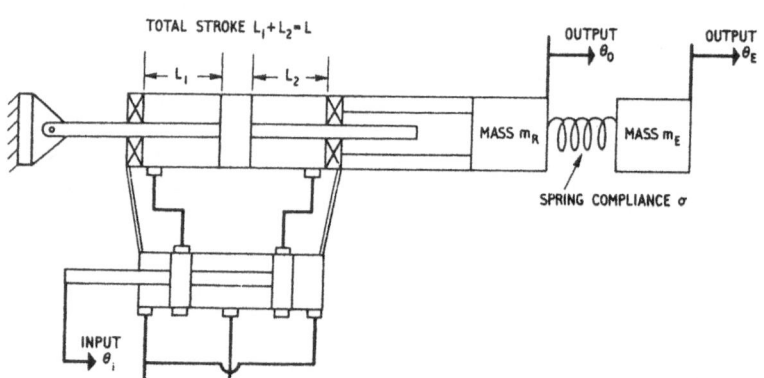

Fig. 24.1 System diagram.

and

$$\frac{\theta_E}{\theta_0} = \frac{1}{1 + \sigma m_E s^2} \tag{24.2}$$

Combining the above equations with eqn (11.12) yields:

$$\frac{\theta_0}{y} = \frac{K(1 + \sigma m_E s^2)}{s\left\{\sigma \lambda m_E m_R s^4 + \dfrac{\sigma K m_E m_R s^2}{S} + [\lambda(m_E + m_R) + \sigma m_E]s^2 \right.}$$
$$\left. + \dfrac{K(m_E + m_R)}{S}s + 1\right\} \tag{24.3}$$

and

$$\frac{\theta_E}{y} = \frac{K}{s\left\{\sigma \lambda m_E m_R s^4 + \dfrac{\sigma K m_E m_R s^2}{S} + [\lambda(m_E + m_R) + \sigma m_E]s^2 \right.}$$
$$\left. + \dfrac{K(m_E + m_R)}{S}s + 1\right\} \tag{24.4}$$

The closed loop transfer function when the feedback signal is taken from the elastically mounted mass, as represented by the block diagram of Fig. 24.2, is then:

$$\frac{\theta_E}{\theta_i} = \frac{K}{\left(\sigma \lambda m_E m_R s^5 + \dfrac{\sigma K m_E m_R s^4}{S} + [\lambda(m_E + m_R) + \sigma m_E]s^3 \right.}$$
$$\left. + \dfrac{K(m_E + m_R)}{S}s^2 + s + K\right) \tag{24.5}$$

When feedback is taken from the rigidly mounted mass, as represented by the block diagram of Fig. 24.3, the system transfer function becomes:

$$\frac{\theta_E}{\theta_i} = \frac{K}{\left(\sigma \lambda m_E m_R s^5 + \dfrac{\sigma K m_E m_R s^4}{S} + [\lambda(m_E + m_R) + \sigma m_E]s^3 \right.}$$
$$\left. + K\left[\dfrac{m_E + m_R}{S} + \sigma m_E\right]s^2 + s + K\right) \tag{24.6}$$

By assuming the entire mass to be elastically mounted, the transfer functions can be reduced to a third order differential equation which

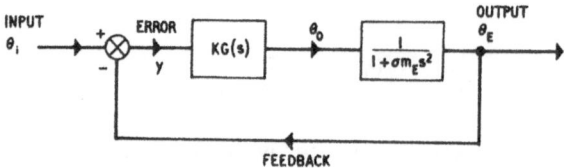

Fig. 24.2 Block diagram, feedback from mass.

facilities direct comparison with the positional control systems analysed in previous chapters.

Let $m_R + m_E = m$, then

$$\frac{\theta_0}{y} = \frac{K(1 + \sigma m s^2)}{s[m(\lambda + \sigma)s^2 + (Km/S)s + 1]} \tag{24.7}$$

and

$$\frac{\theta_E}{y} = \frac{K}{s[m(\lambda + \sigma)s^2 + (Km/S)s + 1]} \tag{24.8}$$

When feedback is taken from θ_E, i.e. from the mass, the system transfer function becomes:

$$\frac{\theta_E}{\theta_i} = \frac{K}{m(\lambda + \sigma)s^3 + (Km/S)s^2 + s + K} \tag{24.9}$$

and when feedback is taken from the output of the actuator θ_0, the equivalent system transfer function then becomes:

$$\frac{\theta_E}{\theta_i} = \frac{K}{m(\lambda + \sigma)s^3 + Km[(1/S) + \sigma]s^2 + s + K} \tag{24.10}$$

Since the corresponding closed loop system transfer function for the rigidly mounted system analysed in Chapter 11 is given by:

$$\frac{\theta_0}{\theta_i} = \frac{K}{\lambda m s^3 + (Km/S)s^2 + s + K} \tag{24.11}$$

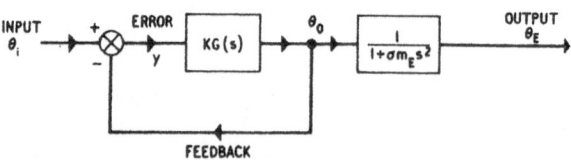

Fig. 24.3 Block diagram, feedback from actuator.

the elastic mass system is dynamically equivalent to a rigid mass system having a total actuator compliance $\lambda + \sigma$.

By comparing eqns (24.9) and (24.10) with eqn (24.11), it can furthermore be seen that when feedback is taken from the elastically mounted mass, the elastic and rigid systems have an identical output stiffness, S, whereas, when feedback is taken directly from the actuator, the output stiffness of the elastic system has to be modified to

$$\frac{1}{(1/S) + \sigma}$$

The following equations are applicable when the transfer function is expressed in terms of the natural frequency and damping factor of the system.

For both types of elastically mounted mass system:

$$\text{Effective natural frequency, } \omega_E = \frac{1}{\sqrt{[(\lambda + \sigma)m]}} \qquad (24.12)$$

$$= \frac{1}{\sqrt{(1/\omega_0^2 + 1/\omega_m^2)}} \qquad (24.13)$$

where $\omega_0 =$ the liquid spring natural frequency, and $\omega_m =$ the mechanical natural frequency.

When feedback is taken from the actuator, the system damping factor is given by the expression:

$$\zeta = \left[\frac{m}{A^2} (R_V + A^2 K\sigma) \right] \left(\frac{\omega_E}{2} \right) \qquad (24.14)$$

When feedback is taken from the mass,

$$\zeta = \frac{1}{2} \frac{m}{A^2} R_V \omega_E \qquad (24.15)$$

A comparison with the damping factor applicable to a rigidly mounted mass system, given in eqn (12.11), shows that the equivalent pressure factor for an elastically mounted mass system with feedback taken from the actuator is given by:

$$(R_V)_E = R_V + A^2 K\sigma \qquad (24.16)$$

or

$$(R_V)_E = R_V + C^2 K\sigma \qquad (24.17)$$

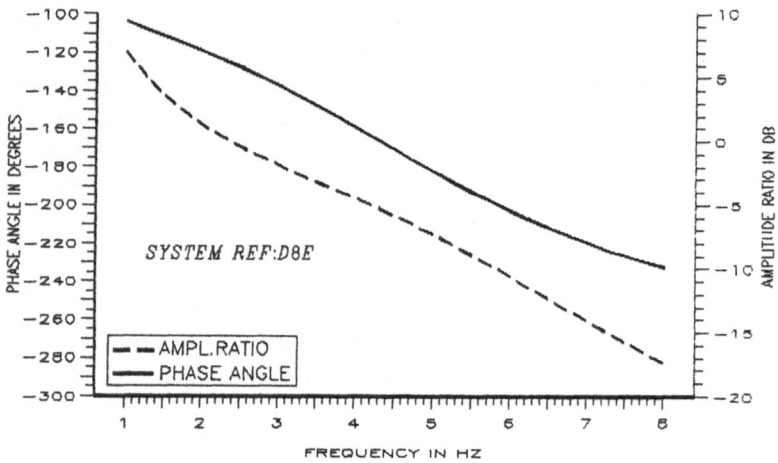

Fig. 24.4 Open loop Bode diagram, feedback from mass.

for a cylinder- and motor-driven system respectively. When feedback is taken from the elastically mounted mass, the damping factors and hence the pressure factors of elastically and rigidly mounted mass systems are identical.

The case studies previously conducted for rigidly mounted mass systems will now be extended to cover elastically mounted mass systems. Parameter print-outs for two alternative systems with identical specifications, but with

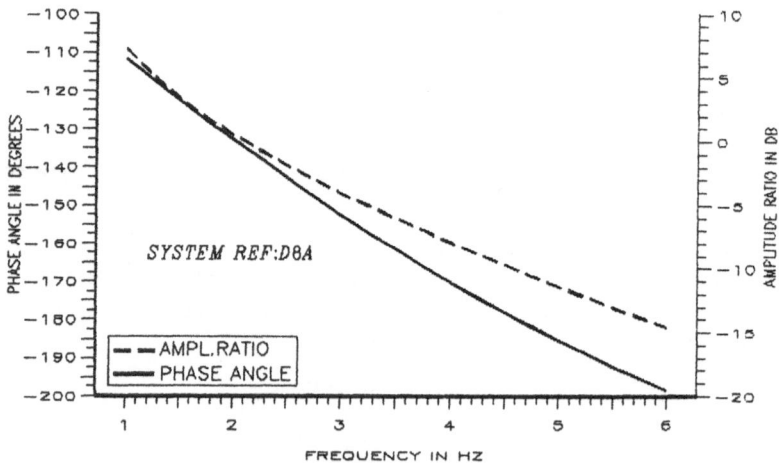

Fig. 24.5 Open loop Bode diagram, feedback from actuator.

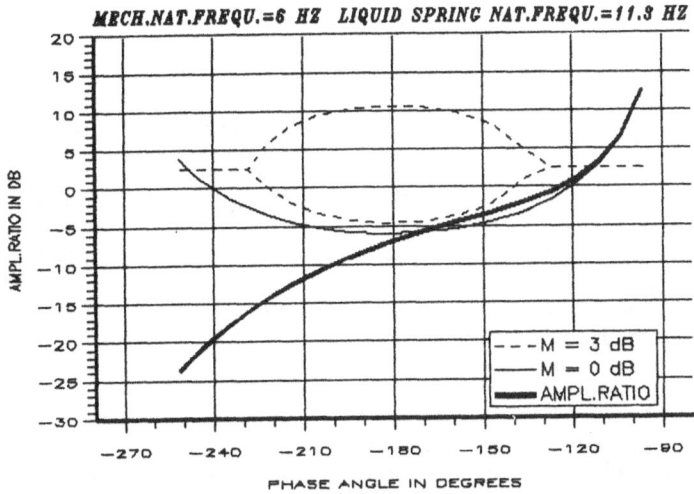

Fig. 24.6 Nichols chart, feedback from mass.

feedback taken from the elastically mounted mass in system ref. D8E and from the actuator in system ref. D8A, are given and corresponding performance curves plotted in Figs 24.4 to 24.9. A summary of performance parameters is given in Table 24.1.

The following general conclusions can be drawn:

(1) Hydraulic transmission natural frequencies are not affected by feedback mode of elastic mass system.

(2) Hydraulic transmission damping factors of rigid and elastic mass systems are identical when feedback is taken from the mass.

(3) Hydraulic transmission damping factor of elastic mass system is increased when feedback is taken from the actuator.

(4) Dynamic performance of elastic mass system is adversely affected when the mass is outside the feedback loop, i.e. when feedback is taken from the actuator.

(5) Loop gains, and hence steady-state performance, for elastic mass systems with feedback from either the mass or the actuator are comparable.

It is of interest to note that the mathematical model of the elastic mass system with feedback taken from the actuator is similar in structure to that of the flow feedback system analysed in Chapter 12. This can clearly be seen by comparing the block diagram of Fig. 12.3 with Fig. 24.3, and system

Table 24.1 Summary of performance parameters

System parameter	Rigid mass	Elastic mass	
		Feedback from mass	Feedback from actuator
Liquid spring natural frequency	11·76 Hz	11·76 Hz	11·76 Hz
Mechanical natural frequency	—	6·00 Hz	6·00 Hz
Hydraulic transmission natural frequency	11·76 Hz	5·34 Hz	5·34 Hz
Hydraulic transmission damping factor	0·55	0·55	0·93
Loop gain[a]	26·98 s^{-1}	13·94 s^{-1}	14·84 s^{-1}
System bandwidth	10·67 Hz	5·26 Hz	4·13 Hz
System natural frequency	6·19 Hz	3·16 Hz	2·52 Hz
System damping factor	0·43	0·41	0·38
Step response time	0·051 s	0·099 s	0·122 s
Time at maximum overshoot	0·089 s	0·174 s	0·214 s

[a] Based on 45° phase margin and 7 dB gain margin stability criteria.

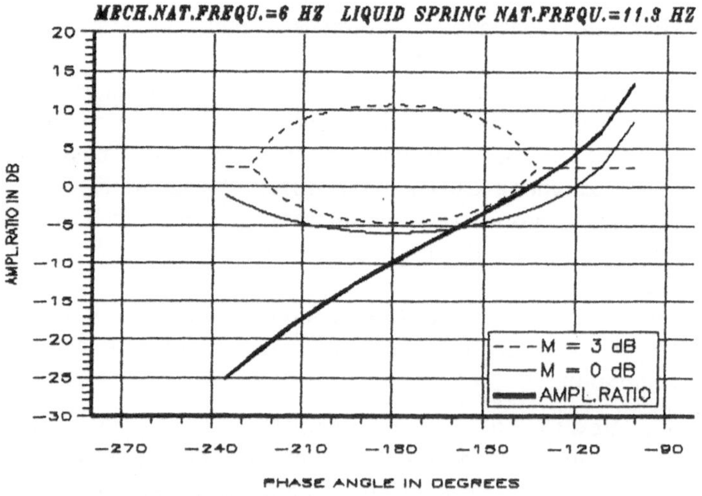

Fig. 24.7 Nichols chart, feedback from actuator.

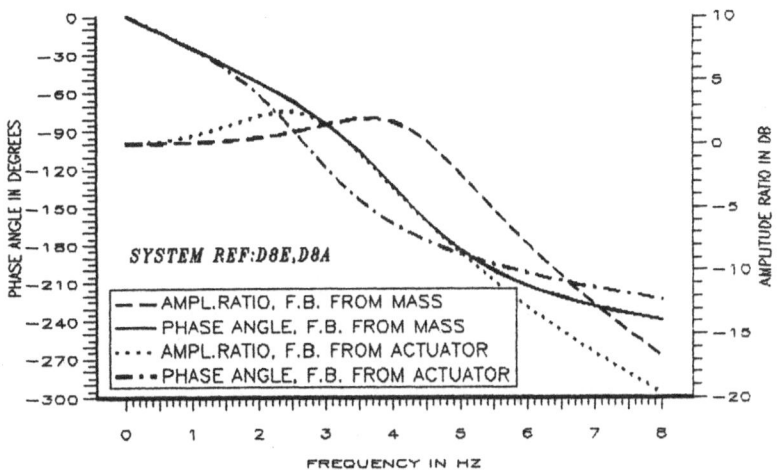

Fig. 24.8 Closed loop Bode diagram.

transfer functions Figs 12.4, 12.8 and 12.9 with eqns (24.6) and (24.10). Both systems give rise to transfer functions with non-constant coefficients if the loop gain, K, is treated as a variable quantity, requiring analytical computerized optimization methods, as outlined in Chapter 13 and described by algorithm 'B', shown in qualitative and quantitative form in Figs 13.8 and 13.9 respectively.

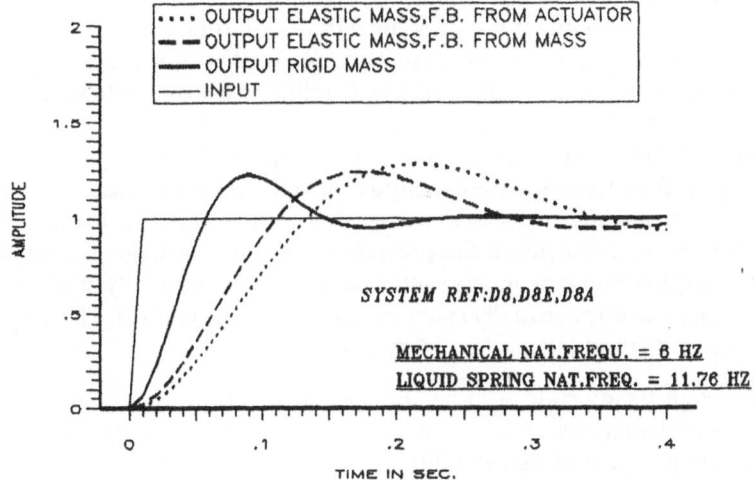

Fig. 24.9 Transient response comparison.

25

The Flow Feedback Option

True flow or velocity valve-controlled systems can only be realized by one of three methods:

(1) Pressure-compensated proportional control valve.
(2) Closed loop system with feedback from actuator.
(3) Closed loop flow feedback system.

The major disadvantages of method (1) are the rather limited degree of load and temperature compensation, 5% being a fairly typical figure, interaction between services supplied from a common power source, and the difficulty of operating overrunning loads under controlled conditions.

Method (2) produces good results but installed cost can be high in comparison with methods (1) and (3), particularly if several actuators are supplied from a common power source, thus requiring separate feedback transducers for each of the actuators. A further drawback can be unfavourable environmental operating conditions which preclude the use of feedback transducers in the vicinity of the actuators.

Method (3) has been briefly described in Chapters 5 and 6, and systems employing flow feedback were analysed in subsequent chapters.

Although method (3) depicts a true flow control system, it does not precisely control velocity in the presence of actuator leakage. In systems employing hydraulic cylinders, actuator leakage is normally negligible, making flow control virtually synonymous to velocity control. The salient features of method (3) can be summarized as:

(1) High steady-state accuracy independent of pressure and temperature variations.
(2) High degree of repeatability.
(3) Ability to control both opposing and overrunning loads.

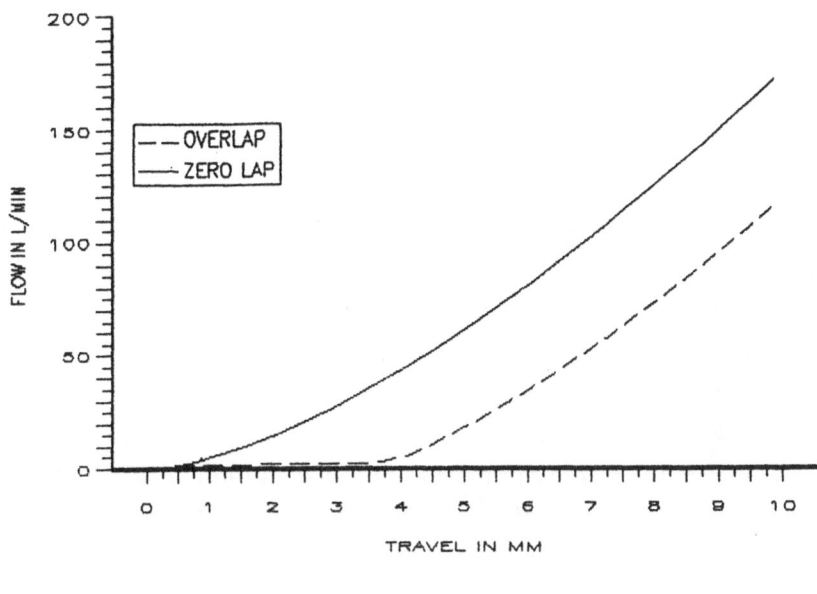

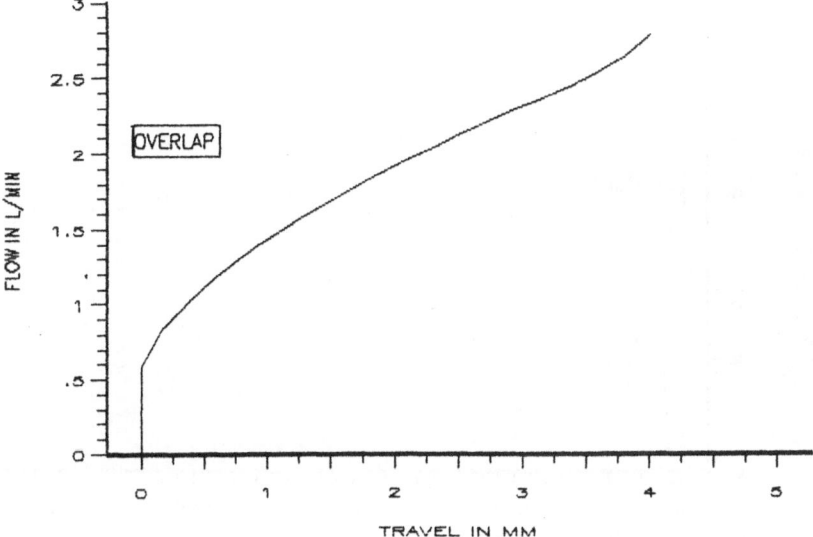

Fig. 25.1 Flow sensor flow–travel characteristics.

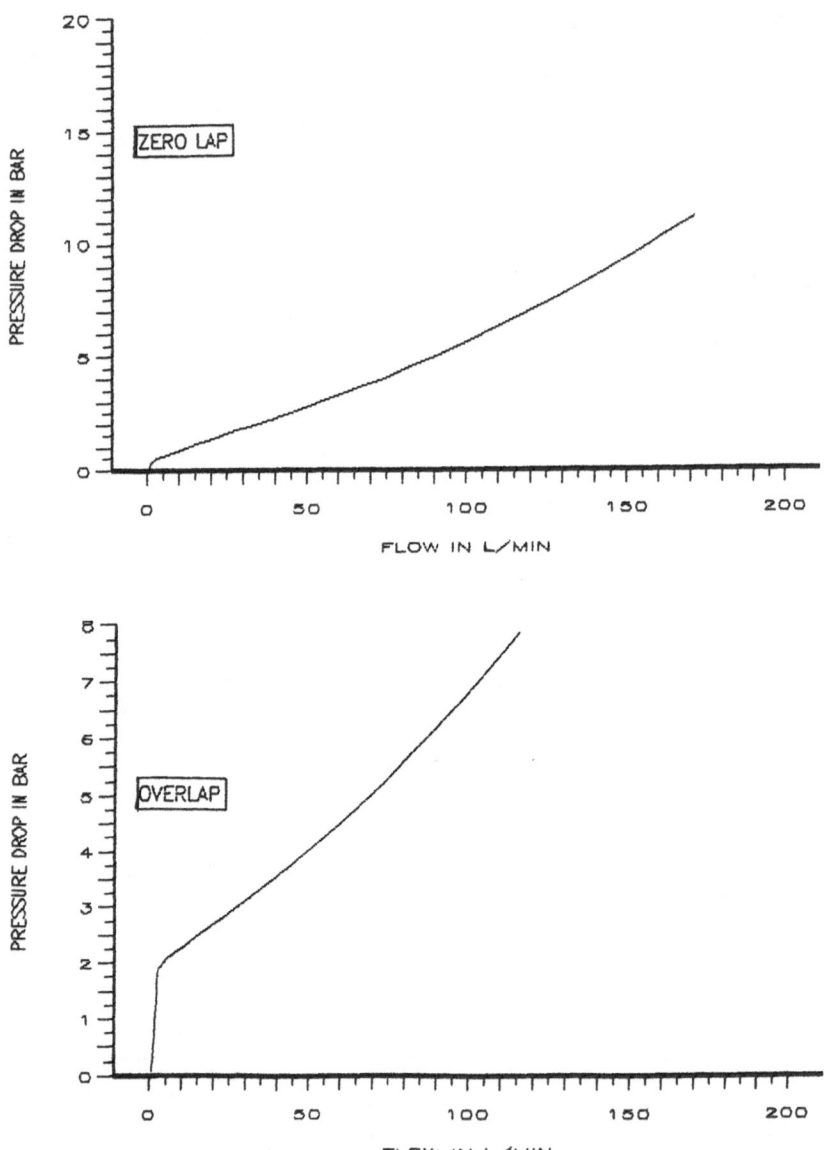

Fig. 25.2   Flow sensor pressure–flow characteristics.

(4) Simultaneous interaction-free operation of services fed from a common power source.
(5) Input/output characteristics of system unaffected by control valve spool configuration.
(6) Closed loop permits use of lower grade control valves and simplified drive amplifier, thus providing cost-effective system solution.
(7) Variable loop gain of electro-hydraulic systems facilitates performance optimization.

Typical flow-travel and pressure-flow characteristics of the flow sensor shown in Fig. 6.3 are plotted in Figs 25.1 and 25.2. The dotted line depicts a sensor with dual gain characteristics, providing turn-down ratios exceeding 1000:1.

A circuit diagram of a bidirectional flow feedback system is shown in Fig. 25.3 and a system suitable for controlling single-acting cylinders, e.g. lift controls, in Fig. 25.4. The power efficiency of the lift control system is very high, since the power-up operating mode uses bleed-off flow control, as described in Chapters 5 and 23, and in the gravity-down operating mode the fixed displacement pump is off-loaded. The bi-directional flow feedback loop ensures highly repeatable velocity, acceleration and deceleration control in both the power-up and gravity-down operating modes.

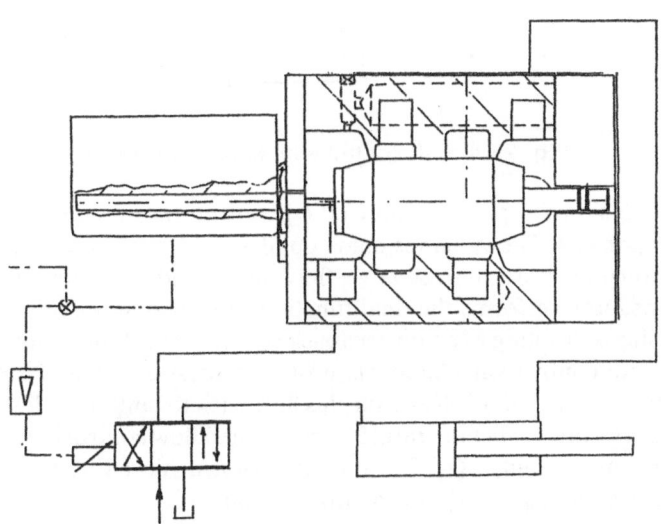

Fig. 25.3 Bidirectional flow feedback loop.

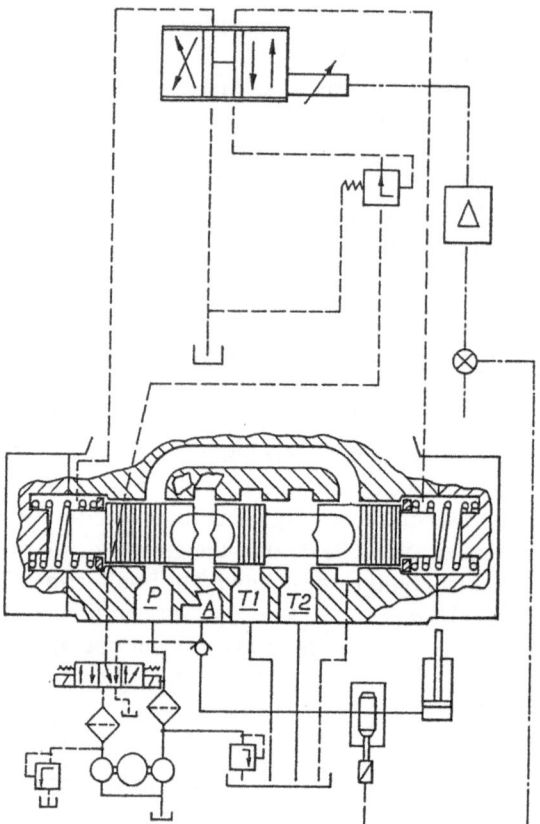

Fig. 25.4   Lift control with flow feedback.

Fig. 25.5 shows a pressure–flow (*P-Q*) control system comprising low cost components. The system is made up of a fixed displacement pump, a proportional pressure control valve, flow sensor and drive amplifier. The flow transducer performs the dual function of providing a pressure–flow relationship and acting as a flow feedback sensor. In the flow control mode the pressure control valve controls upstream pressure, *P*, and acts as a bleed-off valve by spilling off any surplus flow back to tank. The closed loop provides pressure- and temperature-compensated flow control in respect of the flow demand signal, Q_D. Pressure control is achieved by limiting the pressure demand signal, P_D, to the drive amplifier.

In applications where the actuator is allowed to stall, the control valve can saturate, i.e. move to its maximum setting. This can adversely affect

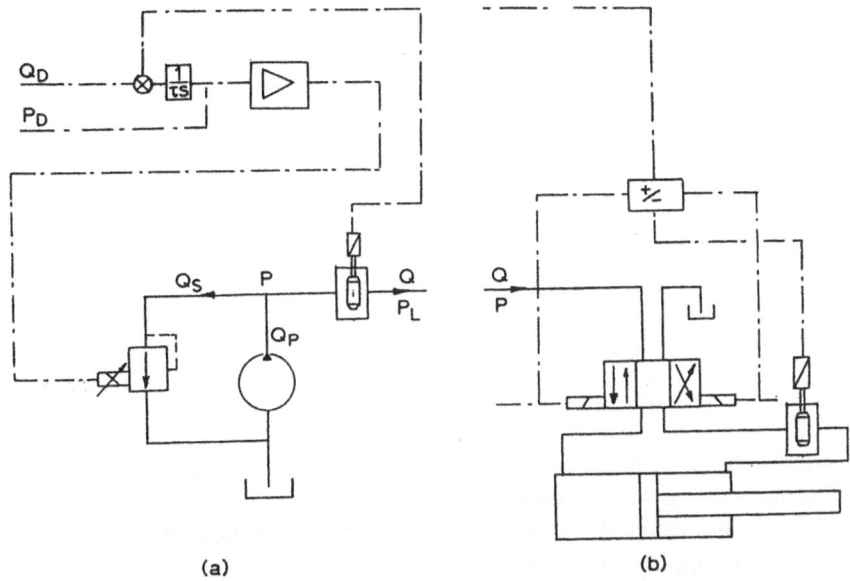

Fig. 25.5 Pressure–flow (P–Q) control with flow feedback.

dynamic performance, particularly during a reversal of direction. This condition can be avoided by introducing an additional spool position feedback loop. The spool position loop should, however, only be used as an alternative and not as an additional feedback loop, since eliminating the free integrator can have an adverse effect on system performance. The introduction of an electronic circuit which selects the larger of the two error signals as the operative input signal to the drive amplifier, thus automatically switching to spool position feedback in case of valve saturation or flow sensor signal failure, will therefore provide enhanced dynamic response and fail-safe characteristics for the system.

26

Non-Linearities

Non-linearities affecting system performance are hysteresis, backlash, dead zone and coulomb friction. In closed loop systems, backlash inside the loop, specifically any backlash associated with feedback transducer drives, should be eliminated to avoid limit cycling. In velocity control systems, a dead zone around the null region is usually deliberately introduced, by using overlapped valves, in order to prevent actuator movement at zero demand signal. In closed loop position control systems employing on–off (or bang-bang) control valves, dead zone has a stabilizing effect, whereas in proportional position control systems, dead zone would give rise to limit cycling. Idealized characteristics of an on–off control system are shown in Figs 26.1 and 26.2. In systems of this type, stability and response are a function of three parameters, i.e. actuator velocity, dead zone and time delay. On–off position control systems can only be used for relatively crude point-to-point positioning.

All control valve characteristics include hysteresis of varying degree which produces a steady-state error in closed loop systems. In some instances this error can be reduced by applying a superimposed dither frequency to the valve input signal, thereby decreasing valve hysteresis, or by increasing the electrical gain of the system, since the system error is inversely proportional to the electrical gain. As the loop gain is the product of the electrical gain and the hydraulic gain, care must be taken not to exceed the limiting loop gain.

Coulomb friction is present in hydraulic cylinder seals, although it is only one of the components constituting seal friction characteristics. The basic lubrication mechanism of seals is shown in Fig. 26.3. It can be seen that the friction force versus velocity curve is non-linear, particularly at low cylinder velocities. A comparison between experimental and theoretical results is shown in Fig. 26.4 for polymer seals and in Fig. 26.6 for piston ring

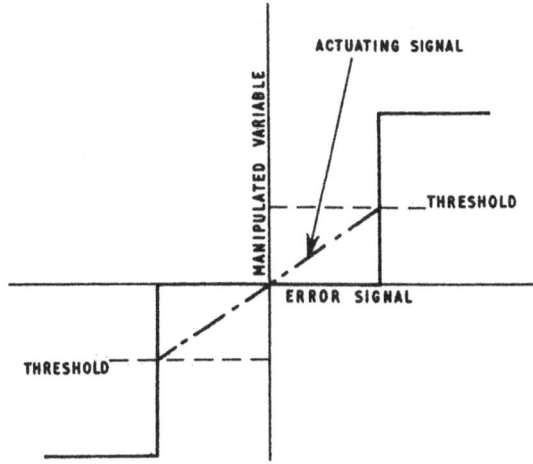

Fig. 26.1 Idealized on–off controller characteristics.

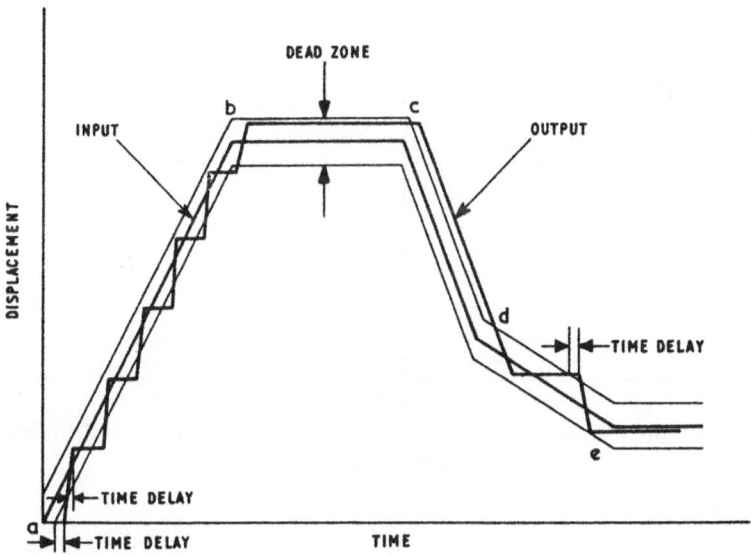

Fig. 26.2 Idealized on–off closed loop control system characteristics.

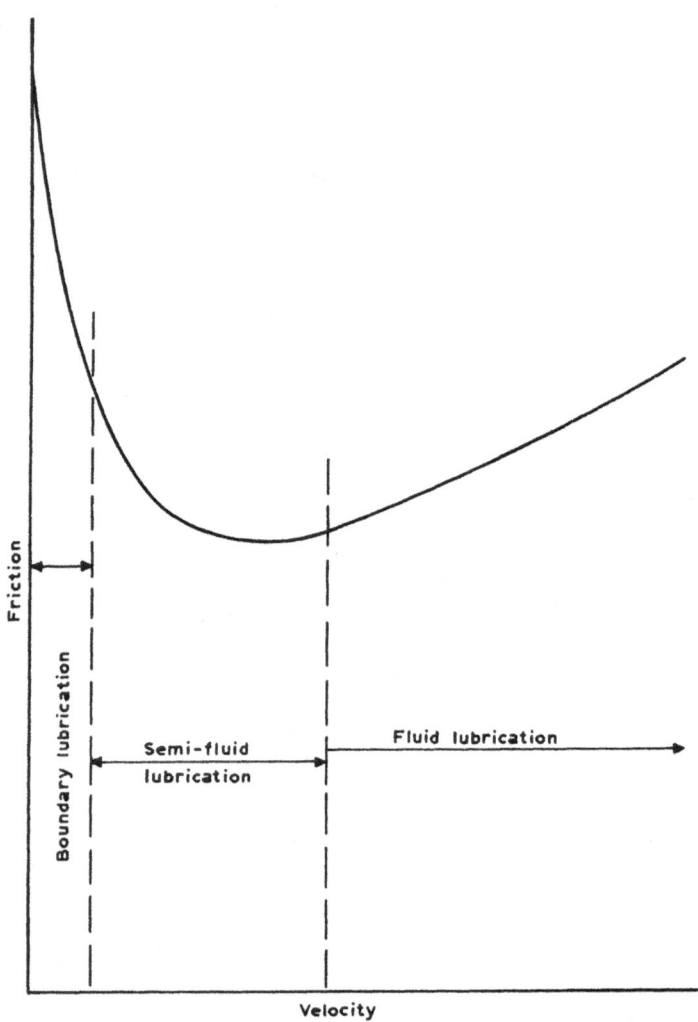

Fig. 26.3   Lubrication mechanism of sliding surfaces (courtesy of Bowden and Tabor).

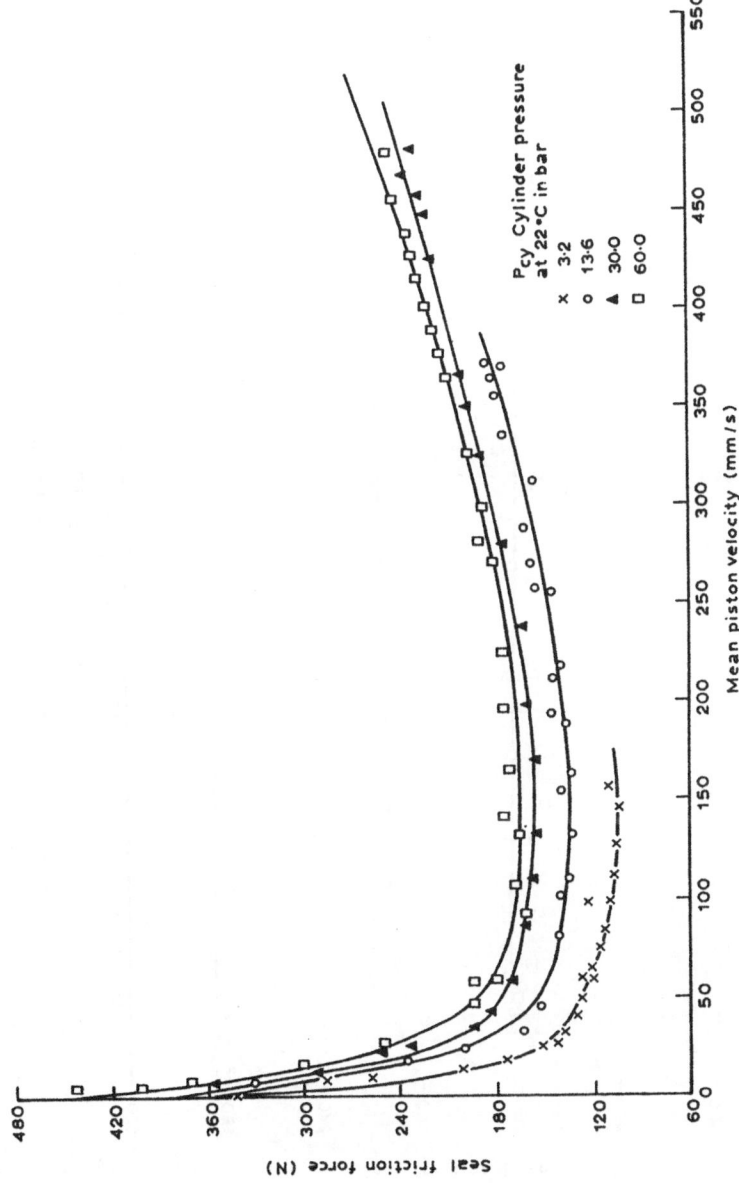

Fig. 26.4 Cylinder friction characteristics for polymer seals.

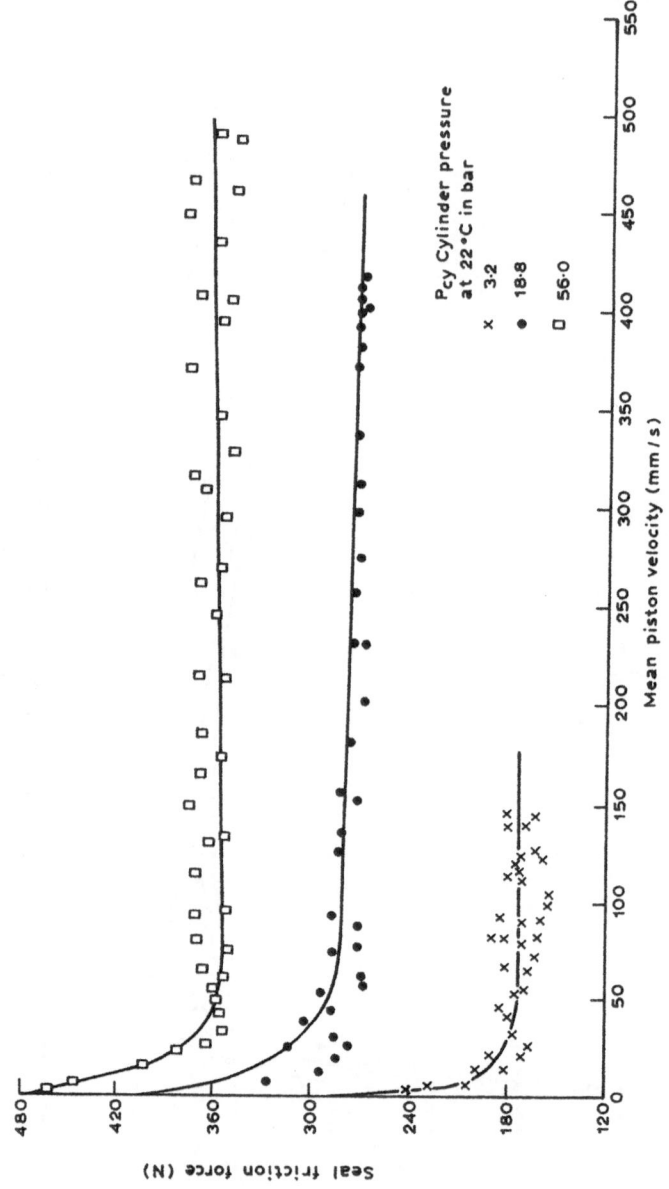

Fig. 26.5   Cylinder friction characteristics for cylinder seals.

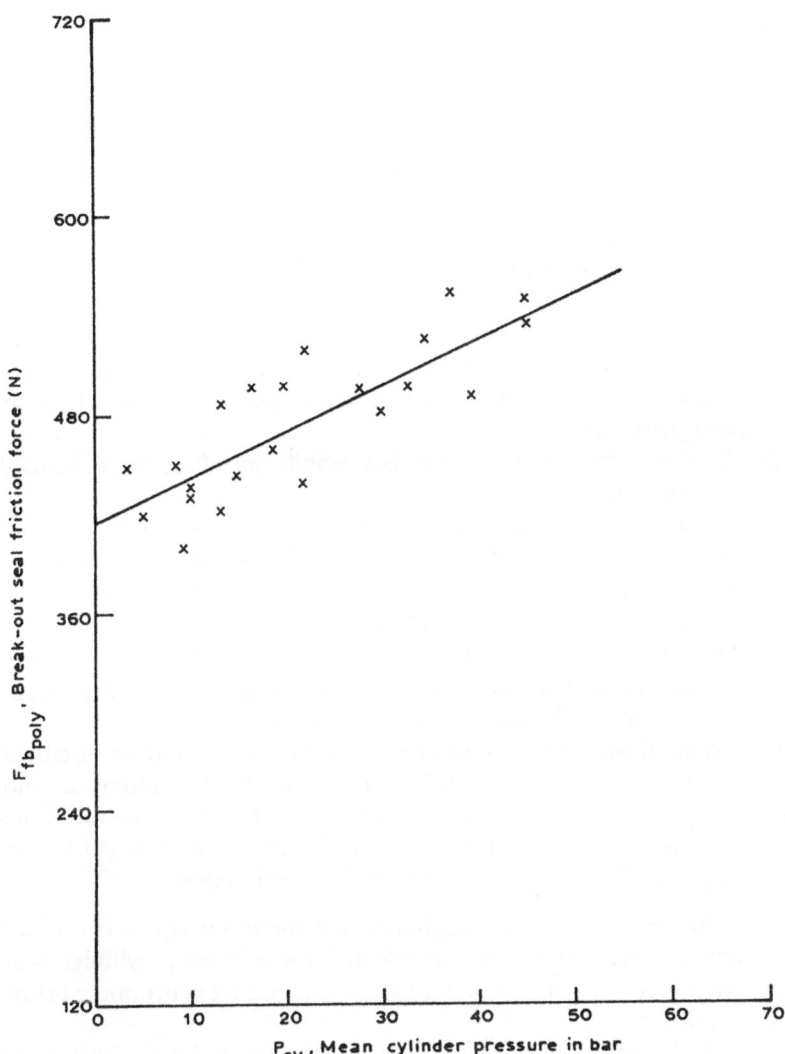

Fig. 26.6 Break-out friction characteristics for piston ring seals.

seals. The theoretical results were obtained by means of a non-linear analytical method. Typical experimental results for break-out friction forces are plotted in Fig. 26.5. The above curves were taken from a thesis covering a comprehensive investigation into cylinder seal characteristics. The following conclusions can be drawn:

(1) Seal friction characteristics of both polymer and piston ring seals are largely non-linear in character.
(2) Seal friction force can be split into three components:
   (i) Coulomb friction.
   (ii) Friction with negative damping coefficient.
   (iii) Friction with positive damping coefficient.
(3) Break-out friction for both types of seal has a linear relationship with working pressure at a given fluid temperature.
(4) Seal friction characteristics of piston ring seals are significantly affected by running-in time, whereas polymer seals are only marginally affected.
(5) Coulomb friction has a marked stabilizing effect on positional control systems.
(6) Negatively damped friction forces present at low cylinder velocities have a destabilizing effect, and can give rise to slip–stick and judder.
(7) Viscous damping is present at higher velocities for polymer seals, but its stabilizing effect is only marginal.
(8) Seal friction has a significant stabilizing effect on closed loop hydraulic positional control systems, and many such systems would be unstable in the absence of cylinder seal friction.
(9) Comparison between non-linear and linearized analyses of closed loop control systems, laboratory and field evaluation and correlation with theoretical results have shown that the following guidelines can be adopted when a linearized small perturbation approach is applied to a positional control system:

In the vicinity of null, cylinder seal friction is equivalent to an actuator damping factor of 0·5, and for a moving cylinder, seal friction prevents the actuator damping factor from dropping below a value of 0·2.

It should be borne in mind that in an electro-hydraulic control system the actuator damping factor is only one of many parameters influencing system performance and that close correlation between linearized system analysis and actual performance has been recorded over many years.

27

Steady-State System Analysis

A block diagram of a typical position control system is given in Fig. 27.1, and an equivalent normalized system block diagram in Fig. 27.2. The free integrator inside the loop eliminates any residual positional error, which leaves three causes giving rise to steady-state errors, i.e. actuator velocity, actuator loading and valve hysteresis.

27.1 VELOCITY ERROR

The velocity error in a position control system is directly proportional to the actuator velocity and inversely proportional to the loop gain, or

$$\varepsilon_V = v/K \qquad (27.1)$$

where v is in length units per second, ε_V in length units, and the loop gain, K, in s^{-1}.

In a closed loop velocity or flow control system containing a free integrator, there is a corresponding acceleration error, which is directly

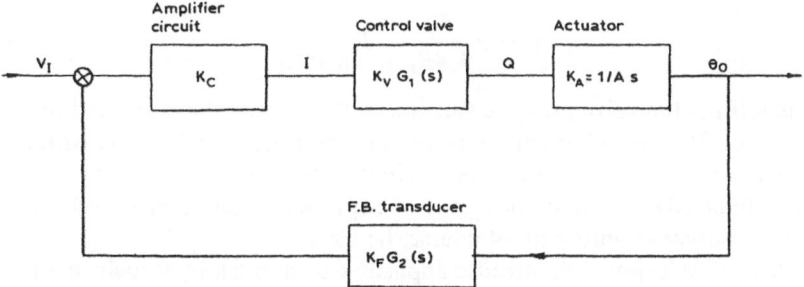

Fig. 27.1 System block diagram; $K_C K_F = K_E$; $K_V K_A = K_H$; $K_E K_H = K$.

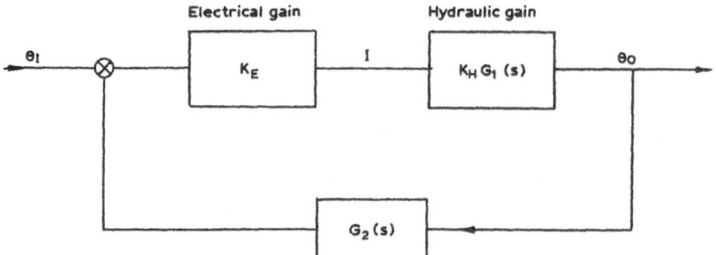

Fig. 27.2   Normalized system block diagram; $K_E K_H = K$.

proportional to actuator acceleration and inversely proportional to the loop gain, where the acceleration error is in length units per second, the acceleration in length units per second per second and the loop gain in s^{-1}.

27.2 HYSTERESIS ERROR

The positional error due to valve hysteresis is given by the expression:

$$\varepsilon_H = \frac{I_H}{K_E} = I_H \frac{K_H}{K} = I_H \frac{K_V}{AK} \tag{27.2}$$

where I_H is the hysteresis of the valve expressed in command units, normally either mV or mA, K_V, the valve flow gain, in cc/s per command unit, and A, the effective cylinder area, in cm². The hysteresis error referred to the actuator ε_H will then be expressed in cm.

27.3 LOAD ERROR

The positional error due to actuator loading is given by the expression:

$$\varepsilon_L = F \frac{K_H}{KAK_P} = F \frac{K_V}{KA^2 K_P} \tag{27.3}$$

where K_P is the valve pressure gain and F the external load referred to the actuator. If K_P is in bar per command unit and the load F in kPa with the other parameters in the units applied in the previous section, then the load error referred to the actuator, ε_L, will be expressed in cm. Alternatively SI or other consistent units can, of course, be used.

The above expressions are also applicable to closed loop velocity control systems where the load error, ε_L, will be expressed in velocity units.

In flow feedback systems, as described and analysed in Chapter 12, the actuator is outside the loop, leading to additional load errors in the presence of actuator shunt leakage. From the equations $Q_S = cP_L$, $Q_S + Q = Q_0$ and

$$P_L = Q\frac{D}{A^2} + \frac{F}{A}$$

$$Q = \left(Q_0 - \frac{cF}{A}\right)\frac{1}{1 + (cD/A^2)} \tag{27.4}$$

Load errors are then increased by the ratio Q_0/Q.

Similarly for elastically mounted mass systems covered in Chapter 24, an additional load error occurs if the elastically mounted mass is outside the feedback loop, i.e. when feedback is taken from the actuator. The additional load error is then given by the product of load and spring compliance, or the total load error

$$\varepsilon_L' = \varepsilon_L + \sigma F \tag{27.5}$$

27.4 CONCLUSIONS

Although the above equations were derived for systems with a linear output they can also be applied to systems with a rotary output by substituting motor displacement for cylinder area and output torque for cylinder loading, providing that consistent units are used.

An examination of the equations describing steady-state errors clearly shows that the critical parameter affecting steady-state accuracy is the loop gain, K, since all steady-state errors are inversely proportional to the loop gain. This also applies to closed loop pressure control systems. A further conclusion which can be drawn is the beneficial effect of a high electrical gain, K_E, which can be achieved by keeping the hydraulic gain, K_H, as low as possible, e.g. by matching valves to the output requirements of the system.

A computer print-out of steady-state performance parameters of typical closed loop position and velocity control systems is shown below. This shows that velocity and acceleration errors are the predominant steady-state errors for position and velocity controls respectively, and that, owing to the relatively high output stiffness of most hydraulic control systems, load errors are normally insignificant compared with other steady-state and dynamic errors.

STEADY-STATE ERRORS
Special Features

VALVE PARAMETERS
Flow Gain = ·7 L/min per command signal
Pressure Gain = 60 bar per command signal
Hysteresis = 5 command signal units
Enter YES (Y) for update?

POSITION CONTROL
Load Error @ 12 kN = 1·470 939E-02 mm
Hysteresis Error = ·288 798 5 mm/sec
Velocity Error @ 53·051 74 cm/sec = 20·628 47 mm

VELOCITY CONTROL
Load Error @ 12 kN = 1·470 838E-02 mm/sec
Hysteresis Error = ·288 798 5 mm/sec
Acceleration Error @ 53·051 74 cm/sec^2 = 20·628 47 mm/sec
To obtain Amplifier Gain, enter Feedback Transducer Gain in Volt/cm
for position control and Volt per cm/sec for velocity control? ·5
Amplifier Gains = 346·262 2 command signal units per volt

STEADY-STATE ERRORS
Special Features

PASSIVE NETWORK
VALVE PARAMETERS
Flow Gain = ·7 L/min per command signal
Pressure Gain = 60 bar per command signal
Hysteresis 5 command signal units
Enter YES(Y) for update?

POSITION CONTROL
Load Error @ 12 kN = 3·079 741E-03 mm
Hysteresis Error = 6·047 058E-02 mm
Velocity Error @ 53·051 74 cm/sec = 4·319 328 mm

VELOCITY CONTROL
Load Error @ 12 kN = 3·079 741E-03 mm/sec
Hysteresis Error = ·6·047 058E-02 mm/sec
Acceleration Error @ 53·051 74 cm/sec^2 = 4·319 328 mm/sec

To obtain Amplifier Gain, enter Feedback Transducer Gain in Volt/cm
for position control and Volt per cm/sec for velocity control? ·5
Amplifier Gain = 1653·697 command signal units per volt

STEADY-STATE ERRORS
Special Features
Elastically Mounted Mass (F.B. from actuator)

VALVE PARAMETERS
Flow Gain = ·7 L/min per command signal
Pressure Gain = 60 bar per command signal
Hysteresis = 5 command signal units
Enter YES(Y) for update?

POSITION CONTROL
Load Error @ 12 kN = 4·247 134 mm
Hysteresis Error = ·499 103 3 mm
Velocity Error @ 53·051 74 cm/sec = 35·650 24 mm

VELOCITY CONTROL
Load Error @ 12 kN = 2·541 912E-02 mm/sec
Hysteresis Error = ·499 103 3 mm/sec
Acceleration Error @ 53·051 74 cm/sec^2 = 35·650 24 mm/sec
To obtain Amplifier Gain, enter Feedback Transducer Gain in Volt/cm
for position control and Volt per cm/sec for velocity control? ·5
Amplifier Gain = 200·359 3 command signal units per volt

STEADY-STATE ERRORS
Special Features

VALVE PARAMETERS
Flow Gain = 2 L/min per command signal
Pressure Gain = 21 bar per command signal
Hysteresis = 2 command signal units
Enter YES(Y) for update?

FLOW CONTROL
Torque Error @ 500 Nm = ·053 991 rad/sec
Hysteresis Error = 5·095 969E-02 rad/sec
Acceleration Error @ 19·999 99 rad/sec^2 = 1·719 743 rad/sec
To obtain Amplifier Gain, enter Feedback Transducer Gain in Volt per
L/min? ·05
Amplifier Gain = 116·323 command signal units per volt

28

Applications

Hydraulic and electro-hydraulic control systems have been applied over a wide range of industries. Applications can be subdivided into four categories: (1) industrial, (2) mobile, (3) marine and (4) aerospace and defence.

(1) Industrial
The most diverse range of applications occurs in the industrial category, which can further be broken down into the following sub-sections:

Machine Tools	Plastics
Process Control	Material Handling
Research & Test	Entertainment

Some typical examples will help to illustrate the versatility of hydraulically driven control systems.

Machine Tools: Honing machines   Gear Cutters
Broach machines   Spline grinders   Horiz. grinders
Tube benders   Power saws   Pipe cutters
Flying shears   Machining centres
Plastics: Injection moulding   Extruders
   Blow moulders   Thermoplastic press
Process Control: Metal forming   Die casting
Wire reinforcement   Welding machine   Glass processing
Food processing   Wire wrapping   Tilting furnace
Coil handling   Cont. casting   Walking beam furnace
Oven control   Billet feeder   Rolling mill
Wood processing

Material Handling: Conveyor drives   Transfer machines
Travelling crane   Brick & concrete block stacker
Block tilter   Transporter (cement, steel, salt, coal)
Robots
Research & Test: Wave generator   Gear box test rig
Flight simulator   Electricity pylon test rig
Fatigue test rig   Performance simulator
Entertainment: Stage control   Flying bedstead
Special effects

(2) Mobile
Earth moving equipment Cranes   Tractor drives
Salting vehicle   Drive control   4-wheel steering
Variable suspension   Train tilt control

(3) Marine
Steering control   Variable pitch hub control
Variable pitch thruster Transfer at sea
Cable laying   Bucket dredger   Barrier control
Winch drives   Hovercraft control   Sail Trainer
Stabilised sector scanner

(4) Aerospace and Defence
Flying controls   Emergency generator   Ram air turbine
Flaps, slats, undercarriage   Radar antennae drive
Missile control   Anti-aircraft turret   Tank controls
Gun controls

The above listing is by no means exhaustive but gives an indication of the wide range of hydraulic control system applications. Some applications which do not readily fall into any of the listed categories, e.g. lift controls, have been omitted. Hydraulic drives have to compete with electric and pneumatic drives in the market place, where performance and cost are the obvious criteria of selection. Sometimes, however, the choice of drive system is carried out on a more subjective basis, i.e. the preference and familiarity of the application engineer with a specific discipline, which can inhibit the selection of the optimum and most cost-effective solution. The tendency towards multi-disciplinary training and education should help to obviate this problem.

29

National and International Standards

The purpose of this section is to highlight the national and international standards which are most relevant to hydraulic control systems employing proportional control components.

(1) BS 4062 ≡ ISO 4411 *Valves for hydraulic fluid power systems.*
(2) BS 5995 ≡ ISO 6404 *Methods of test for electrohydraulic servo valves.*
(3) BS 6697 *Methods of test for electrohydraulic proportional control valves.*
(4) BS 6494: Part 2 *Specification for four-port and five-port servo valves.*
(5) BS 2917 ≡ ISO 1219 *Specification for graphical symbols used on diagrams for fluid power systems and components.*
(6) ISO 5598 *Fluid power system and components vocabulary.*
(7) BFPA/P49 *Guidelines to electrohydraulic control systems.*

The majority of the above documents were initially prepared by the British Trade Association BFPA (formerly AHEM), converted into a British national standard by BSI, and ultimately revised and issued as an international standard by ISO.

30

Introduction

The simulation package 'Hydro Analyst' is the culmination of over twenty years' continuous work of adapting the mathematical models described in Part 1 to the latest state of the art in Information Technology. The Windows' based 'Hydro Analyst' supersedes the MS Dos based 'Hydrasoft' programs.

The programs contain several special features, e.g.

Automatic component selection from an extensive data base covering all major hydraulic equipment manufacturers.

Automatic Loop Gain optimisation.

Dynamic performance parameters as a function of independent variables under optimised and non-optimised operating conditions.

Comprehensive graphic facilities for single and multiple plots.

Ability to generate any duty cycle, including continuous input functions.

System enhancement, e.g.. P.I.D., passive networks, flow feedback and multiple feedback loops.

Comprehensive interactive HELP file.

All calculations are carried out in SI units, input and output values are entered and displayed in practical units.

Two alternative program versions, one in metric and one in US units, are available. The metric demonstration version is supplied with the textbook.

The US demonstration version can be downloaded from the Flotron Ltd website: www.flotron.co.uk

The programs have been used ,world wide , by Hydraulic equipment manufacturers, end users, consultants and Universities with an excellent correlation record.

31

Structure

The Hydro Analyst program group contains four interactive programs:
  Flow Control
  Pressure Control
  Power Efficiencies
  Help
The Power Efficiency file can also be accessed from the Flow Control file.

Flow Control

The program is arranged in a modular structure comprising eight full screen (1 to 8) and four partial screen (9 to 12) forms as enumerated below.
  1. Cylinder
  2. Motor
  3. Components
  4. Options
  5. Frequency Domain
  6. Time Domain
  7. Power Efficiencies and Dissipation
  8. Graphics Display
  9. Summary
  10. Examples
  11. File
  12. Help

Pressure Control

The program comprises five full screen (1 to 5) and three partial screen (6 to 8) forms:
  1. Hydraulic System
  2. Components
  3. Frequency Domain
  4. Time Domain
  5. Graphics Display
  6. Examples
  7. File
  8. Help

Help

A comprehensive interactive file tailored to the specific requirements of individual modules and including a search mode and 'Hot Keys'.

Power Efficencies and Dissipation

The program comprises two full screen (1 to 2) forms and one partial screen form (3).
   1. Power
   2. Graphics Display
   3. Help

General

All modules and forms are accessed via the Menu Bar.
In the non-tutorial mode, modules and forms can be accessed in any chosen order, whereas in the tutorial mode, the order of access is preset.

32

Modules

32.1 Cylinder and Motor (Flow Control)

16 alternative hydraulic circuit configurations are available,(Fig.1 & Fig.2)..
All relevant input data are entered into the textboxes provided, and dependant
variables, e.g. flow rate, load pressure, valve pressure drop are automatically entered by the program.
This module performs two functions:
    1.Hydraulic system integrity check.
    2.Hydraulic system Transfer function derivation.
If the system fails the integrity check,message boxes containing warnings and
recommended remedial actions will be displayed.

32.2 Components

This form contains a database of components from all major hydraulic equipment
manufacturers and includes both steady-state and dynamic performance parameters. Components can be selected either manually or automatically.
Steady-state errors and amplifier gains are also calculated and displayed.

32.3 Options

Two types of options are catered for:
    1. Customised
    2. Multiple Variable
The 'Customised' categories are:
    1.Gearing between output and motor
    2. Linear output from motor
    3.Multiple cylinders, i.e. cylinders connected in parallel
    4.Shunt leakage pressure control
The 'Multiple Variable' categories are:
    1.Inclined cylinder
    2. Load sensing
    3. Constant horsepower

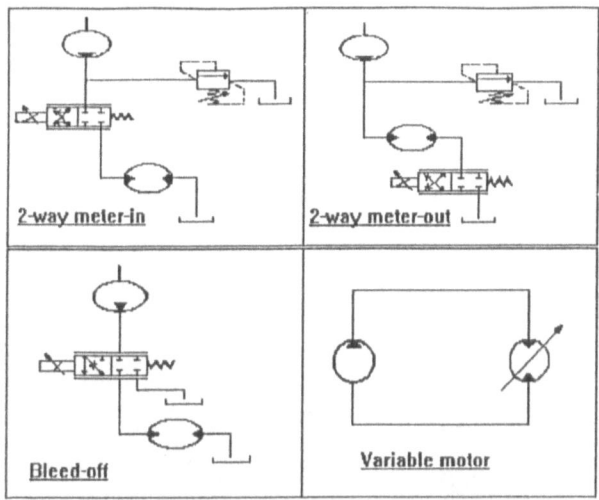

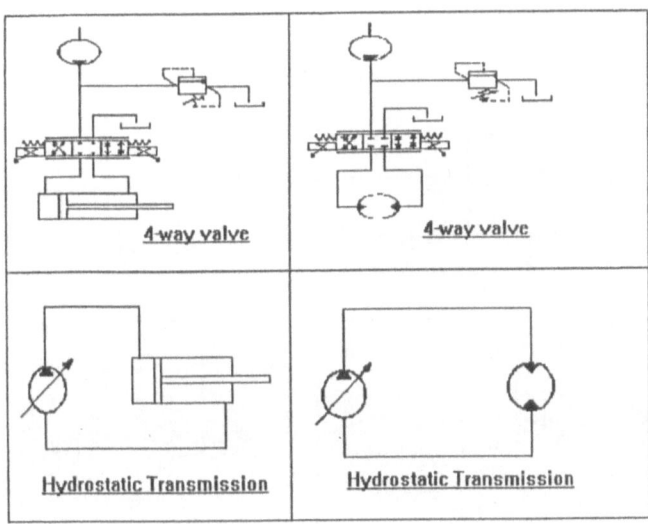

Fig. 1 Hydraulic Circuit Diagrams

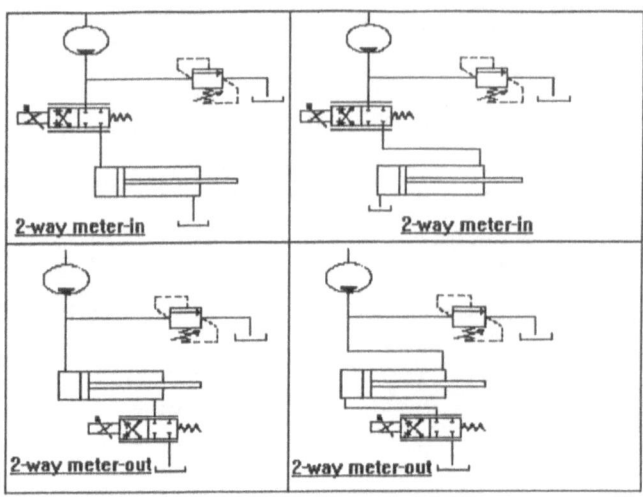

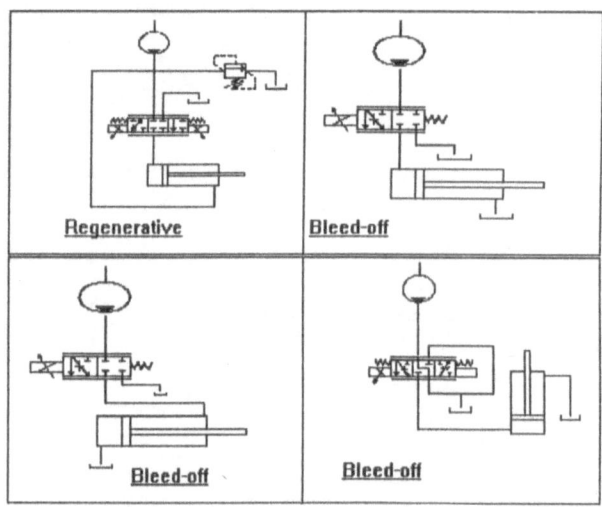

Fig. 2 Hydraulic Circuit Diagrams

The Multiple Variables form the link between independent and dependent variables and these can be summarised as:

Multiple Variable	Setting	Independent Variable	Dependent Variable
Inclined cylinder	Angle	Mass ref.to cylinder	Corresp. Load
Load sensing			
(Pressure match)	Valve press. drop	Load pressure	Supply pressure
(Power match)	Compensator press.setting	"	"
Constant h.p.	Const. power Max. flow rate Max. pressure	Flow rate	Supply pressure

32.4 Frequency Domain

Five basic tasks are performed:
1. Establishing system feedback mode:
    1.1 Open Loop
    1.2 Position feedback
    1.3 Velocity feedback
    1.4 Flow feedback
    1.5 Force feedback
    1.6 Pressure feedback
    1.7 Multiple feedback
2. Loop gain setting for closed loop systems:
    2.1 Manual
    2.2 Semi-automatic
    2.3 Automatic
3. Derivation of system transfer function
4. Stability check
5. Determining system frequency response parameters:
    5.1 Natural frequency
    5.2 Damping factor
    5.3 Band width
An additional task of this module is to generate frequency based graphs. These are:
1. Open and closed loop Bode Diagrams
2. Nyquist Diagram
3. Nichols Chart
4. Stability Margins
5. Automatic Looping under optimised conditions

6. Automatic Looping under non-optimised conditions

32.5   Time Domain

Time response parameters corresponding to the frequency response parameters determined in the Frequency module are displayed. In its basic form these are the response of the system to a step demand and can be summarised as:

Response time to reach 90 % of final value
Maximum overshoot
Time taken to reach maximum overshoot
Settling time, i.e. time taken to settle within 5 % of final value

Graphs covering the following input demands are generated:
   1. Step demand
   2. Ramp-step response parameters
   3. Triangular response parameters
   4. Large step response (valve saturated or 'bang-bang' valves)
   5. Generated duty cycle (up to 8 transitions)
   6. Continuous function generator:
      6.1 Sinusoidal
      6.2 Triangular
      6.3 Reverse Triangular
      6.4 Square
      6.5 Reverse Square
      6.6 Saw tooth
      6.7 Inverted Saw tooth

32.6   Power Efficiencies and Dissipation

Power Efficiencies and Dissipation for five alternative pump configurations can be displayed or plotted:
   1. Fixed displacement pump at constant supply pressure
   2. Pressure-compensated variable displacement pump
   3. Fixed displacement pump with load sensing (pressure match)
   4. Variable displacement pump with load sensing (power match)
   5. Fixed displacement pump with bleed-off

32.7 Graphics Display

The number of graph types available in the three functional programs are:
    Flow Control: 22
    Pressure Control: 14
    Power Efficiencies: 6 (4 line graphs and 2 bar charts)

All line graphs are either single or multiple plots.

All graphs can be displayed and printed as full screen or sized as partial screen plots.

32.8 Summary

The partial screen display lists system characteristics and can be called at any stage of the analysis to identify the current position.

32.9 Examples

12 typical examples are provided in the Flow Control program and 3 examples in the Pressure Control program.
For demonstration disks, data cannot be entered in the textboxes and input data can therefore only be entered via an example file.

32.10 File

The usual File commands are included in the Cylinder, Motor and Hydraulic System forms, i.e.
    Save
    Save As
    Open (Not accessible on demonstration disks)
    Print
    New
    Exit

Note that File names starting with WE are reserved for example files.
Files incompatible with the chosen program will automatically be rejected, e.g. a Flow Control data file will not be accepted by the Pressure control program and a Pressure Control data file will not be accepeted by the Flow control program.

32.11 Help

The Help file can either be accessed from the icon in the program group, or from any form or module of the three working programs.

This file can be called from the Menu bar or via 'Hot Keys' by clicking the cursor on the chosen item and entering the F1 key.

32.12 Hydraulic System (Pressure Control)

The hydraulic system comprises a dead-headed actuator, a pressure source,e.g. a pump or an accumulator, and a pressure control valve, either manually pre-set or electrically modulated. Due to the trapped oil volume and the compressibilty of the fluid, specified by the Bulk Modulus, the system acts essentially as a liquid spring system, which can be identified by a first order transfer function.
A typical hydraulic circuit diagram is shown in Screen S13. Although a cylinder is shown in this diagram, the analysis is equally applicable to a dead-headed motor.
All relevant input data are entered into the textboxes provided, and parameters defining hydraulic system dynamics are automatically computed and entered by the program.

32.13 Screens

S1:   Cylinder Module
S2:   Motor Module
S3:   Components Module
S4:   Frequency Domain Module
S5:   Time Domain Module, Continuous Function Generator
S6:   Time Domain Module, Generated Duty Cycle
S7:   Frequency Domain Module, Open Loop System
S8:   Frequency Domain Module, Optimised Auto Looping-f(mass)
S9:   Frequency Domain Module, Optimised Auto Looping-int. network
S10: Graphics Module, Graph Options
S11: Options Form, Constant Power Pump
S12: Cylinder Module, Constant Power Pump
S13: Pressure Control System
S14: Power Efficiency and Dissipation

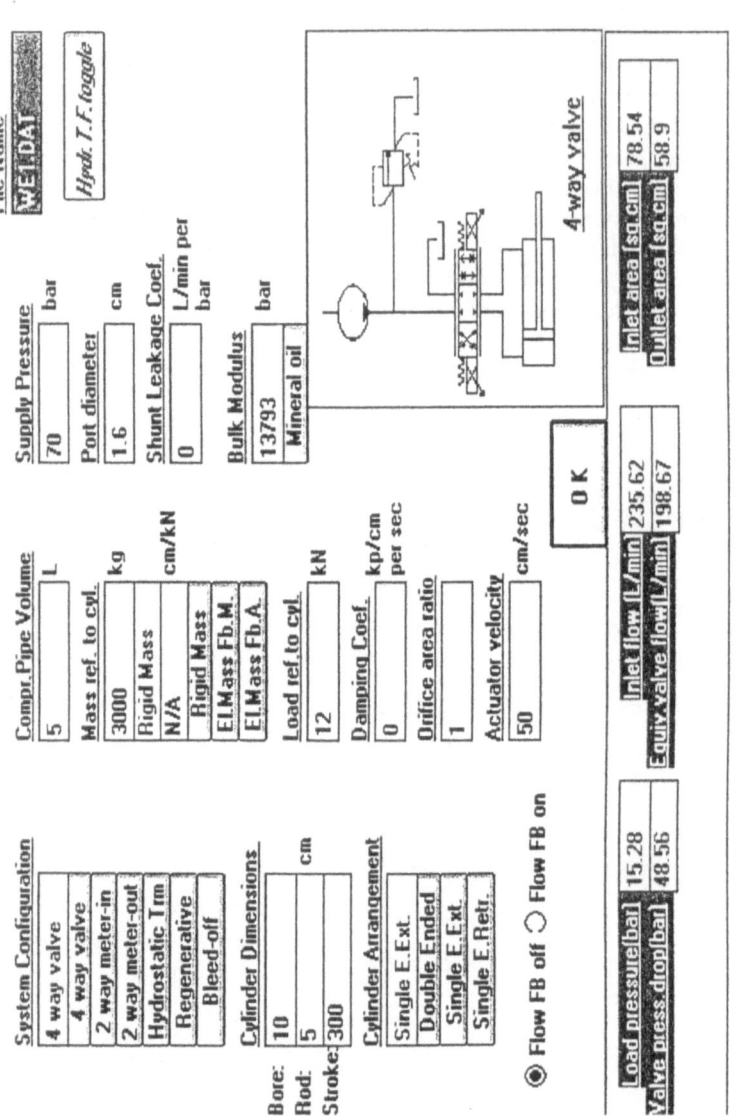

Screen S1 Cylinder Module

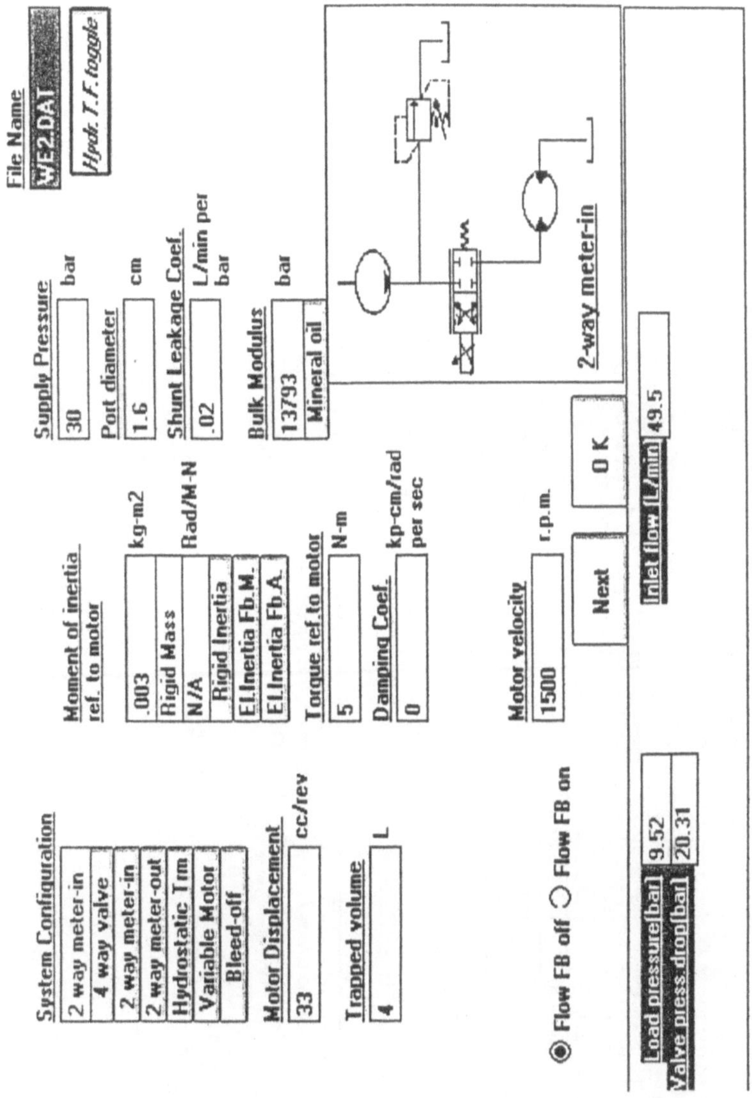

File Name
WE2.DAT

Hpdr. T.F. toggle

Supply Pressure
30 bar

Port diameter
1.6 cm

Shunt Leakage Coef.
.02 L/min per bar

Bulk Modulus
13793 bar
Mineral oil

Moment of inertia ref. to motor
.003 kg-m2

Rigid Mass   N/A   Rad/M-N
Rigid Inertia
El.Inertia Fb.M.
El.Inertia Fb.A.

Torque ref. to motor
5 N-m

Damping Coef.
0 kp-cm/rad per sec

Motor velocity
1500 r.p.m.

System Configuration
2 way meter-in
4 way valve
2 way meter-in
2 way meter-out
Hydrostatic Trm
Variable Motor
Bleed-off

Motor Displacement
33 cc/rev

Trapped volume
4 L

● Flow FB off ◯ Flow FB on

Load pressure(bar)   9.52
Valve press. drop(bar)   20.31

Next   O K   Inlet flow (L/min)   49.5

2-way meter-in

Screen S2 Motor Module

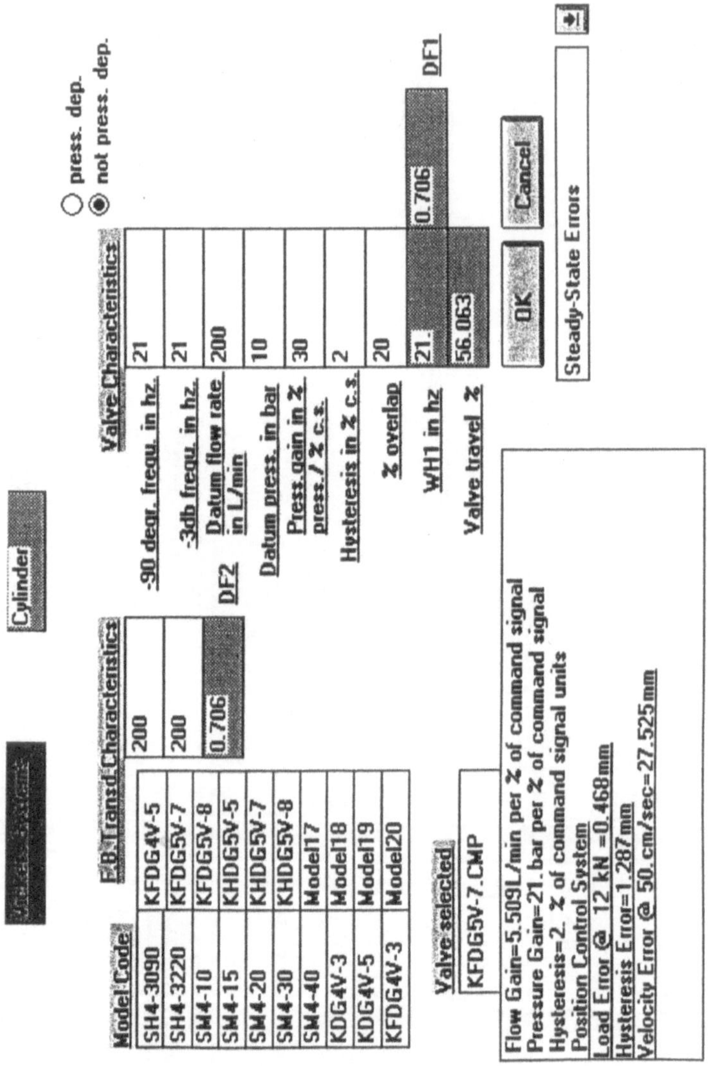

Valves Systems
Cylinder
Model Code
SH4-3090
SH4-3220
SM4-10
SM4-15
SM4-20
SM4-30
SM4-40
KDG4V-3
KDG4V-5
KFDG4V-3
KFDG4V-5
KFDG5V-7
KFDG5V-8
KHDG5V-5
KHDG5V-7
KHDG5V-8
Model17
Model18
Model19
Model20
F.B. Transd.Characteristics
200
200
0.706
DF2
Valve selected
KFDG5V-7.CMP
Valve Characteristics
-90 degr. frequ. in hz.
21
-3db frequ. in hz.
21
Datum flow rate in L/min
200
Datum press. in bar
10
Press.gain in % press./% c.s.
30
Hysteresis in % c.s.
2
% overlap
20
WH1 in hz
21.
0.706
Valve travel %
56.063
DF1
press. dep.
not press. dep.
OK
Cancel
Steady-State Errors
Flow Gain=5.509L/min per % of command signal
Pressure Gain=21.bar per % of command signal
Hysteresis=2. % of command signal units
Position Control System
Load Error @ 12 kN =0.468mm
Hysteresis Error=1.287mm
Velocity Error @ 50.cm/sec=27.525mm

Screen S3 Component Module

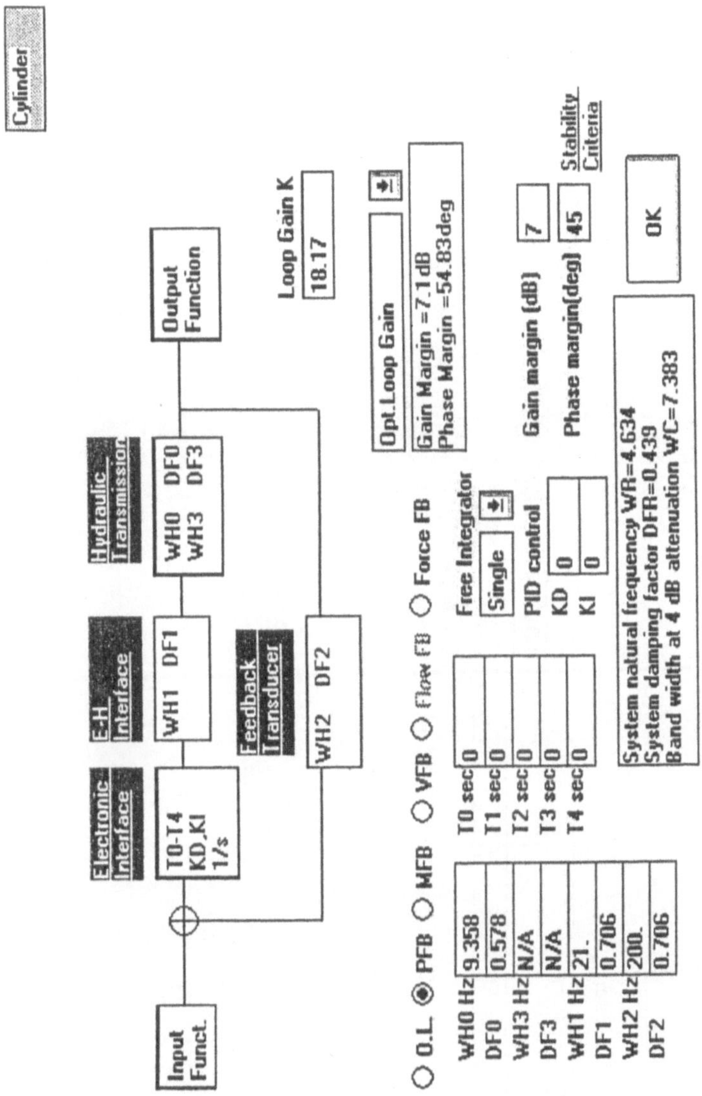

Screen S4 Frequency Domain Module

Screen S5   Time Domain Module

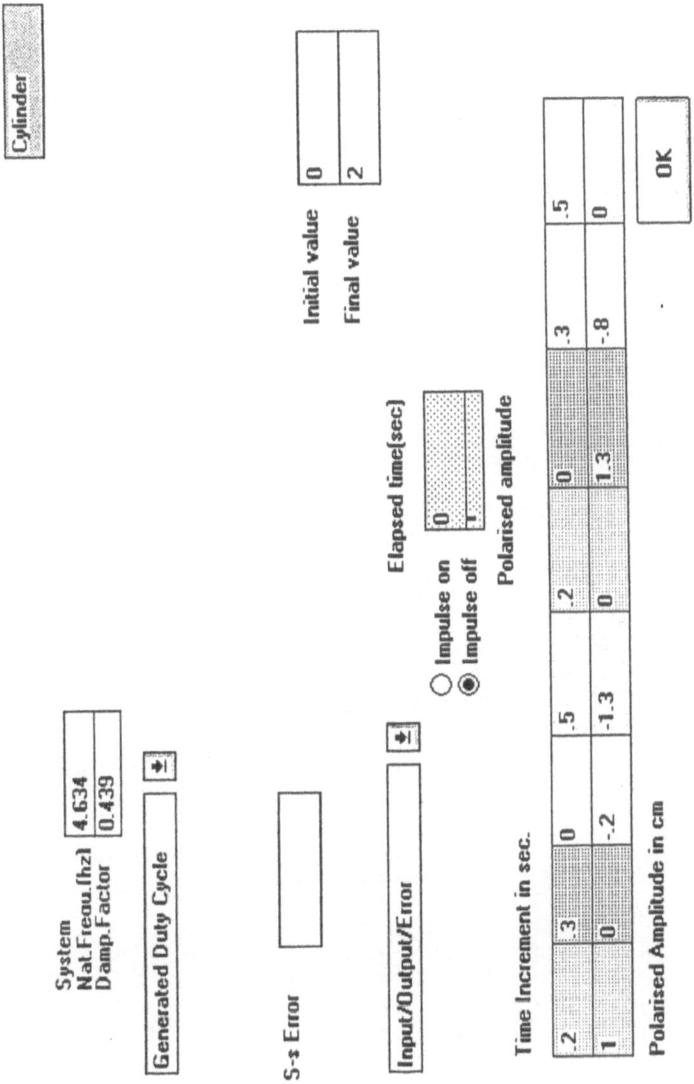

Cylinder

System
Nat.Frequ.(hz) 4.634
Damp.Factor 0.439

Generated Duty Cycle

S-$ Error

Input/Output/Error

Time Increment in sec.

.2	.3	0	.5	0	.2	0	.3	.5
1	0	-.2	-1.3	0	0	1.3	-.8	0

Polarised Amplitude in cm

Initial value 0
Final value 2

Elapsed time(sec)

○ Impulse on
◉ Impulse off

Polarised amplitude

OK

Screen S6   Time Domain Module

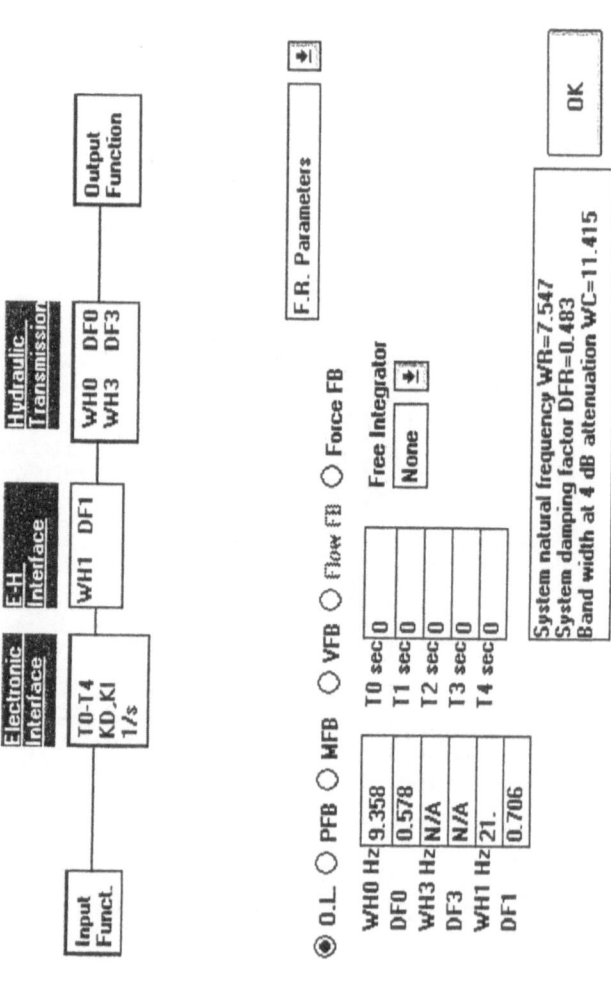

Screen S7 Frequency Domain Module

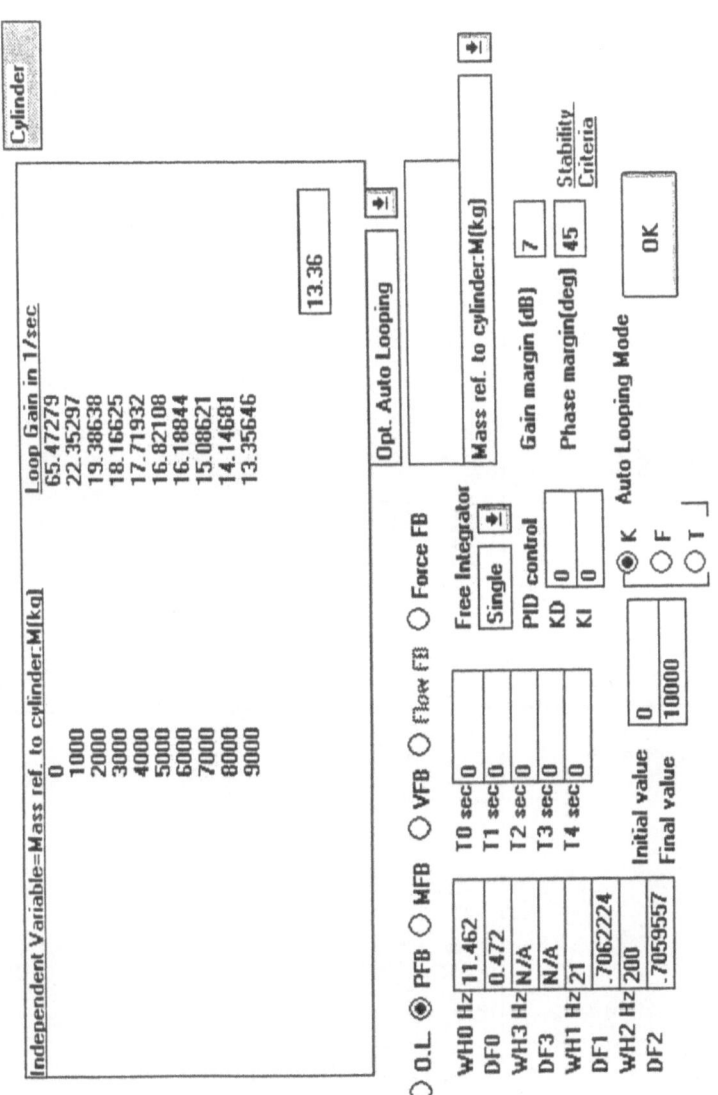

Screen S8 Frequency Domain Module

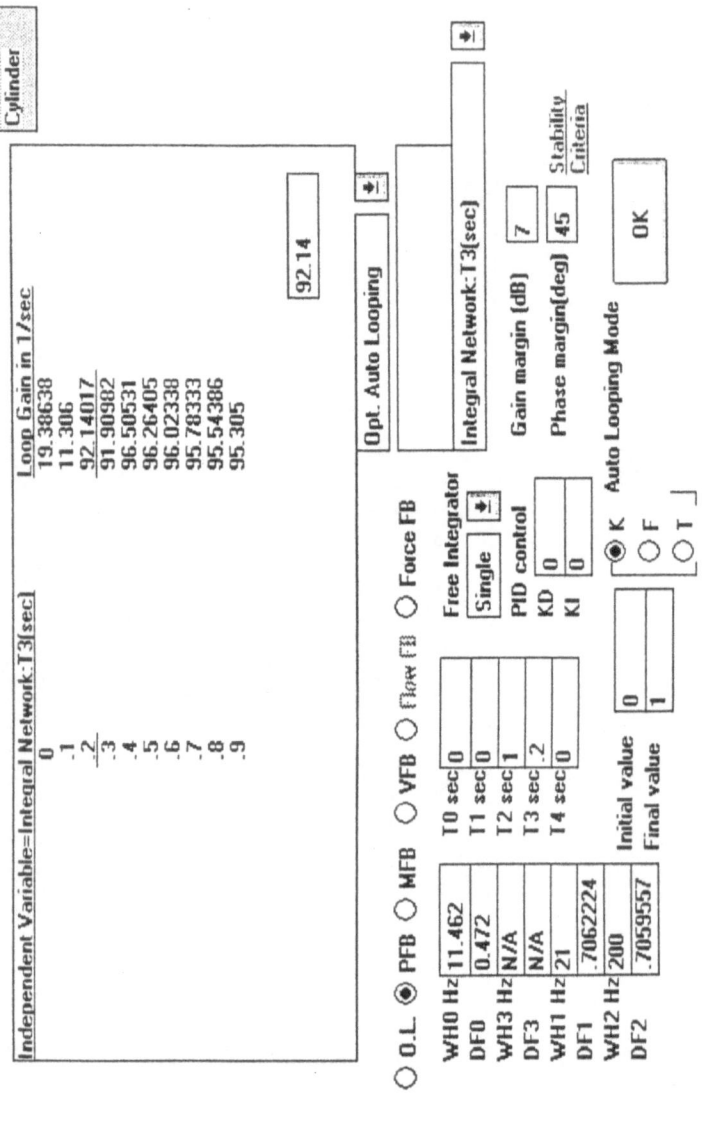

Screen S9 Frequency Domain Module

WE1.DAT

GRAPHICS DISPLAY

Const. Power Pump[Click]
Shunt Leakage Pressure Control
Bode Diagram[Click]
Nichols Chart
Nyquist Diagram
Opt.Auto Looping[K]
Opt.Auto Looping[F]
Opt.Auto Looping[T]
Non-Opt.Auto Looping[F]
Non-Opt.Auto Looping[T]
Transient Response
Displacement Profile
Ramp-Step Response Parameters
Triang. Response Parameters
Large Step Response[Click]
Velocity/Displacement Profile[Click]
Velocity Derivative
Accelerating Loads
Stability Margins

Reset Values   OK

Graph   □ Variable Thickness

Screen S10 Graphics Module

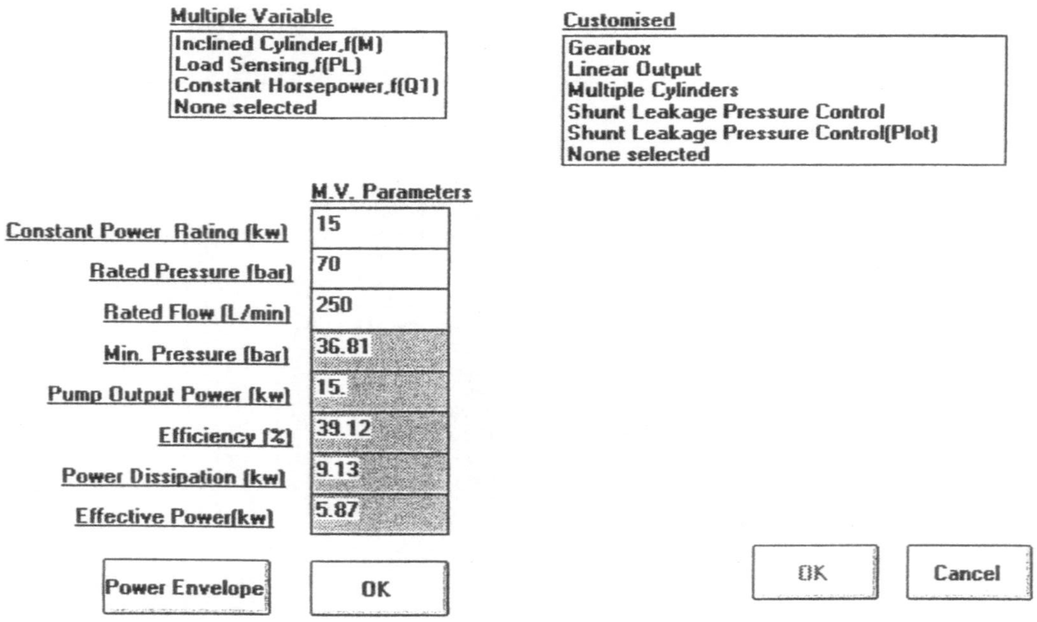

Multiple Variable

Inclined Cylinder.f(M)
Load Sensing.f(PL)
Constant Horsepower.f(Q1)
None selected

Customised

Gearbox
Linear Output
Multiple Cylinders
Shunt Leakage Pressure Control
Shunt Leakage Pressure Control(Plot)
None selected

M.V. Parameters

Constant Power Rating [kw]	15
Rated Pressure [bar]	70
Rated Flow [L/min]	250
Min. Pressure [bar]	36.81
Pump Output Power [kw]	15
Efficiency [%]	39.12
Power Dissipation [kw]	9.13
Effective Power[kw]	5.87

Power Envelope

OK

OK

Cancel

Screen S11 Options Form

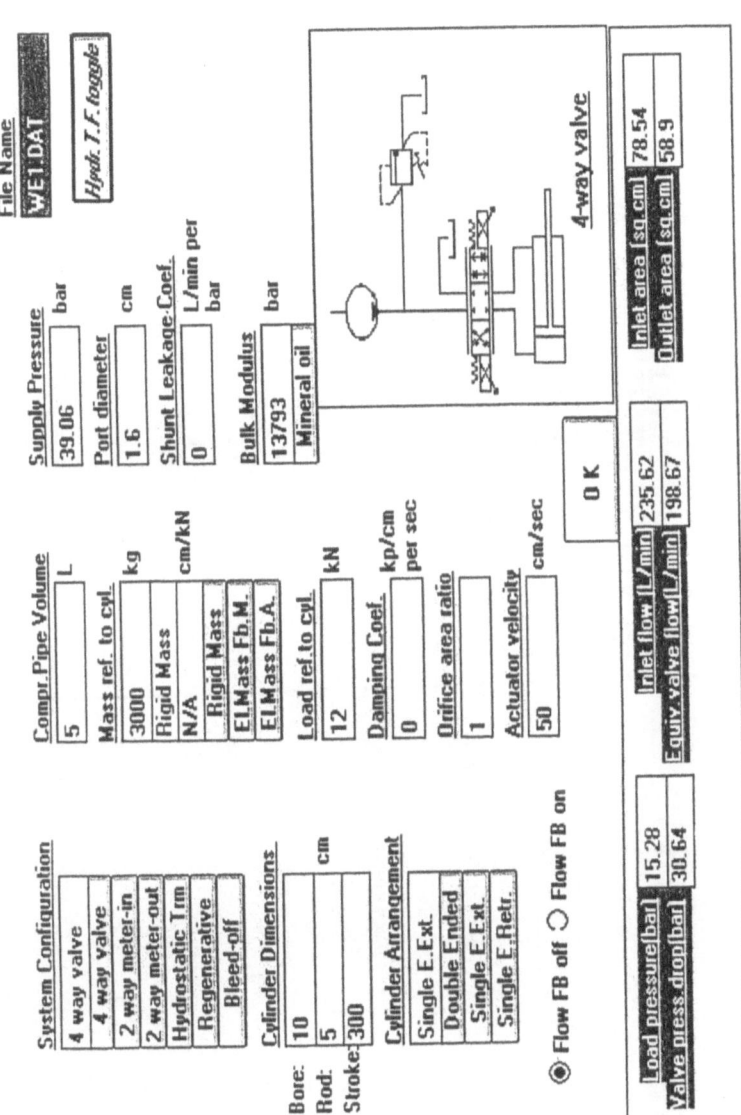

Screen S12 Cylinder Module

File Name

WE2P.DAT

Control Valve Override [bar per L/min]

.125

Effective cylinder area [cm^2]

Bulk Modulus [bar]

13793

Mineral oil

Trapped Volume [L]

20

Shunt Leakage Coef. [L/min per bar]

.2

Time constant T0 [sec]

0.0106

Liquid Spring Response Time [sec]

0.0244

OK

Pressure Control

Screen S13 Pressure Control

Supply Pressure [bar]	70
Valve Pressure setting [bar]	15
Pump Flow rating [L/min]	300
Actual Flow rate [L/min]	235.62
Load Pressure drop [bar]	15.28

FIXED DISPLACEMENT PUMP
Efficiency=17.144%   Power Dissipation=28.362kW

PRESSURE MATCH
Efficiency=39.633%   Power Dissipation=8.938kW

PRESSURE COMP. PUMP
Efficiency=21.829%   Power Dissipation=21.016kW

POWER MATCH
Efficiency=50.462%   Power Dissipation=5.761kW

BLEED-OFF
Efficiency=78.54%   Power Dissipation=1.603kW

Effective Output Power = 5.868 kW

OK   Cancel

GRAPHICS

Efficiency_f(P)   Bar Chart, Efficiency

Efficiency_f(Q)   Bar Chart, Dissipation

Dissipation_f(P)   Plot all

Dissipation_f(Q)

Screen S14 Power Efficiency and Dissipation

33

How to Get Started (Browsing)

An easy way to familiarise yourself with the programs is to to browse through the forms and modules as outlined in the following procedure.

1. Insert set-up disk in drive 'a', go to RUN and enter 'a:setup' to initiate installation procedure.For the full working program you will be asked to insert two additional disks. Follow the instructions ; a 'Readme' screen will be displayed and subsequently the program group 'HYDRO ANALYST' will be set up.

2. For all programs, other than demonstration programs, insert the dongle in the parallel printer port and install the dongle driver by (double-)clicking the appropriate icon. This is a 'one-off' procedure for the chosen computer. The system is now operational.

3.(Double-)click the 'Flow Control' icon

4. Select 'Module' from menu to display 'Hydraulic System' then click on 'Cylinder' to display blank Cylinder form..

5. Select any 'System Configuration' and observe corresponding hydraulic circuit diagram and input prompts.Note that 'OK' button is disabled (greyed out), and will only be enabled once all input textboxes have been filled in.

6. Select, in turn, system configurations 2 way meter-in, 2 way meter-out, Regenerative and Bleed-off, and observe effect of changing 'Cylinder Arrangement' from 'Single ended Extension' to 'Single ended Retraction'. Note automatic return to single ended extension when Regenerative system is selected.

7. From menu bar select 'Module', 'Hydraulic System', 'Motor' and repeat procedure outlined in para. 5.

8. From menu select 'Components', then choose 'Component Supplier' to display valve model codes included in component data base.
Double-click any model code to display steady-state and dynamic valve characteristics.

9 Click arrow on bottom right hand Options box to display the four available options.

> Steady-State Errors
> Automatic Component Selection
> Update Mode
> Amplifier Gain

10.From menu bar select 'Module', 'Frequency Domain'.
Enter alternative feedback mode options and note changes to system block diagram.

11. From menu select 'Module', 'Time Domain'.
Click arrow on 'MODE' box to display 8 alternative operating modes.

12. Click arrow on 'PLOTS' box to display 6 types of plot.

13. From menu choose 'Module', 'Graphics Display'.
Click on 'Graph' button for Listing of available graph types.

14. From menu bar choose 'Options'.
5 customised and 3 Multiple Variable options will be listed.

15. Now return to cylinder form by selecting 'Module', Hydraulic System', 'Cylinder'.

16. Repeat steps 5 to 14 in conjunction with the 'Help' screen.
Individual items can be specified by making use of the 'Hot Key' F1.

34

Guided Tour (Tutorial Mode)

In the tutorial mode the sequence of events is preset and in addition explanatory messages are displayed.

1. Return to 'Module', 'Hydraulic System', 'Cylinder'.
When the Cylinder form is displayed, click on 'Tutorial', 'On'

2. Select 'Examples' from the menu bar and choose one of the listed examples, e.g. 'Cylinder System', then click 'OK' button to display hydraulic input parameters and hydraulic circuit diagram.

3. Click 'OK' button for listing of dependent hydraulic parameters at the bottom of the screen. Transfer function of hydraulic transmission can be viewed by clicking 'Hydr. T.F. toggle' in top right-hand corner.

4. Click on 'Next' buttton when message will be displayed.
Click on 'OK' button to display Component Data form.

5 Steady-state and dynamic characteristics for the selected valve are listed. Click on 'OK' button to confirm selection. To illustrate Automatic Component Selection, choose 'Component Supplier', 'Vickers' from the menu bar. The model codes contained in the Vickers data base will be listed. Click on bottom right-hand 'Options' arrow and select 'Automatic Component Selection' to highlight suitable model codes. To select valve, double-click model, e.g.KFDG5V-7 and 'OK' selection.

6. Click on 'Next' button when a message will be displayed.
Click on 'OK' button to display Frequency Domain form. It can be seen that the feedback mode has not yet been entered.To confirm this, click on 'Summary', which can be called on at any time.'OK' summary and select 'PFB' (positional feedback) Option button.The loop gain K shown is the optimised and stored value.

7. Click on 'Next' button to display message box.
Click on 'OK' button for Component Data form, then click on arrow of bottom right-hand 'Options' box and select 'Steady-State Errors'.
Flow Gain, Pressure Gain and Hysteresis for the given operating conditions and the corresponding Steady-State Errors, i.e. Load, Hysteresis and Velocity errors will be listed.For most positional feedback systems, the predominant error is the Velocity Error.

8. Click on 'Next' button to display message box and then click on 'OK' button for 'Graphics Display'. Click on 'Graph' button for listing of available graphs. The plots will show the dynamic system performance in both the frequency and time domains.

9. Select 'Bode Diagram' and click 'OK' button, or alternatively double-click 'Bode Diagram'. A message box will appear, enter 'YES' to display Frequency domain form.

10. The system block diagram and all relevant parameters are displayed. For a positional control system, a single free integrator has been entered. Initial and final plot values have been set to 0.1 and 100 hz. Any of the listed values can be overridden if required.

11. Click on 'OK' button to generate and display 'Closed Loop Bode Plot'.Click on graph to toggle between Amplitude Ratio and Phase Angle.
To illustrate how to change the plotted frequency range, click on 'Graph' button, then 'Reset Values' and 'YES' to display Frequency form. Change 'Final value' from 100 to 10, and click on 'OK' button to display new plot.
The amplitude plot shows that the system reaches a maximum amplitude ratio of about 1 dB at 4.5 hz, and tha t the frequency band width at -4 dB is 7.4 hz
The phase angle plot shows that system natural frequency, i.e.at 90 degrees phase lag, occurs at 4 hz.

12. Click on 'Graph' button to show listing of graphs and select 'Nichols Chart' and click 'OK' button for message box, then click on 'YES' to display Frequency domain form. Note that suitable initial and final frequency values have been automatically entered. To generate and plot graph click on 'OK' button'.
The loop gain for this example had previously been optimised to satisfy the stability criteria of gain margin > 7 dB and phase margin > 45 degrees. It can be seen that in this case the gain margin is the critical criterion (7 dB), whereas the phase margin is 55 degrees.

13. To plot a Nyquist diagram, repeat the procedure outlined in para. 12 and select 'Nyquist Diagram' from the list.

14. To establish system transient response parameters in the time domain, select 'Module', 'Time Domain'.from the menu bar. System frequency identification is shown in the top left-hand corner. Click right-hand arrow on the 'Mode' box' and select 'T.R. Parameters', then click 'OK' button to show listing of transient response parameters.

15. To examine system dynamic performance in the time domain in more detail, return to the 'Graphics Display' screen and select 'Transient Response' from the graph listing. Clicking the right-hand arrow of the 'MODE' box lists the following input mode options:

        T.R. Parameters
        Continuous Function Generator
        Ramp-step response parameters
        Triangular response parameters
        Step demand
        Generated Duty Cycle
        Large Step response

Note that the 'PLOTS' option has already been set for a closed loop positional feedback system to 'Input/Output/Error'.

16. Select 'Step demand' as the input mode. A suitable operating range for the given example of 0.5 sec has been entered. The entered values can, of course, be overridden, if required.
To generate and plot the step response curve, click the 'OK' button.
The plot confirms the transient response parameters obtained in para. 14.

35

Worked Examples

15 worked examples, 12 applicable to Flow Control and 3 to Pressure Control, have been chosen to represent a cross-section of typical electro-hydraulic control system applications.

The Flow Control examples are:
> we1.dat: Valve controlled Cylinder System
> we2.dat: Valve controlled Motor System
> we3.dat: Valve controlled Regenerative Cylinder System
> we4.dat: Hydrostatic Transmission with Cylinder
> we5.dat: Valve controlled Bleed-off Cylinder System
> we6.dat: Valve controlled Cylinder System with elastically mounted mass
> we7.dat: Valve controlled Cylinder System with PID Control *
> we8.dat: Valve controlled Motor System with PID Control *
> we9.dat: Valve controlled Cylinder System with Integral Network *
> we10.dat:Valve controlled Motor System with flow feedback & gearing *
> we11.dat: Variable Motor System
> we12.dat: Valve controlled Multiple Cylinder System

The Pressure Control examples are:
> we1p.dat: Pressure & Force Control with low trapped volume
> we2p.dat: Pressure & Force Control with medium trapped volume
> we3p.dat: Pressure & Force Control with high trapped volume

Note that examples marked with an asteric are closed loop systems, all others can be either closed or open loop.

A brief outline of each example, emphasising any special features, will now be given.
we1.dat: Valve controlled Cylinder System

This is a general purpose control system utilising a 4-way meter-in/meter-out proportional flow control valve. This hydraulic circuit configuration enables both opposing and assisting loads to be controlled.
The following feedback options are applicable:
> Open loop
> Positional feedback
> Velocity feedback
> Force feedback
> Multiple feedback

It should be noted that for Force control an elastic member has to be interposed between the cylinder and the output. The stiffness of this member is entered when the steady-state option of the 'Component' form has been selected.

The Loop gain displayed on the 'Frequency' form is the optimised value, compatible with the set stability criteria of Gain margin > 7 dB and phase margin > 45 degrees. By choosing 'MODE', 'Stability Margins', 'OK', it can be seen that the gain margin is the critical criterion.

we2.dat: Valve controlled Motor System

This is a general purpose motor control system utilising a 2-way meter-in proportional flow control valve.This hydraulic circuit configuration cannot control negative, i.e. overrunning loads without the provision of a counter-balance valve.
The dynamic performance characteristics of a meter-in circuit are inferior to those of an equivalent meter-in/meter-out circuit. This is due to the increased oil compliance of the meter-in circuit by a factor of 4:1, resulting in a hydraulic transmission natural frequency reduction of 2:1. This can be illustrated by observing the effect on the natural frequency WH0 when changing the 'System Configuration' to '4 way valve', 'OK'. (with the 'Hydr. T.F. toggle' enabled)
Power dissipation of a meter-in valve is less than that of a meter-in/meter-out valve.

we3.dat: Valve controlled Regenerative Cylinder System

This configuration is only applicable to an extending asymmetrical cylinder.operated by a 3-way flow control valve. Normally a differential cylinder with a 2:1 area ratio would be employed.
The flow from the annular cylinder chamber is fed back to the valve inlet, thus reducing the pump flow rating.
Both opposing and assisting loads can be accommodated.
On retraction, the system should be treated as a '2 way meter-out' configuration.

we4.dat: Hydrostatic Transmission with Cylinder

This is a very power-efficient system configuration since no flow throttling is involved. Infinitely variable flow control is achieved by varying the swash angle of a variable displacement pump. For positional or reversible directional control of the actuator, an over-centre swash angle mechanism is required.
Note that the 'Supply Pressure' textbox has been replaced by a 'Pump leakage coef.' textbox.
Since the valve characteristics listed in the Component form are not applicable, the dynamic characteristics of the swash angle control system should be entered as a second order transfer function in the Frequency Domain, identified by the

natural frequency WH1 and the damping factor DF1.

we5.dat: Valve controlled Bleed-off Cylinder System

This system configuration is particularly suitable for applications with a limited flow range operating over a wide load range, e.g. a lift control system.
Note that the 'Supply Pressure' textbox has been replaced by a 'Pump Flow' textbox and 'Valve flow' has been added to the list of dependent hydraulic variables. Since there is no restriction between the pump outlet and actuator, the pump operating pressure is directly proportional to the applied load, giving high power efficiency. A bleed-off system can only deal with opposing loads.

In a typical lift control system, the power-up mode would comprise a vertically mounted extending or retracting cylinder subjected to an opposing load. To descend, the system configuration would change from 'Bleed-off' with an opposing load to '2-way meter-out' with an assisting, i.e. negative , load with the 'Supply Pressure' set to zero. In the gravity-down and stationary modes, the pump would normally be off-loaded.

we6.dat: Valve controlled Cylinder System with elastic mass

Athough in the majority of electro-hydraulic control systems the mass can be regarded as rigidly connected to the actuator, there are some applications where elasticity of the mass mounting has to be taken into account.
Typical examples are flying controls where the elasticity of the control surfaces, i.e. aelerons, tailplane and rudder, can be regarded as elastically mounted mass systems. A typical industrial application would be a machine tool control incorporating long leadscrews.
The program allows for two alternative feedback options; positional feedback from the actuator or positional feedback from the elastically mounted mass.
The effect on the hydraulic transmission transfer function can be illustrated by enabling the 'Hydr. T.F. toggle' and switching from 'El.Mass Fb.M.' to 'El.Mass Fb.A.' Although the hydraulic natural frequency WH0 remains at 4.608 hz, the hydraulic damping factor DF0 increases from 0.472 to 0.817.
In the worked example, the natural frequencies of the mechanical compliance WHEMM and the liquid spring mass compliance WHOIL are 5.033 hz and 11.462 hz respectively.

we7.dat: Valve controlled Cylinder System with PID Control

PID Control, frequently referred to as a Three-term Controller, can be used in some applications to enhance dynamic system performance.
In the worked example, referring to the 'Frequency Domain' module, a derivative term KD of 0.5 and an integral term KI of 0.5 have been entered in conjunction

with a Loop Gain K of 10 1/sec.

Comparing the dynamic performance characteristics of the worked example
we1.dat with the enhanced system we7.dat, the following conclusions can be
drawn:

	-we1.dat	we7.dat
1.Hydraulic nat. frequ. WH0:	9.358 hz	9.358 hz
2.System nat. frequ. WR:	4.634 hz	8.574 hz
3.System band width WC:	7.383 hz	10.15 hz
4.Ratio WR/WH0:	0.495	0.91
5.Ratio WC/WH0:	0.79	1.08
6.Step response time:	0.068 sec	0.054 sec
7 Overshoot:	21.54 %	2.338 %

The improved dynamic performance of the enhanced system is self-evident.

we8.dat: Valve controlled Motor System with PID Control

A comparative study of the motor driven systems, one without and one with PID
control, reveals an improved dynamic performance in line with the results obtained for the cylinder driven systems. The results are summarised below:

	-we2.dat	we8.dat
1.Hydraulic nat. frequ. WH0:	8.961 hz	8.961 hz
2.System nat. frequ. WR:	3.938 hz	6.038 hz
3.System band width WC:	6.34 hz	8.534 hz
4.Ratio WR/WH0:	0.439	0.674
5.Ratio WC/WH0:	0.707	0.952
6.Step response time:	0.078 sec	0.06 sec
7 Overshoot:	29.29 %	11.566 %

we9.dat: Valve controlled Cylinder System with Integral Network

Integral type passive networks are used for the purpose of enhancing steady-state
performance. In the worked example, values of 1 sec and 0.2 sec have been entered as the respective time constants T2 and T3 in the Frequency Domain module. Consequently the maximum permissible Loop Gain K has been increased by
a ratio of 4.6:1 to 83 1/sec.
Since steady-state errors, i.e. Load Error, Hysteresis Error and Velocity Error,
are all inverseley proportional to the Loop Gain, these errors will be reduced by
a factor of 4.6:1. This can be seen by selecting the 'Steady-state Errors' Option of
the 'Components' form

Selecting the T.R. Parameters' Mode in the 'Time Domain' Module shows that dynamic system performance is virtually unaffected.

we10.dat:Valve controlled Motor with flow feedback & gearing

Incorporating a flow sensor in the feedback loop provides a flow control system which is unaffected by load, pressure and temperature variations.
Flow feedback introduces an addditional second order term to the hydraulic transfer function.The additional natural frequency WH3 and corresponding damping factor DF3 are 8.961 hz and 0.01 respectively
This is illustrated by enabling the 'Hydr. T.F. toggle' on the 'Motor' module.
Selecting 'Options' from the menu bar and then clicking on 'Gearbox' lists the Output parameters.Clicking the 'OK' button reverts to the 'Motor' module.

we11.dat: Variable Motor System

This system configuration is similar to the Hydrostatic Transmission, we4.dat in that no flow throttling is involved, producing high power efficiency.
In this case, however, a variable displacement motor is used driven by a fixed displacement pump.Varying the swash angle of the motor will vary its output torque which, in turn, will control the output velocity.
The hydraulic transmission is identified by a first order transfer function.
For this example the applicable time constant T0 = 0.0004 sec as listed in the Frequency Domain module.
As for the Hydrostatic Transmission, the dynamic characteristics of the swash control system should be entered as a second order transfer function in th Frequency Domain, identified by the natural frequency WH1 and the damping factor DF1.

we12.dat: Valve controlled Multiple Cylinder System

This is an example of several cylinders connected in parallel and controlled by a single control valve.
Output parameters can be displayed by selecting 'Options' from the menu and then clicking 'Multiple Cylinders'. For this application, 11 cylinders of 25 cm bore, 16.5 cm rod and 280 cm stroke are connected in parallel.
Clicking the 'OK' button, reverts to the Cylinder module, which lists an equivalent single cylinder of 82.916 cm bore, 54.724 cm rod and 280 cm stroke.
The total flow ratings are listed among the dependent hydraulic variables.
Performance characteristics of the control valve are displayed in the Component form.

we1p.dat: Pressure & Force Control
we2p.dat: dto.
we3p.dat: dto

These three application examples are outlined in chapter 32, para.12
'Hydraulic System (Pressure Control).

36

Loop Gain Selection

We shall select the Valve controlled Cylinder System, worked example we1.dat, to illustrate the three alternative methods of Loop Gain selection, i.e

Manual
Semi-automatic
Automatic

The chosen feedback option will be Positional feedback.

Procedure:
1 .Select 'Cylinder System' from the Worked Examples listing
2. Cylinder screen with File Name: WE1.DAT will be displayed.
3. Click on 'OK' button to list dependent hydraulic variables.
4. From menu bar select: 'Module', 'Frequency Domain'
5. Click on feedback option 'PFB' to select positional feedback.
   The previously optimised Loop Gain K of 18.166 1/sec is displayed.
6. Delete the listed loop gain; this will disable the 'OK' button

36.1 Manual

The objective is to select a loop gain compatible with the stability criteria applicable to an adequately damped system, e.g. Gain Margin > 7 dB and Phase Margin > 45 degrees. This is obviously going to be a laborious, trial and error based procedure:
1. Enter an arbitrary value in the 'Loop Gain K' textbox, say, 100.
2. Click on right-hand arrow 'MODE' and select 'Stability Margins'.
3. Click on 'OK' button to display:
   Stability Margin = -7.71dB
   Phase Margin = -65.11 deg
the negative signs indicating a negatively damped unstable system.
   4. Reduce Loop Gain to , say, 50, and click 'OK' button, the negative signs still indicating a negatively damped unstable system.
   5. Reduce Loop Gain to 25 and click 'OK' button. System now stable but inadequately damped.
   6. Repeat above procedure, adjusting the Loop Gain value until the condition Gain Margin > 7 dB and Phase Margin > 45 degrees has been satisfied.
The Loop Gain wil then be in the region of 18 1/sec.

36.2   Semi-automatic

In this mode, we shall generate a graph, plotting the gain margin and phase margin as a function of the loop gain to enable us to choose an appropriate value for the loop gain.

Procedure:

1. From the menu bar select: 'Module', 'Graphics Display', then click the 'Graph' button to display list of graph options.

2. Select 'Stability Margins' and click 'OK' button: message box will be displayed: Click on 'YES'

3. The Frequency Domain form will appear with all relevant system parameters listed.The independent variable for this plot is the 'Loop Gain K(1/sec)' with the initial and final value set at 10 and 100 1/sec. Any of the listed values can be changed if so required.

4. To plot the graph, click the 'OK' button.

The graph will now be generated and displayed. The plot shows that the Gain Margin is the critical criterion at a loop gain of 18 1/sec.

36.3   Automatic

This is the preferred procedure:

1. In the Frequency Domain, click right hand 'MODE' arrow and select 'Opt.Loop Gain'. Stability Criteria, i.e. Gain margin = 7dB and Phase margin = 45 degrees will be displayed.

2. Click 'OK' button to initiate loop gain optimisation. Since optimisation is arrived at by means of an iterative algorithm, accuracy limiting tolerances have to be imposed, in this case +/- 0.2 dB for gain margin and +/- 2 degrees for phase margin. The optimised Loop Gain, actual gain and phase margins, and system dynamic performance characteristics are now displayed. For this system configuration, the gain margin is the critical criterion.

3. If required, the above procedure can be repeated at different Stability Criteria settings.

37

Applying a Duty Cycle

We shall again select the Valve controlled Cylinder System we1.dat, to illustrate the application of different types of input functions as outlined in chapter 32.5, 'Time Domain'.
Both open and closed loop systems will be investigated.

Procedure:

1. Repeat steps 1 to 5 detailed in chapter 36 'Loop Gain selection procedure'.We have now entered a positional feedback system.

2. From the menu bar, select 'Module', 'Graphics Display', then click on 'Graph' button. List of available graph options appears.

3. Select 'Transient Response' and click on 'OK' button;when message box appears, click on 'YES' to display Time Domain module.. Note that the system identification has been entered in the top left hand corner as a second order transfer function.

4.Click on 'MODE' right hand arrow to reveal Mode Options, then choose 'Cont.Funct.Generator'. The input function default parameters of 1 cm amplitude, 10 hz frequency and zero offset are listed as are the plot initial and final time values compatible with the default parameters.

5. Click 'CONT.FUNCTION GENERATOR' arrow and select required wave form, say, 'Sinusoidal'. Note that an 'Input/Output/Error' Plot has been entered.

6.Click 'OK' button to generate and plot graph. Since the input frequency had been set at approximately twice the system natural frequency, the output wave form is greatly attenuated.

7 To achieve better response characteristics, we shall now change the input frequency to, say, 5hz.

8. Click on 'Graph' and 'Reset Values' for message box, then 'YES' to revert to Time Domain.Re-select 'CONT.FUNCT.GENERATOR' from MODE box.

9. Alter the input frequency to 5 hz; note automatic change of Final value to 0.6 sec. Click 'OK' button to plot new graph. The change in input frequency has a dramatic effect on the output amplitude, which now actually overshoots the input amplitude by a small amount.

10. The above procedure can now be repeated for other wave forms.

11. An alternative method to redraw a graph is to select 'Clear graphs' from the 'File' menu and click the 'OK' button instead of the 'Reset Values' command.

12. We shall now generate a customised duty cycle:
Repeat steps 2 and 3. Click on 'MODE' right hand arrow to reveal Mode Options, then choose 'Generated Duty Cycle'. The input data screen catering for 8
transitions now appears.

13. Enter the arbitrary eight transition duty cycle shown in Screen S6.
For convenience use the tab key to fill in the table. Note that the 'Final value' automatically updates as you enter the time increments.

14. Click 'OK' button to generate and plot graph. It can be sen that, for the
given values, the output follows the input fairly closely.

15 To illustrate the effect of system dynamics on the transient response,
we shall now reduce the system natural frequency to , say, 2 hz and replot the
graph. The new plot shows that the transient output characteristics have changed
considerably.

16. To investigate the transient response of an open loop system we shall
now proceed as follows:
Repeat steps 1 to 4 detailed in chapter 36, 'Loop Gain selection procedure'.

17. Click on feedback option 'O.L.' to enter an open loop system. Note the
change in the block diagram.

18. Click the 'MODE' right hand arrow and select 'F.R. Parameters' and
then 'OK' to display the dynamic system characteristics.

19. To generate a continuous wave form, proceed as outlined in steps 2 to
4. Note that the input amplitude is now a velocity set at a default value of 1
cm/sec. For a more realistic input velocity change the value to , say, 20 cm/sec.

20. Click 'CONT.FUNCTION GENERATOR' arrow and select required
wave form. Note that for an open loop system, a 'Velocity profiles' Plot has been
automatically entered.

21. Click 'OK' button to generate and plot velocity profile graph.

22. To generate alternative plots, i.e.
      Displacement Profile
      Velocity/Displacement Profile
      Accelerating Loads
select required plot from Graph List and and follow previously outlined procedure. Note that 'Velocity Derivative' is not applicable to an open loop system.

23. The above procedure, steps 20 to 22, can now be repeated for other
wave forms.

38

Automatic Looping

The objective of Automatic Looping is to establish performance trends or, more specifically, to investigate the sensitivity of the system to input parameter variations. Automatic Looping can be performed under either optimised or non-optimised conditions. Under optimised conditions, the loop gain is adjusted to provide optimum system dynamic performance over the entire operating range. Under non-optimised conditions, the loop gain remains at a constant value. In the Flow Control program version, 24 independent variables are listed, while in the Pressure Control version, 10 independent variables are available.

38.1 Optimised Auto Looping

To demonstrate the procedure, we shall again use the worked example we1.dat: 'Cylinder System' from the 'Examples' menu.

1. Enter example 'Cylinder System'. The Cylinder module will be displayed
2. Select 'Frequency Domain' from menu, then click on 'PFB' feedback option.
3. Click 'MODE' arrow and choose 'Opt.Auto Looping'. 'Variable' box will appear.
4. Click 'VARIABLE' arrow and select 'Mass ref. to cylinder:M(kg)'
5. Enter 'Initial value': 0 and 'Final value': 10000, then click on 'K' option. A blank screen will appear.
6. Click on 'OK' button to initiate computation.
7. To plot graph,enter 'Module', Graphics Display' from menu. Stability Boundary plot will be displayed.
8. To plot corresponding Frequency Response parameters, click on 'Graph' button, select 'Opt.Auto Looping(F)' and 'OK'. When Frequency module appears, choose 'F' option and 'OK'
9. To plot Transient Response parameters,repeat step 8, but this time selecting 'Opt.Auto Looping(T)' and 'T' option.
10. Since the mass is now set at the final value of 10000 kg, you can revert to the original operating conditions by returning to the Cylinder module and clicking the 'OK' button.
11. Above procedure can be repeated for any of the other listed independent variables, eg to investigate optimum time constants for an integral passive network, select 'Opt.Auto Looping(K)' from graph listing and 'Integral Network:T3(sec)' from 'VARIABLE' option. Note that default Initial and Final values are automatically entered.

12.Click on 'K' button and then 'OK' for computation of Stability Boundary and step 7 for display of graph. Note that the preferred value of time constant T3 is underlined.

13.Carry out steps 8 and 9 to obtain Frequency Response and Transient Response parameters.

38.2   Non-Optimised Auto Looping

Under non-optimised auto looping, the loop gain remains constant, hence the 'K' option and Stability Boundary are not applicable. Frequency and Transient Response parameter plots can be obtained using the procedure outlined above. Optimised and non-optimised Frequency and Transient Response parameters have been allocated separate file names and are therefore stored as four accessible graphs while the program is running.

39

System Enhancement

Several methods can be applied to enhance both steady-state and dynamic performance characteristics of electro-hydraulic control systems.

Five alternative methods are described in this section. Input shaping is the only method applicable to both open and closed loop systems, all other methods are only applicable to closed loop systems.

39.1 Passive Network (Integral or Phase Advance)

Phase advance networks are used to enhance dynamic system performance whereas integral networks enhance steady-state performance with only marginal effect on dynamic response.

Worked example we9.dat is a typical positional control system incorporating an integral network.Improvement of steady-state performance parameters of 5:1 are achievable for many system configurations.

39.2 Three-term Controller (PID Control)

PID (Proportional-Integral-Derivative) control has been widely used in process control applications, predominantly pneumatic, to enhance system performance. It can also be applied to some electro-hydraulic control systems to enhance dynamic performance, although not all system configurations will benefit from it.

The effect of PID control on two typical positional control systems is demonstrated in worked examples we7.dat and we8.dat for a cylinder and motor controlled system respectively.

The results show that applying PID control improves dynamic response by a factor of approximately 2:1, with a marked reduction of overshoot, i.e. from 21.5% to 2.3% in one case and from 29.3% to 11.5% in the other case.

39.3 Multiple Feedback

The most commonly applied multiple feedback arrangement in electro-hydraulic control systems is a minor velocity feedback loop in conjunction with a major positional feedback loop.

To illustrate the effect on system performance, we shall apply a minor velocity feedback loop to the valve controlled Cylinder System we1.dat.

Procedure:

    1. Enter the first example 'Cylinder System' and 'OK' the Cylinder module.

    2. Select 'Components' from the menu and 'OK' Valve selected.

3. Select 'Frequency Domain' from the Module menu to display system paramers and click 'MFB' button.A message box will appear,click OK' buttton.

4. The system block diagram now shows the additional feedback loop and all parameters listed refer to the inner, or minor, velocity feedback loop.Note that the free integrator has changed to 'None' and that the Loop Gain has been reduced to 1.1.Rather less stringent stability criteria have been set in establishing the Loop Gain, i.e. the Gain margin has been reduced to 5dB.
The dynamic parameters WH2 and DF2 refer to the velocity feedback transducer, or tacho.

5. Double-click the 'MFB' button to establish the major feedback loop.
A message box will appear. Enter WH2 and DF2 applicable to the position feedback transducer, then select 'Opt.Loop Gain' and 'OK'. The major positional feedback loop has now been established.

6. Select 'Module' and 'Time Domain' from the menu, then select 'T.R.Parameters' and 'OK' The Time Response parameters will be displayed in the top right hand corner.

7. We can now compare the performance parameters of the original single feedback loop and the new double feedback loop systems:

	Feedback loop	
	Single	**Double**
1.Hydraulic nat. frequ. WH0:	9.358 hz	9.358 hz
2.System nat. frequ. WR:	4.634 hz	7.586 hz
3.System band width WC:	7.383 hz	10.592 hz
4.Loop Gain:	18.17 1/sec	18.67 1/sec
5.Ratio WR/WH0:	0.495	0.81
6.Ratio WC/WH0:	0.79	1.13
7.Step response time:	0.068 sec	0.051 sec
8.Overshoot:	21.54 %	7.87 %

The comparison shows a marked improvement of dynamic performance.
Since the loop gain of the two systems is virtually identical, steady-state performance is unaffected.

39.4 Input Shaping

As an example of input shaping, we shall apply a ramp-step input to example we1.dat in order to reduce the step response overshoot.

Procedure:

1. From the 'Graphics Display' module select 'Ramp-Step Response Parameters' and 'OK'. After message box, Time Domain module will be displayed. Note that system transfer function and initial and final plot values have been entered.

2. Click 'OK' button to display 'Ramp-Step Response Parameters'. The plot shows that after a ramp time of approximately 0.2 sec, we reach the point of diminishing returns for the overshoot, which has been reduced from 21.5% for a step to around 5% for a ramp-step demand.

3. To plot the transient response to a ramp-step demand, choose 'Transient Response' from the graph listing. When the Time Domain module is displayed, select 'Generated Duty Cycle' from the 'MODE' options, and enter 0.2 sec as the first Time Increment and ,say, 0.3 sec for the second Time Increment, then change the second Polarised Amplitude to 0.

4. Click 'OK' button to display plot, which shows that the overshoot has been reduced to 5%.

An alternative method of reducing overshoot would be to introduce a negative impulse of half of the step demand after 0.5 sec.
Procedure:

1. From graph listing select 'Transient Response', then 'Reset Values, to display Time Domain. Choose 'Generated Duty Cycle, and enter 0 for first time increment and 0.3 for second time increment with 0 for second Polarised Amplitude..For negative impulse, enter 0.05 as 'Elapsed time(sec)' and -.5 as 'Polarised amplitude'. Click on 'Impulse on' button.

2. Click 'OK' button to display graph which shows that overshoot has been virtually eliminated

39.5 Adaptive Control

In an adaptive control system, the loop gain is automatically optimised by varying the amplifier gain accordingly.
A typical example of an adaptive control system was described in chapter 38 para 38.1, 'Automatic Looping, Optimised'.

.Relevant Screen is S8.
Relevant graphs are G8, G9 and G10.

40

Graphics

Graphs are generated in one of the following modules:
Frequency Domain
Time Domain
Options
Power Efficiencies and Dissipation
All graphs are subsequently plotted in the Graphics Display module.
All graphs listed in this module are allocated a specific file name and all plotted graphs are stacked. The graph selected from the list will be transferred to the top of the stack. All graphs will be retained as long as the program is running.
Graphs can be saved by selecting 'Copy' from the File menu. This command will copy the file to the Clipboard and Paintbrush, where the graph can be annotated and saved as a Bitmap(BMP) file.

40.1 Sizing of Graphs

Graphs can be sized by changing the format from full to partial screen and dragging the outer limits of the form to the required size.
This facilitates producing multiple graphs, as shown in graph G31, which combines graphs G4 and G5, and in graph 32, which combines graphs G6 and G7.

40.2 Editing Graphics

To edit graphs, select 'Module', Graphics Display', then 'Toolbar', 'on'.
The toolbar will now appear on any graph displayed.
Click the toolbar; the screen 'Graph Control' will appear. This screen contains 15 options. Click on any of the listed textboxes to display the corresponding screen. Note that the 'Overlay' screen option is applicable to graphs incorporating two vertical axes and, in the listed graphs, refers to the right-hand axis.

For more detailed information, refer to the 'Graphics Server' Operating Manual.

41

Description of Graphs

Most of the graphs listed are applicable to worked example we1.dat.

G1: Power Envelope, Constant Power Pump
G2: Power Efficiency & Dissipation, Constant Power Pump
G3: Shunt Leakage Pressure Control
G4: Closed Loop Bode plot, Amplitude Ratio
G5: Closed Loop Bode plot, Phase Angle
G6: Nichols Chart
G7: Nyquist Diagram
G8: Stability Boundary with variable mass
G9: Optimised Frequency Response parameters with variable mass
G10 Optimised Transient Response parameters with variable mass
G11: Stability Boundary with integral network
G12: Optimised Frequency Response parameters with integral network
G13: Optimised Transient Response parameters with integral network
G14: Non-optimised Frequency Response parameters with variable mass
G15: Non-optimised Transient Response parameters with variable mass
G16: Ramp-Step Response parameters
G17: Triangular Response parameters
G18: Large Step Response, velocity and displacement
G19: Large Step Response, Accelerating Loads
G20: Transient Response, Generated duty cycle
G21: Transient Response, Sinusoidal input
G22: Velocity Derivative, Sinusoidal input
G23: Velocity-Displacement, Sinusoidal input
G24: Accelerating Load, Sinusoidal input
G25: Velocity Profile, Sinusoidal input, Open Loop System
G26: Displacement Profile, Sinusoidal input, Open Loop System
G27: Ideal Velocity-Displacement, Sinusoidal input, Open Loop System
G28: Actual Velocity-Displacement, Sinusoidal input, Open Loop System
G29: Accelerating Load, Sinusoidal input, Open Loop System
G30: Stability Margins, Closed Loop Position Control System
G31: Multiple Graphs
G32: Multiple Graphs
G33: Transient Response, Pressure Control System

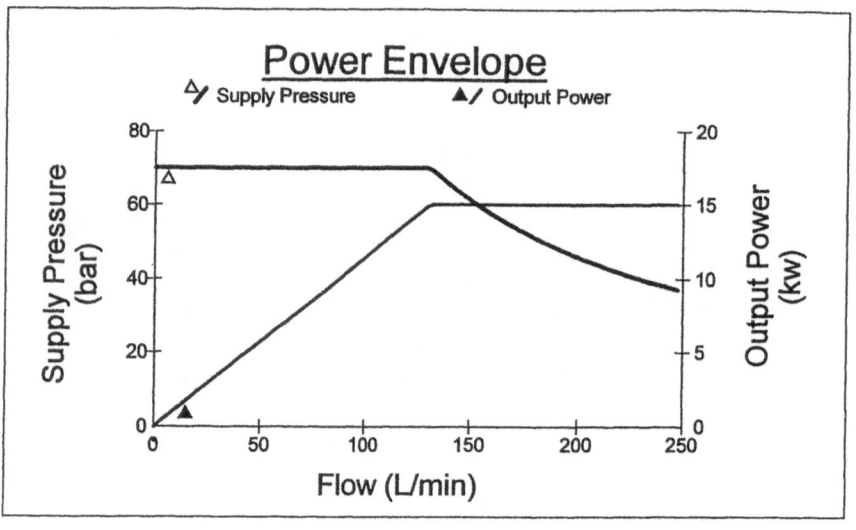

Graph G1 Power Envelope

This plot shows the power and pressure envelope of a variable displacement
pump with power limited to a constant value of 15 kw. This is achieved by reducing the supply pressure above the flow rating corresponding to the limiting
power output of 15 kw.
Screens S11 and S12 are applicable to this system.
The purpose of introducing constant power characteristics is to limit the torque
requirements of the prime mover, and hence to prevent stalling of the pump drive

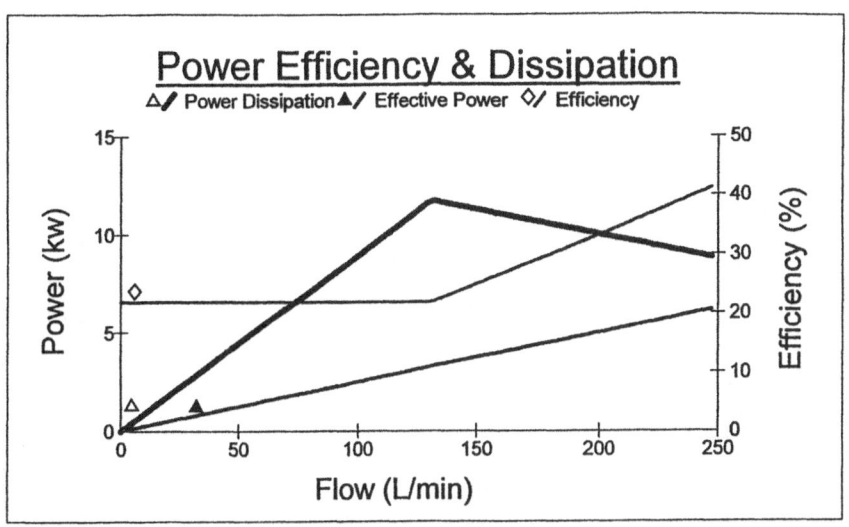

Graph G2 Power Efficiency & Dissipation

This plot shows power efficiency and dissipation of the power unit detailed
above It can be seen that for a given load pressure and flow, power dissipation
is directly proportional to flow and efficiency remains constant up to the cut-off
point. Beyond this point, power dissipation is reduced and efficiency increased.

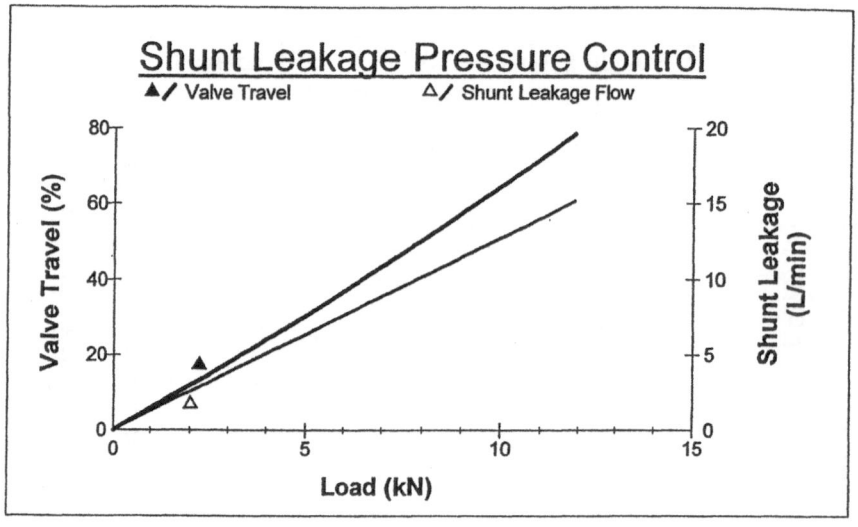

Graph G3 Shunt Leakage Pressure Control

In this mode a flow control valve meters the flow through a shunt to tank.In most cases a laminar flow shunt, i.e. providing linear pressure-flow characteristics, would be used.Such a shunt is viscosity sensitive and the fluid should preferably be maintained at a fairly constant temperature.

The plot shows flow and valve travel characteristics. The controlled load is directly proportional to the flow and,for the chosen valve, an almost linear relationship is established between load and valve travel.

The load pressure can be applied to either a cylinder or motor to control the load or torque of the actuator.

For added accuracy, a force or torque feedback transducer can be used to close the loop.

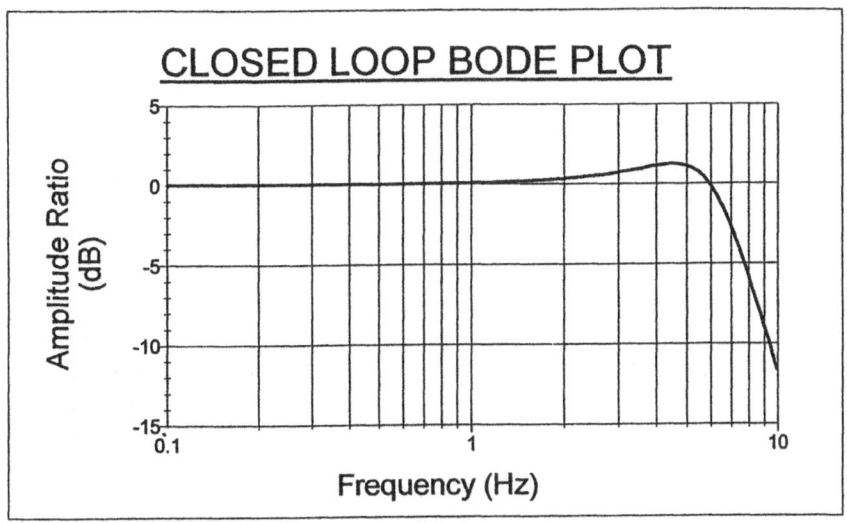

Graph G4 Closed Loop Bode Diagram

Frequency response gives a good indication of the stability and dynamic performance of closed loop systems. In this graph, amplitude ratio is plotted as a function of frequency. It can be seen that the system reaches its maximum amplitude of 1 dB at approximately 4.5 hz. The system band width at -4 dB is 7.4 hz.

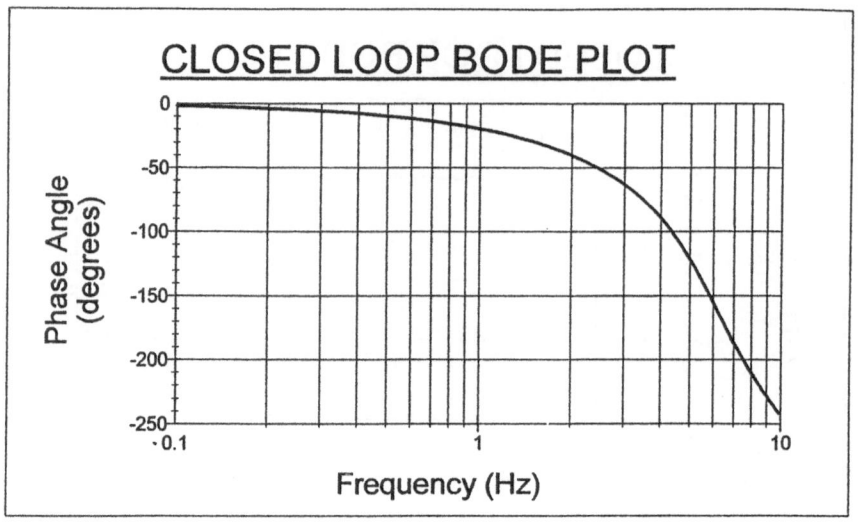

Graph G5 Closed Loop Bode Diagram

In this graph, phase angle is plotted as a function of frequency.
The frequency at 90 degrees phase lag is usually referred to as natural or resonant frequency, in this case around 4 hz. The Frequency Domain Module,Screen
S4 shows that the system transfer function of example we1.dat is represented by
a 7th order differential equation depicted by three 2nd order terms and a free integrator.

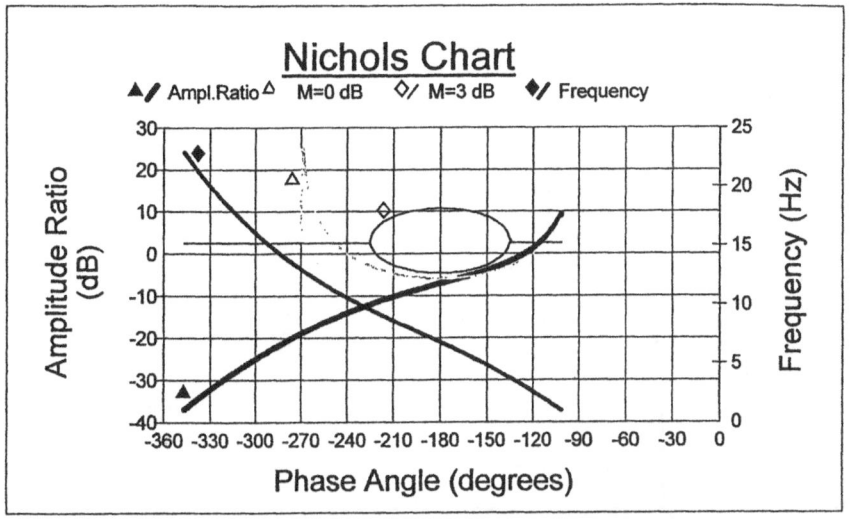

Graph G6 Nichols Chart

A Nichols Chart is an open loop frequency contours plot on a Log modulus versus phase angle chart, which is a convenient method of determining the stability of a closed loop system. The stability of a system can be defined by the gain margin and phase margin criteria. In optimising the loop gain, the gain and phase margin criteria were set at 7dB and 45 degrees, as shown on Screen S4.
By referring to graph G6 and screen S4 it can be seen that for the system analysed, the gain margin is the critical criterion.
The M contours included on the Nichols chart are the closed loop amplitude ratio plots at 0 and 3 dB.

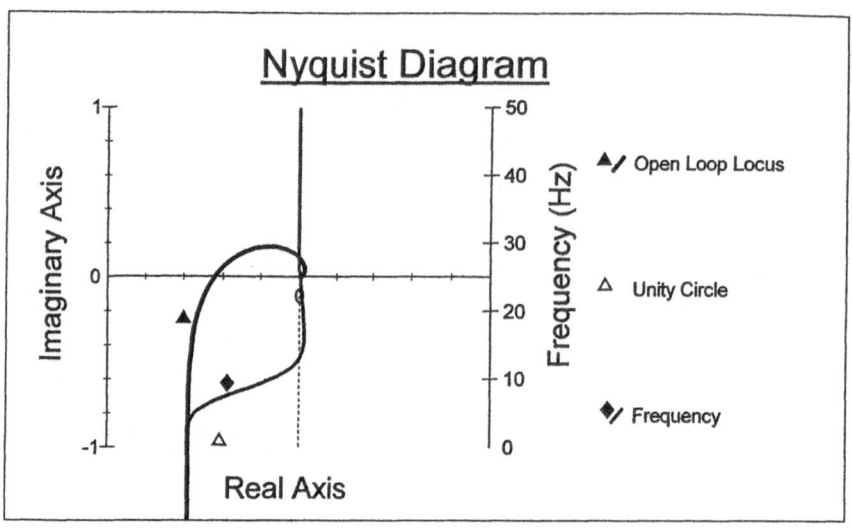

Graph G7 Nyquist Diagram

In this diagram, open loop frequency contours are plotted on a complex plane.
The limiting criterion for absolute stability is the -1 +0j point.
The gain margin is defined as the reciprocal of the modulus at -180 degrees
phase shift. The phase margin is defined as the phase difference between the
phase shift and -180 degrees when the modulus is unity.Alternatively the phase
margin can be defined as the angle subtended by the negative real axis and the
intersection of the open loop frequency contour with the unity circle.
In practice the Nichols Chart is a more convenient method and is therefore the
preferred option.

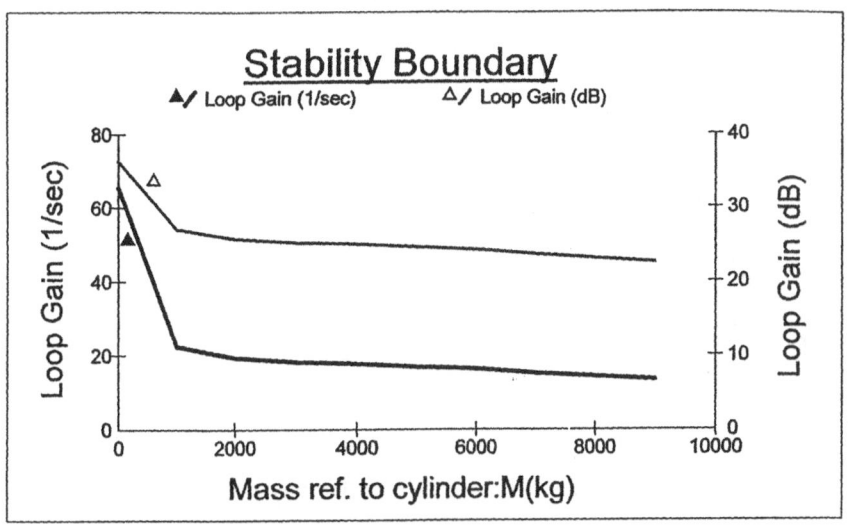

G8 Stability Boundary with variable mass

The steps to be taken to plot this graph are outlined in section 10.1.
Corresponding values of the two variables, in this case mass and loop gain, are displayed on the Frequency Domain Screen S8. The loop gain at zero mass is determined by the dynamic characteristics of the conrol valve.

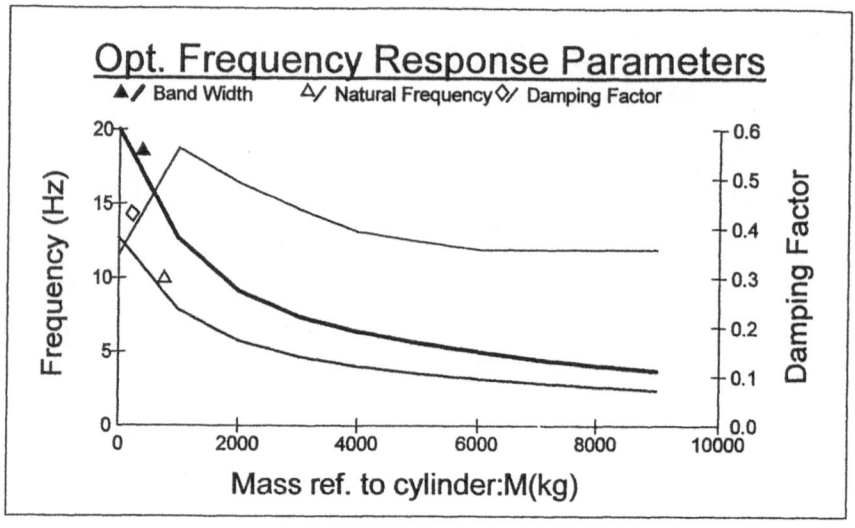

G9 Optimised Frequency Response parameters

This graph shows the effect of varying mass on the Natural Frequency, Damping Factor and Band Width of the system.

The plot shows that both the band width and natural frequency are considerably reduced with increasing mass, whereas the damping factor is only marginally affected, particularly at mass values greater than 4000 kg.

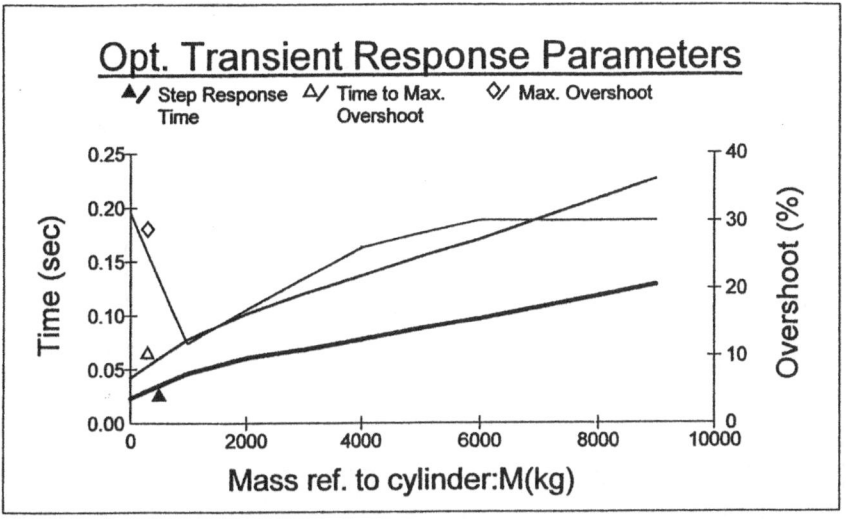

G10 Optimised Transient Response parameters

The results of the previous graph are reflected in the Time Response.
Step response times increase considerably over the total range; overshoot is only
marginally affected at mass values above 4000 kg.

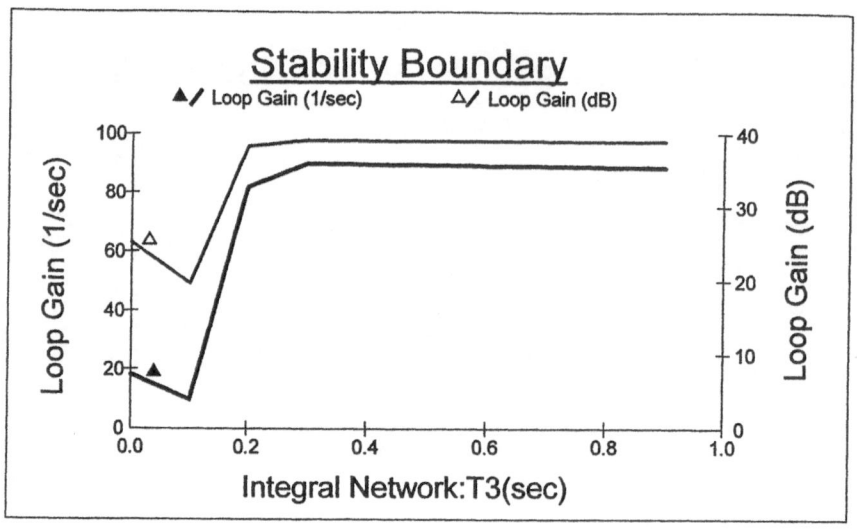

G11 Stability Boundary with integral network

It can be seen from this plot and the corresponding print-out in the Frequency
Domain Screen S9 that the optimum value for the time constant T3 is 0.2 sec
with the time constant T2 set at 1 sec.
A system incorporating a passive network is described in section 7, worked example we9.dat and in section 11, paragraph 11.1.

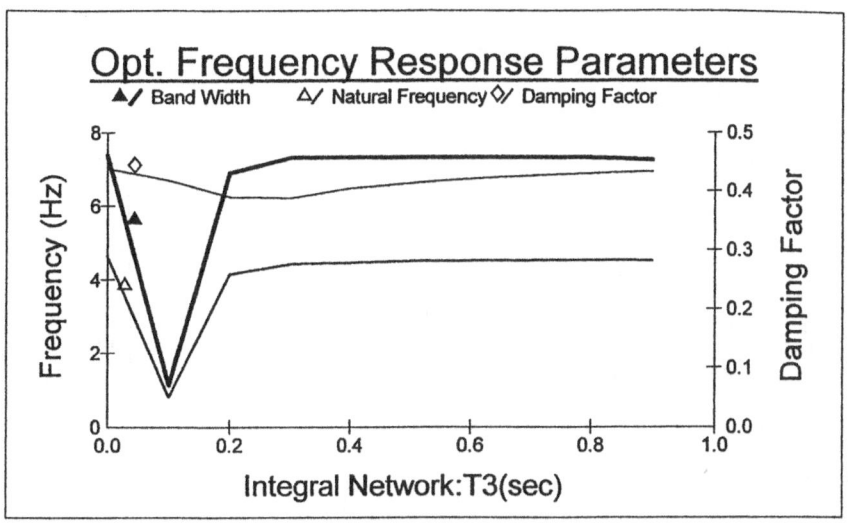

G12 Optimised Frequency Response parameters

This graph shows that an integral network does not affect Frequency Response if the time constants of the network are set at the optimum value of 0.20 and 1 sec.

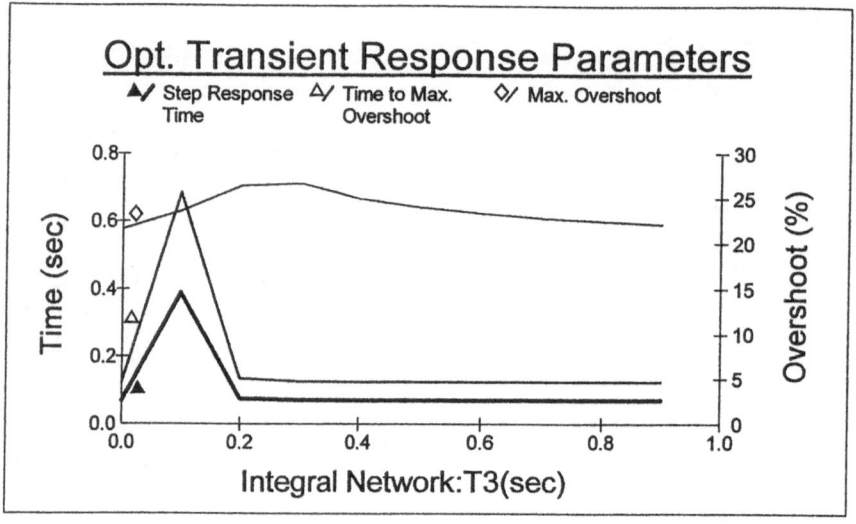

G13 Optimised Transient Response parameters

In line with the results shown in graph G12, Transient Response parameters are not affected at optimum network time constant settings.

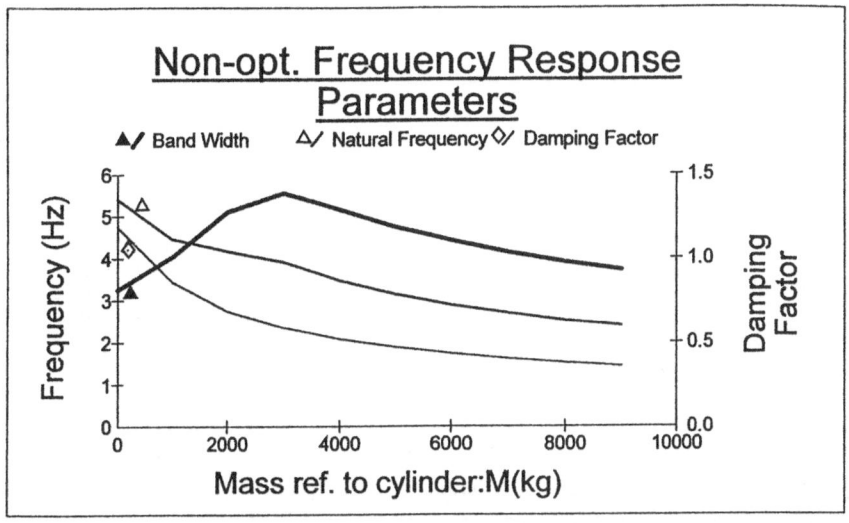

G14 Non-optimised Frequency Response parameters

To maintain stability over the entire operating range, the loop gain has to be set at the lowest value corresponding to the largest mass, as shown in Graph G8. Particularly at masses below 4000 kg the optimised system has considerably better dynamic performance. Above 4000 kg performance ratings of the optimised and non-optimised systems are virtually identical.

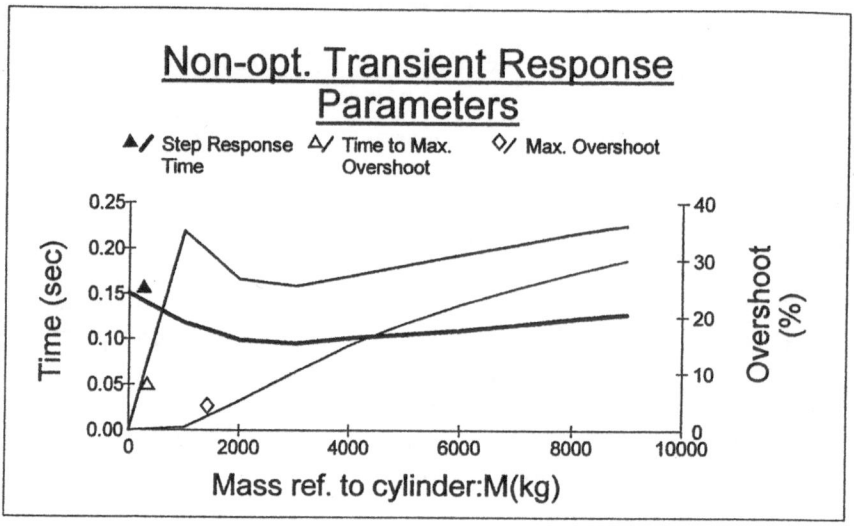

G15 Non-optimised Transient Response parameters.

A comparison of Graph G10 and G15 shows that while step response times vary considerably over the operating range for the optimised system, response times for the non-optimised system remain almost constant.

At a mass of 2000 kg, step response times for the optimised and non-optimised systems are 0.05sec and 0.10 sec respectively. Since the loop gain is the same for both systems at maximum mass, transient response times are identical at 9000 kg.

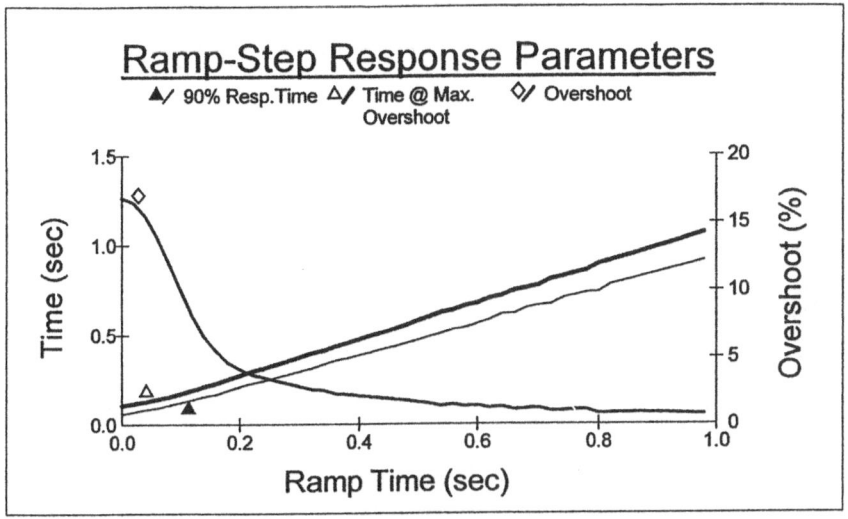

G16 Ramp-Step Response parameters

A convenient method of reducing overshoots is to apply a ramp to the input signal. The graph shows that overshoot can be drastically reduced, in this case from 21.5 % to 2 %. Overshoots are reduced at the expense of response time; it is therefore prudent to strike a compromise. We reach a point of diminishing returns around 0.2 sec ramp time when the overshoot has been reduced to 5 %

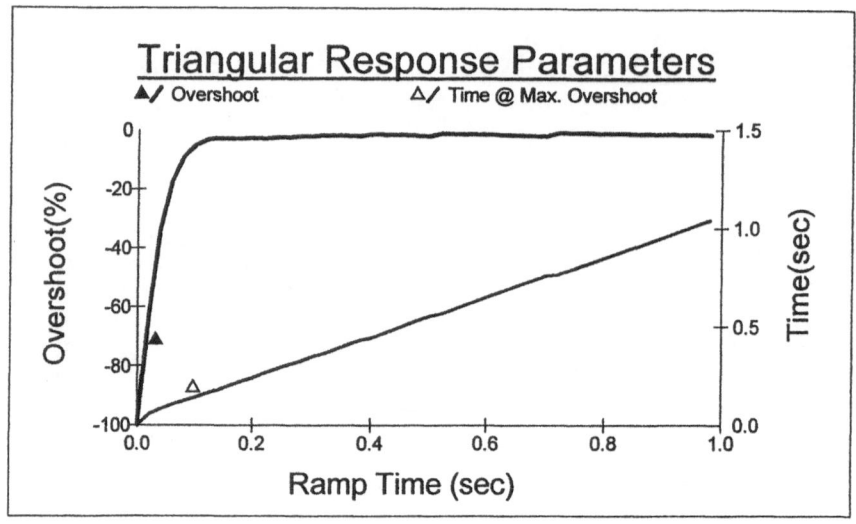

G17 Triangular Response parameters

This graph shows the effect of transient response to a single triangular demand signal. At ramp times of less than 0.1 sec, the system is unable to reach the required amplitude. At ramp times greater than 0.2 sec the system will respond to the demand signal without overshoot. We can infer from this that system we1.dat is able to follow a triangular wave of up to 2.5 hz frequency.

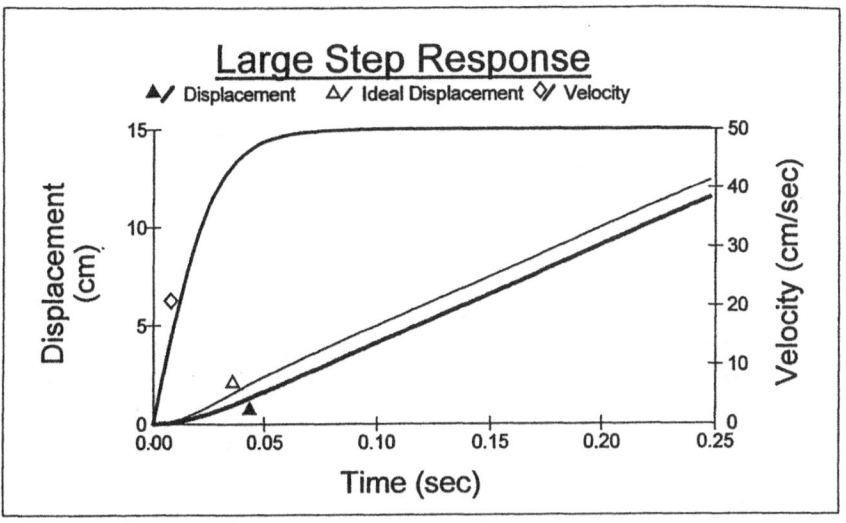

G18 Large Step Response

This graph illustrates the the transient response of the system to a step demand under valve saturated condition. It also applies to an on-off or 'bang-bang valve. The condition of valve saturation often occurs when a large mass or inertia has to be moved at insufficient supply pressure. The plot takes the throttling action of the valve into account.

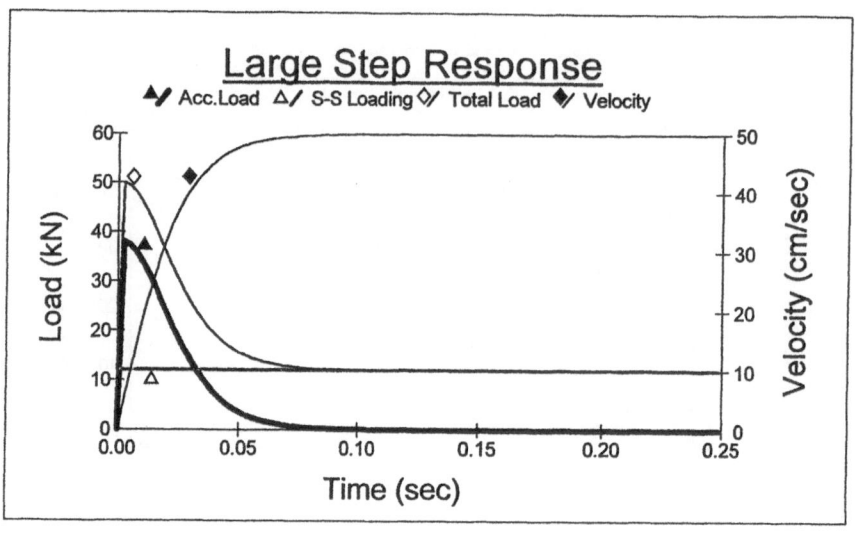

G19 Large Step Response

When moving a large mass, the maximum available accelerating force is limited by the supply pressure. As the velocity increases, the acceleration decreases until the steady-state velocity has been reached.

For the system transient response shown, the initial accelerating load is 38 kN. Steady-state velocity is reached after 0.11 sec.

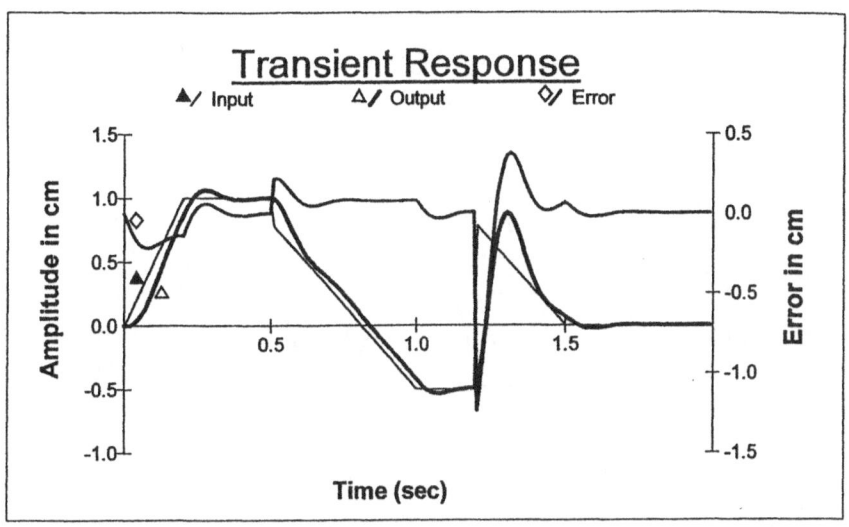

G20 Transient Response.

The steps taken to generate a customised duty cycle consisting of 8 transitions are outlined in section 9. The input, output and error profiles corresponding to the duty cycle tabulated in Screen S6 are plotted in this graph.

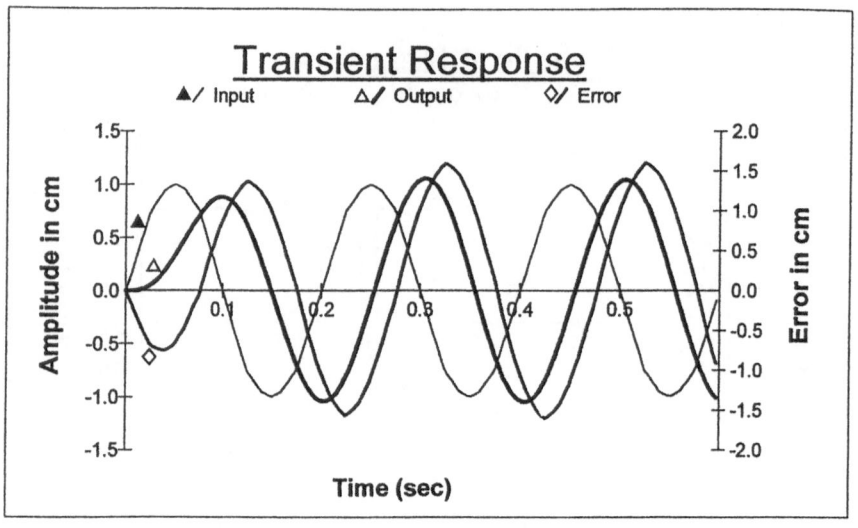

G21 Transient Response

The steps taken to generate the transient response of a postional control system
to a continuous duty cycle are outlined in section 9.
The graph shows the response to a sinusoidal input of 1 cm amplitude and 5 hz
frequency.

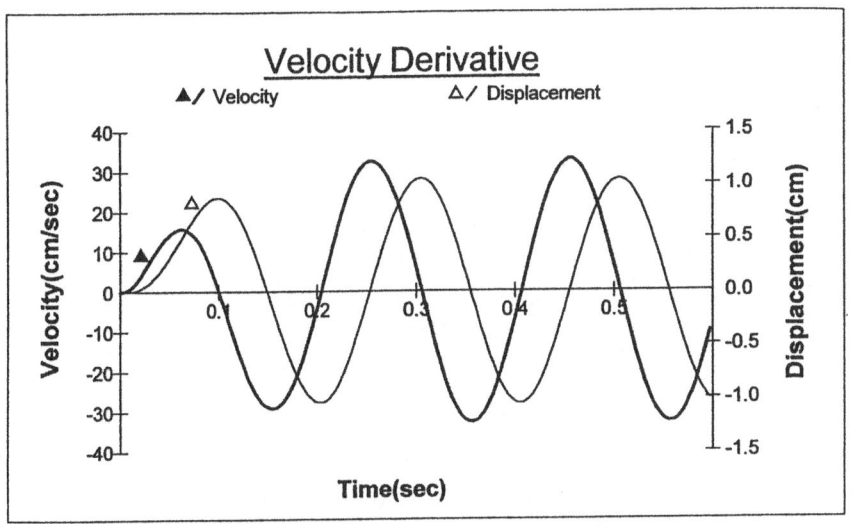

G22 Velocity Derivative

The velocity derivative corresponding to the duty cycle generated by the continuous function generator, as shown in graph G21, is a sine wave, peaking at 32 cm per sec velocity.

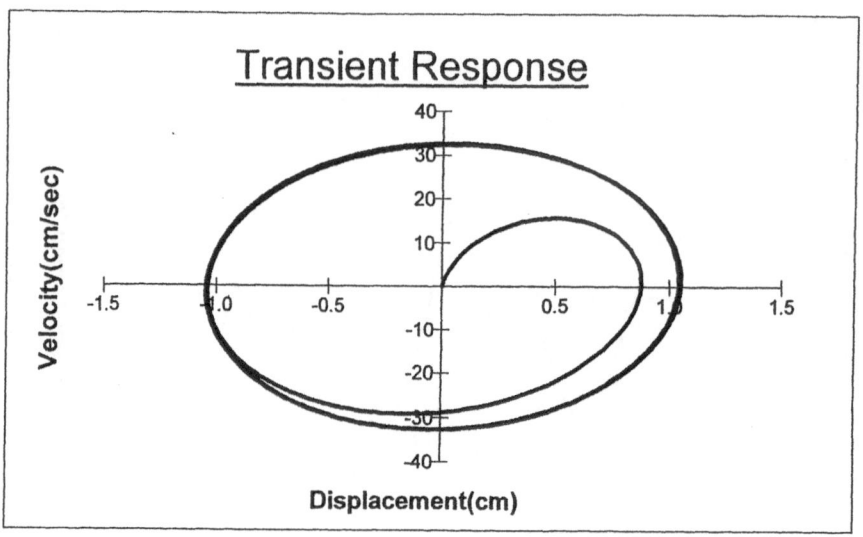

G23 Velocity-Displacement

The velocity-displacement plot of a sinusoidal input takes the form of a continuous circle. The output plot for the duty cycle shown in graph G21 is plotted in this graph. The amplitude peaks at 1.1 cm and the velocity at 32 cm per sec.

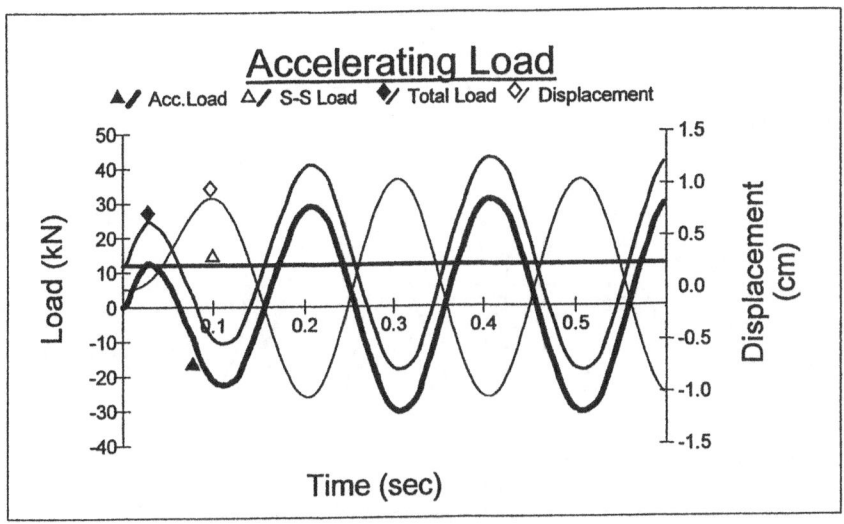

G24 Accelerating Load

This graph shows that accelerating forces of 30 kN are required to meet the duty cycle plotted in graph G21
The accelerating force is superimposed on the steady-state load of 12 kN.

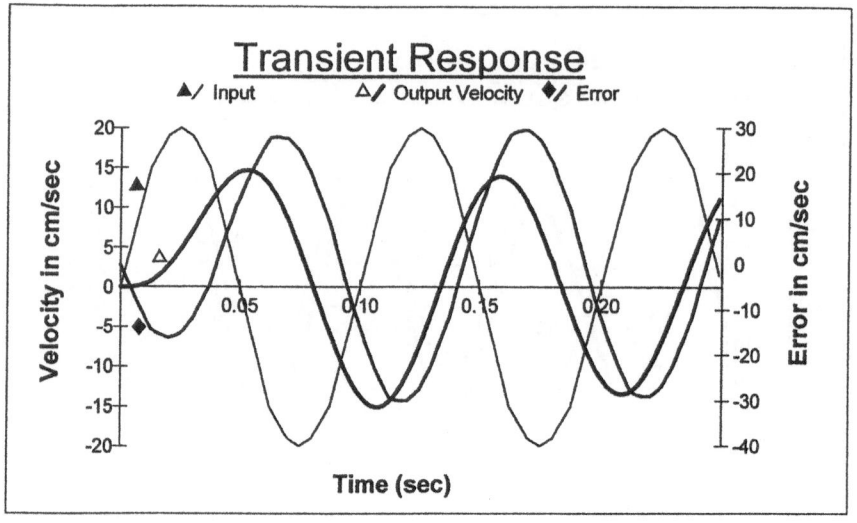

G25 Velocity Profile

This is the velocity profile of an open loop system subjected to a sinusoidal duty cycle of 10 hz frequency and 20 cm per sec. amplitude. The plot shows that the output velocity attenuates to 15 cm per sec.

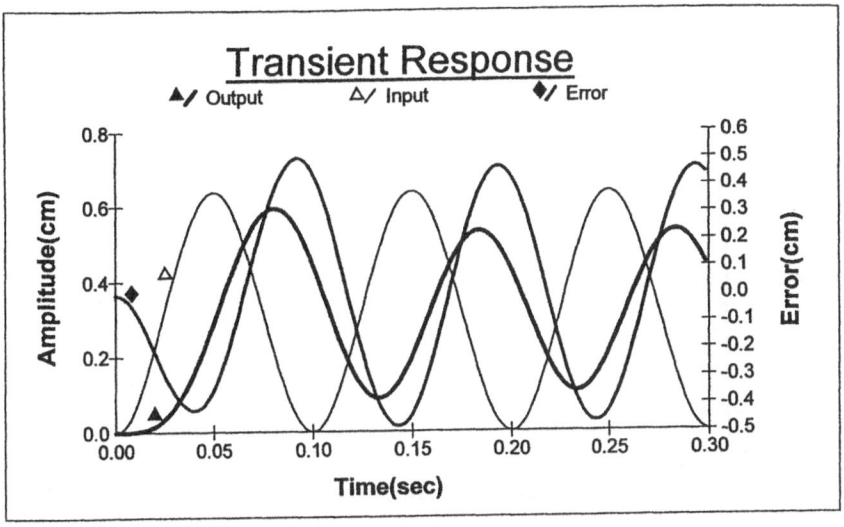

G26 Displacement Profile

The displacement profile corresponding to the velocity profile G25 is plotted in this graph. The displacement profile is a sinusoidal wave settling to an operating range of 0.53 cm maximum to 0.1 cm minimum.

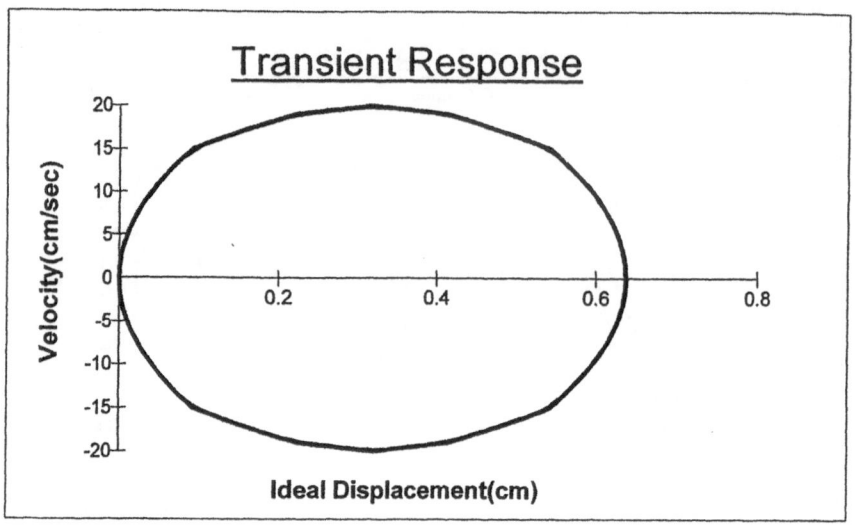

G27 Ideal Velocity-Displacement

A combination of graphs G25 and G26 yields a velocity-displacement profile
which is circular in shape, as shown by the plot of an ideal system.

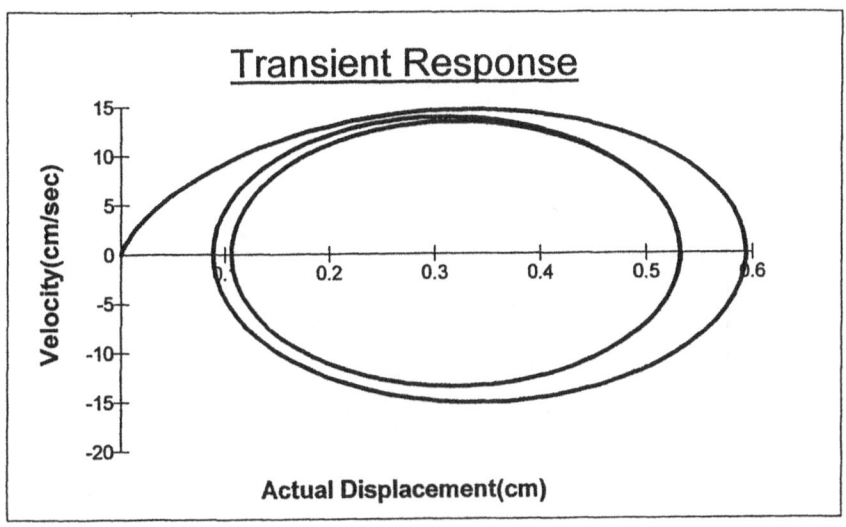

G 28 Actual Velocity-Displacement

The actual velocity-displacement profile corresponding to the ideal profile, as shown in graph G27, shows a number of contours settling to an operating range varying between a maximum of 0.53 cm and a minimum of 0.1 cm.

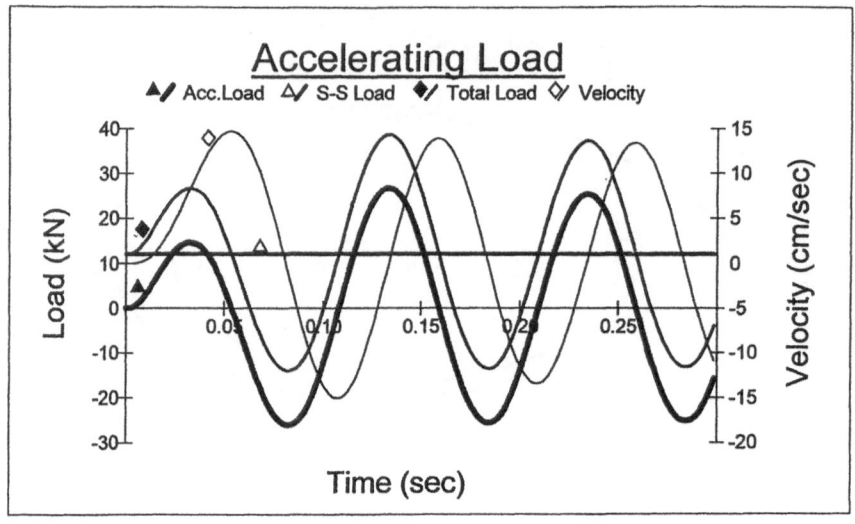

G 29 Accelerating Load

The accelerating force required to sustain the velocity profile, shown in graph G25, is plotted in this graph. This force of 26 kN is superimposed on the steady-state load of 12 kN to give the total load requirement also plotted in this graph. For reference, the velocity profile is also plotted.

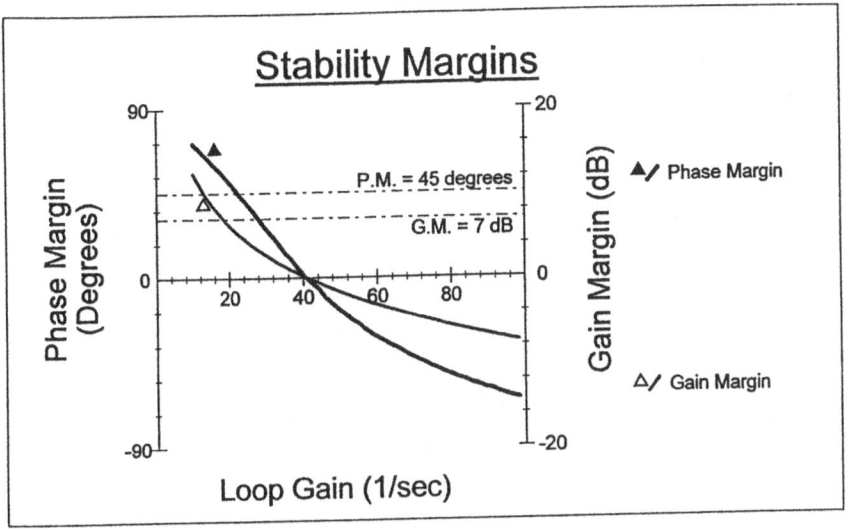

G 30 Stability Margins

In this graph gain and phase margins are plotted as a function of loopgain.
The maximum permissible loop gain is reached when the respective contour in-
tersects the corresponding stability criterion, i.e. 45 degrees for phase margin and
7 dB for gain margin. In this case the gain margin is the crirical criterion at an optimum loop gain of 18 1/sec.
This graph is generated in the Semi-automatic Loop Gain selection described in
section 8 paragraph 8.2.

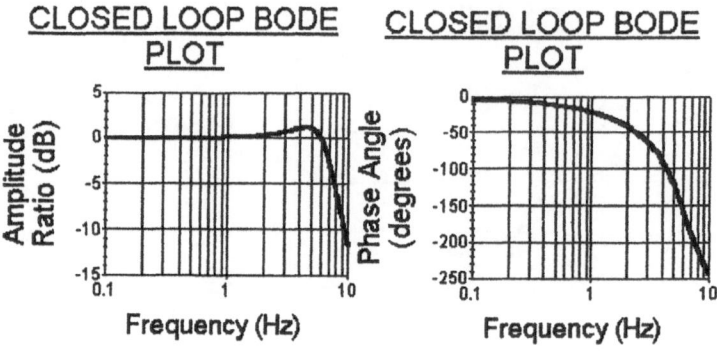

Bode Diagrams

G 31 Multiple Graphs.

The creation of multiple graphs is covered by section 12 'Graphics' under paragraph 12.1 'Sizing of graphs'.

This plot combines graphs G4 and G5.

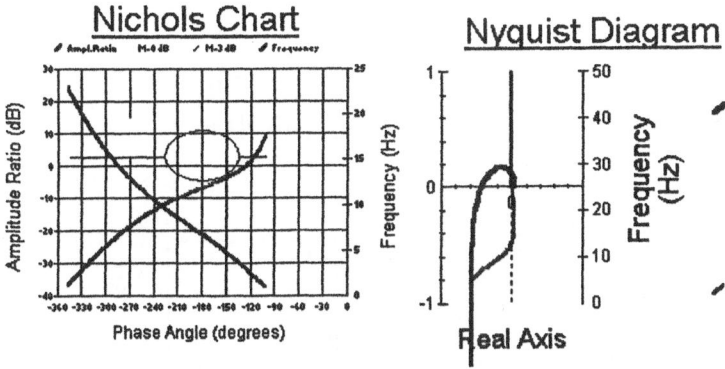

Frequency Response Diagrams

G 32 Multiple Graphs.

The creation of multiple graphs is covered by section 12 'Graphics' under paragraph 12.1 'Sizing of graphs'.

This plot combines graphs G6 and G7.

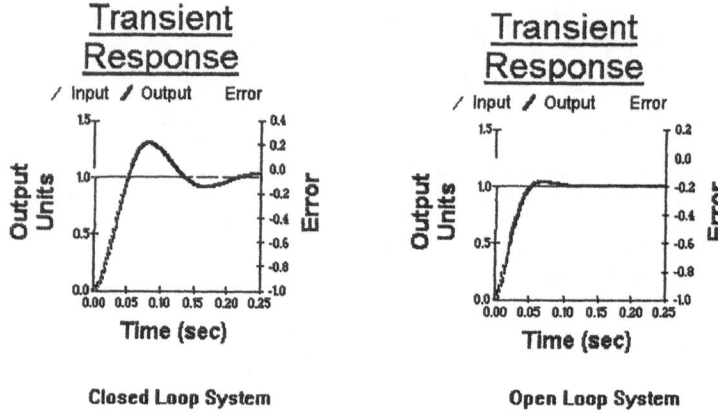

Closed Loop System   Open Loop System

Pressure Control System

G 33 Pressure Control System

The two graphs compare the transient response to a step demand of an open and closed loop system. It can be seen that the open loop system has more favourable dynamic characteristics. The system parameters correspond to the worked example we2p.dat. The only purpose of selecting a closed loop system would be to minimise steady-state errors. For the optimised system the loopgain was set to a value of 34.6 1/sec, thereby limiting the steady-state error to 0.17 bar.

Bibliography

The following books and papers, though not referred to in the text, are recommended for further study.

AHRENDT, W. R., and TAPLIN, J. F., *Automatic Feedback Control*, McGraw-Hill, New York (1951)

AREFIN, K. M. M., 'Seal Characteristics of a Hydraulic Ram and Effects on Servo System Performance, School of Mechanical Engineering, Thames Polytechnic, London, June (1975)

BLACKMAN, P. F., *The Pole-Zero Approach to System Analysis*, Rose Muir Publications, London

BOWDEN, F. P., and TABOR, D., *Friction and Lubrication*, Oxford (1950 & 1967)

BROWN, G. S., and CAMPBELL, D. P., *Principles of Servo Mechanisms*, Chapman and Hall, London (1948)

EVANS, W. R., *Control System Dynamics*, McGraw-Hill, New York

HADEKEL, R., *On the Stability of Flying Controls*, Technical Information Bureau, Ministry of Supply (1954)

HADEKEL, R., 'Hydraulics and Pneumatic Servos', *Automation* (1954–55)

HARPUR, N. F., 'Some Design Considerations of Hydraulic Servos of the Jack Type', *Proc. Conf. Hydraulic Servo Mechanisms* 4 (1953)

JAMES, H. M., NICHOLS, N. B., and PHILIPS, R. S., *Theory of Servo Mechanisms*, McGraw-Hill, New York (1947)

NOTON, G. I., and TURNBULL, D. E., 'Some Factors Influencing the Stability of Piston Type Control Valves', *Proc. Inst. Mech. Engrs*, London (1958)

RAVEN, F. H., *Automatic Control Engineering*, McGraw-Hill, New York (1961)

ROBERT BOSCH GmbH, *Proportional Valves: Theory & Application* (1987)

SHEPHERDSON, M., and WALTERS, R., 'A Stroboscopic Method of Making Frequency Response on Small Electromechanical Devices', *Electron. Eng.* 31, 374, 220 (1959)

THALER, G. J., and BROWN, R. G., *Analysis and Design of Feedback Control Systems*, McGraw-Hill, New York (1960)

VICKERS SYSTEMS LTD, *Proportional Valves* (1968)

WALTERS, R., 'Some Hydraulic Servo Applications', *Proc. Conf. on Oil Hydraulic Power Transmission and Control*, 69 (1961)

WALTERS, R., *Hydraulic Servo Systems Analysis and Synthesis*, Sperry Gyroscope Co. Publication (1964)

WALTERS, R., 'Synthesising Hydraulic Control Systems from Available Components', *Hydraulic Pneumatic Power* **2**, 126 (1965)

WALTERS, R., *Hydraulic and Electro-Hydraulic Servo Systems*, Iliffe Books, London (1967)

WALTERS, R., 'Electrically Modulated Actuator Controls', *4th International Fluid Power Symposium*, Sheffield, April (1975)

WALTERS, R., 'Mathematical Model of a New Actuator Control System', *Pneumatics-Hydraulics '75 Conference*, Gyor, Hungary, September (1975)

WALTERS, R., 'Electrically Modulated Actuator Controls', *International Symposium on Oil Hydraulics*, Hannover, April (1977)

WALTERS, R., 'The Electronic-Hydraulic Interface', *Design Engineering Conference*, Birmingham, December (1978)

WALTERS, R., 'Desk Top Computer Performance Prediction Package', *I.Mech.E.Seminar on Computer Aided Design*, London, November (1983)

WALTERS, R., 'Proportional Control Systems and Applications', *FLUMEX '84 Conf.*, Birmingham, June (1984)

WALTERS, R., 'State of the Art Lecture: Electro-Hydraulic Proportional Control', I.Mech.E., London, April (1986)

WALTERS, R., 'Electro-Hydraulic Proportional Controls', I.Mech.E. Lecture, Oxford University, February (1987)

WALTERS, R., 'Electro-Hydraulic Control Systems', Research Seminar, University of Salford, October (1987)

WALTERS, R., 'Modular Optimised System Simulation (MOSS)', *BHRA 8th International Symposium on Fluid Power*, Birmingham, April (1988)

WALTERS, R., and HARRISON, D., 'An expert Simulation Software Package for Electrohydraulic Control Systems', *6th Fluid Power Workshop, Bath University*, September (1993)

Index

Errata

Hydraulic and Electro-hydraulic Control Systems
Second Enlarged Edition

by R. B. WALTERS

ISBN 978-94-015-9429-5

The corrections are:
Graph G8   page 299: section 10.1 should read: section 38
Graph G11   page 302: section 7 should read: section 21
                    section 11 paragraph 11.1 should read: section 39
                    paragraph 39.1
Graph G20   page 311: section 9 should read: section 37
Graph G21   page 312: section 9 should read: section 37
Graph G30   page 321: section 8 paragraph 8.2 should read:
                    section 36 paragraph 36.2
Graph G31   page 322: section 12 paragraph 12.1 should read:
                    section 40 paragraph 40.1
Graph G32   page 323: section 12 paragraph 12.1 should read:
                    section 40 paragraph 40.1